Douglas William Freshfield

Travels in the Central Caucasus and Bashan

Douglas William Freshfield

Travels in the Central Caucasus and Bashan

ISBN/EAN: 9783741171833

Manufactured in Europe, USA, Canada, Australia, Japa

Cover: Foto ©Klaus-Uwe Gerhardt /pixelio.de

Manufactured and distributed by brebook publishing software (www.brebook.com)

Douglas William Freshfield

Travels in the Central Caucasus and Bashan

TRAVELS

IN THE

CENTRAL CAUCASUS AND BASHAN

INCLUDING

VISITS TO ARARAT AND TABREEZ

AND

ASCENTS OF KAZBEK AND ELBRUZ.

BY

DOUGLAS W. FRESHFIELD.

'Per Alpium juga
Inhospitalem et Caucasum.'
HOR. Epod. 1. 12.

LONDON:
LONGMANS, GREEN, AND CO.
1869.

PREFACE.

THE FOLLOWING PAGES sufficiently explain how the journey described in them was planned and carried out. In the course of our wanderings, we visited two countries, well known, indeed, by name to the general reader, but concerning which vague, and in some respects incorrect impressions are frequently entertained. A truthful traveller may do as good service by destroying illusions as by bringing forward fresh information, and I have felt bound to record our conviction that the belief that there are 'Giant Cities' in Bashan is as unfounded as the still more prevalent idea that all the men in the Caucasus are brave, and all the women beautiful.

Our Syrian travels owed their chief interest to a sudden access of vigour on the part of the Turkish Pashas, which enabled us to visit, with little risk or expense, the remarkable ruins of the Hauran and Lejah, and to form our own opinion as to their date—a question as yet discussed principally by unskilled witnesses, and still awaiting the decision of a competent judge.

The exploration of the passes and glaciers of the Central Caucasus, and the ascent of its two most famous summits, formed the chief aim of our journey, and are the main subject of the present volume. I trust that the record of our adventures in the mountain fastnesses may prove of sufficient interest to draw the attention of our countrymen to a range surpassing the Alps by two thousand feet in the average height of its peaks, abounding in noble scenery and picturesque inhabitants, and even now within the reach of many 'long-vacation tourists.' When the Caucasus, as yet less known than the Andes or the Himalayas, becomes a recognised goal of travel, this work will have fulfilled its object, and will be superseded by the production of some author better qualified, both by literary skill and scientific attainments, to treat of so noble a theme.

The reader will not find in these pages any political speculations, for which so rapid a journey afforded scant opportunity; he may more justly complain of the absence of ethnological details concerning the tribes of the Caucasus. My excuse is, that information filtered through an uneducated interpreter is difficult to obtain and little trustworthy; the subject, moreover, has been fully treated of by German travellers, in works already translated into English, and accessible to those in whom the present account of the natural features of the Caucasian region may raise a wish to learn more of its inhabitants.

The Map of the Central Caucasus is reduced from the Five Verst Map, executed by the Russian Topographical Department at Tiflis, with many corrections suggested by our own experience. The illustrations are derived from various sources; some have been engraved from paintings by a Russian artist resident at Tiflis, others are from photographs or pencil-sketches. Two of the smaller plates are borrowed from a privately-printed work of Herr Radde, our numerous obligations to whom I gladly take this opportunity of acknowledging.

I owe my best thanks to Mr. Edward Whymper for the skill he has shown in dealing with the rough materials placed at his disposal, a task for which his well-known knowledge of mountain scenery eminently qualified him. I have also to thank Mr. Weller for the care he has taken to make the maps accurate and intelligible.

I cannot conclude these few words of preface without bearing grateful witness to the constant encouragement, and very important aid, which I have received from my companions, Mr. A. W. Moore and Mr. C. C. Tucker, in the preparation of the volume now submitted to the public.

CONTENTS.

CHAPTER I.
EGYPT AND PALESTINE.

Introductory—Choosing a Dragoman—Djebel Mokattam—The Nile Steamer—The Mecca Caravan—Sail for Syria—A Poor Traveller—Struck by Lightning—Syrian Sloughs and Storms—The River Kishon—Arrival at Jerusalem—An Idea worked out—"Vive la Mer Morte!"—Jericho—We fall among Thieves—The Jordan Valley—Capture of a Standard-bearer—Ferry of the Jordan 1

CHAPTER II.
BASHAN.

The English Soldier—A Mountain Ride—Es-Salt—Lost on the Hills—The Jubbek—Camp of the Beni-Hassan—Suppressing a Sheikh—The Oak Forests of Gilead—The Tablelands—An Uxorious Sheikh—Deraā—The Roman Road—The Robbers repulsed—Ghusum—Bozrah—Honoured Guests—A Ramble in the Ruins—Kureiyeh—Patriarchal Hospitality—Hebran—A Stone House—Rufu—Ascent of El-Kleib—Suweideh—Kunawat—Noble Ruins—Shuhba—Hades on Earth—Visiting Extraordinary—The Lejah—A Lava Flood—Abireh—Khubab—A Rush to Arms—The Stolen Mule—A Village in Pursuit—Mismiyeh—The 'Giant Cities' are Roman Towns—The Wrath of the Boys—A Friendly Salut—Keawra—Entrance to Damascus 16

CHAPTER III.
LEBANON AND THE LEVANT.

Damascus—Bazaars and Gardens—An Enthusiastic Freemason—Snowstorm on Anti-Lebanon—Baalbec—An Alpine Walk—The Cedars—Return to Beyrout—Cyprus and Rhodes—Smyrna—The Valley of the Mæander—Excavations at Ephesus—Constantinople—The Persian Khan—May-Day at the Sweet Waters—Preparations for the Caucasus . 53

CHAPTER IV.

TRANSCAUCASIA.

On the Black Sea—Trebizonde—Rival Interpreters—Paul—Running a Muck—Batoum—The Caucasus in Sight—Landing at Poti—The Rion Steamer—A Drive in the Dark—Kutais—Count Levaschoff—Splendid Costumes—Mingrelian Princesses—Azaleas—The Valley of the Quirili—A Post Station—The Georgian Plains—Underground Villages—Gori—First View of Kazbek—Tiflis—The Hôtel d'Europe—The Streets—Silver and Fur Bazaars—Maps—German Savants—The Botanical Garden—The Opera—Officialism Rampant—A False Frenchwoman—A Paraclinasis—The Postal System in Russia 74

CHAPTER V.

THE PERSIAN POST-ROAD.

The Banks of the Kur—Troops on the March—A Romantic Valley—Delidschan—A Desolate Pass—The Gokcha Lake—Ararat—Erivan—The Kurds—The Valley of the Araxes—A Steppe Storm—A Dangerous Ford—Nakhitchevan—A Money Question—Djulfa—Charon's Ferry and a Modern Cerberus—A Friend in Need—A Persian Khan—Marand—Entrance to Tabreez—Olim Lazarus 112

CHAPTER VI.

TABREEZ, ARARAT, AND THE GEORGIAN HILL-COUNTRY.

The City—Brick Architecture—The Shah's Birthday—The European Colony—A Market Committee—Return to Djulfa—A Dust Storm—Ford of the Araxes—Aralykh—Start for Ararat—Refractory Kurds—A Moonlight Climb—Failure—A Lonely Perch—Vast Panorama—Tucker's Story—A Gloomy Descent—Return to Erivan—Etchmiadzin—The Armenian Patriarch—A Dull Ride—Hammamly—The Georgian Hills—Digbloghlu—A Moist Climate—Schulaweri—Tiflis again—Moore joins us 141

CHAPTER VII.

THE KRESTOWAJA GORA AND ASCENT OF KAZBEK.

Start for the Mountains—The Pass of the Caucasus—Kazbek Post Station—The Governor—A Reconnaissance in force—Legends—Avalanches—The Old Men's Chorus—Men in Armour—Our Bivouac—A Critical Moment—Scaling an Icewall—The Summit—The Descent—A Savage Glen—A Night with the Shepherds—Return to the Village—Caucasian Congratulations 170

CHAPTER VIII

THE VALLEYS OF THE TEREK, ARDON, AND RION.

PAGE

A Geographical Disquisition—The Upper Terek—Savage Scenery—Ferocious Dogs—Abano—A Dull Walk—Hard Bargaining—An Unruly Train—A Pass—Zaramag on the Ardon—A Warm Skirmish and a Barren Victory—An Unexpected Climb—The Lower Valley—A Russian Road—Tseh—The Ossetes—The Mamisson Pass—Adai Khokh—A Shift in the Scenery—Gurshavi—The Boy-Prince—An Idle Day—View from the Rhododendron Slope—Glola—The Pine-Forests of the Rion—Chiora 205

CHAPTER IX.

THE GLACIERS AND FORESTS OF THE CENTRAL CAUCASUS.

Caucasian Shepherds—A Lovely Alp—Sheep on the Glacier—A New Pass—A Snow Wall—A Rough Glen—The Karagam Glacier—Bivouac in the Forest—An Icefall—A Struggle and a Victory—The Upper Snowfields—The Watershed at last—Chek—A Useful Gully—An Uneasy Night—Glola again—Pantomime—Gebi—Curious Villagers—A Bargain for Porters—Azalea Thickets—The Source of the Rion—Rank Herbage—Camp on the Zenes-Squali—A Low Pass—Swamps and Jungles—Path-finding—The Glen of the Scena—Wide Pasturages—The Nakasagar Pass 245

CHAPTER X.

SUANETIA.

Free Suanetia, Past and Present—Herr Radde's Experiences—Physical Features—Fortified Villages—Jibiani—Picus Savage—A Surprise—Glaciers of the Ingur—Petty Theft—Threats of Robbery—Alarms and Excursions—A Stormy Parting—The Horseman's Home—The Ruined Tower—A Glorious Icefall—Adisch—Sylvan Scenery—The Muchelala—Suni—Ups and Downs—Midday Halt—Latal—A Suanetian Farmhouse—Murder no Crime—Tan Tikunal—A Sensation Scene—The Caucasian Matterhorn—Pari at last—Hospitable Cossacks . . . 292

CHAPTER XI.

FROM PARI TO PÄTIGORSK, AND ASCENT OF ELBRUZ.

A Captive Bear—Moore Harangues the Porters—Camp in the Forest—A Plague of Flies—Lazy Porters—A Nook in the Mountains—Cattle-Lifting—Across the Chain in a Snowstorm—A Stormy Debate—A Log Hut—Baksan Valley—Uruspieh—The Guest House—Village Rewarded—Minghi-Tau—An Idle Day—An Enlightened Prince—Passes in the Karatshai—Tartar Mountaineers—A Night with the Shepherds—A Steep Climb—Camp on the Rocks—Great Cold—On the Snowfield—In a Crevasse—Frigid Despair—A Crisis—Perseverance Rewarded—The Summit—Panorama—The Return—Enthusiastic Reception—The Lower Baksan—A Long Ride—A Tcherkess Village—Grassy Downs—Zountzki—Pätigorsk 337

CHAPTER XII.

PÄTIGORSK AND THE TCHEREK VALLEY.

The Caucasian Spas—Their History and Development—View from Machoucha—The Patients—Essentuky—Kislovodsk—The Narsan—Hospitable Reception—A Fresh Start—A Russian Farmhouse—By the Waters of Baksan—Naltshik—The Tcherek—Camp in the Forest—A Tremendous Gorge—Balkar—A Hospitable Sheikh—The Mollah—Gloomy Weather—A Solemn Parting—Granitic Cliffs—Karaoul—A Mountain Panorama—Sources of the Tcherek—The Saxlerweek Pass—Koschtantau and Dychtau—A Noble Peak—Our Last Camp . 381

CHAPTER XIII.

THE URUCH VALLEY AND RETURN TO TIFLIS.

Woodwl Defiles—Styr Digor—A Halt—We Meet a Cossack—A Rain-storm—Zadolesk—The Gate of the Mountains—Across the Hills and Through the Forest—Tugauria—Novo-Christiansky—A Christian Welcome—A Wet Ride—Ardonsk—A Breakdown on the Steppe—Vladi-kafkaz—A Diligence Drive—The Dariel Gorge—Return to Tiflis—Reflections on the Caucasian Chain—Its Scenery and Inhabitants—Comparison with the Alps—Hints for Travellers 422

CHAPTER XIV.

TRANSCAUCASIA AND THE CRIMEA: HOME THROUGH RUSSIA.

Borjom—Bad Road—Beautiful Scenery—Achaltsich—Across the Hills—Abastuman—A Narrow Valley—The Burnt Forest—Panorama of the Caucasus—Last Appearance of Kazbek and Elbruz—A Forest Ride—Ingilad—Mingrelian Hospitality—A French Baron's Farm—The Rion Basin—Kutais—The Postmaster—Poti—A Dismal Swamp—Soukhoum-Kalé—Sevastopol—The Battlefields—The Crimean Corniche—Bakhchi-Sarai—Odessa—A Run across Russia—A Jew's Cap—The Dnieper Steamboat—Kieff—Picturesque Pilgrims—The Lavra—Sainted Mummies—A Long Drive—Vitebsk—St. Petersburg—Conclusion 465

APPENDIX.

I. The Elbruz Expedition of 1829 497
II. Heights of Peaks, Passes, Towns, and Villages in the Caucasian Provinces 500
III. Catalogue of Plants 502

LIST OF MAPS AND ILLUSTRATIONS.

MAPS.

		PAGE
I. Route Map of the Haurán	. . . To face	16
II. The Caucasian Provinces	. . . ,,	74
III. The Central Caucasus	. . . End of Vol.	

FULL-PAGE ILLUSTRATIONS.

Elbruz from the North	. . . Frontispiece	
Ararat	. . . To face	125
Kazbek from the Post Station	. . . ,,	185
Kazbek from the South	. . . ,,	197

PANORAMAS.

The Caucasus from Pâtigorsk	. . . ,,	281
The Kasbimstau Group	. . . ,,	281

WOODCUTS IN TEXT.

A Georgian Church	95
The Georgian Castle, Tiflis	104
Mountaineers in Ararat	135
An Ossete Village	215
An Ossete	227
Adai Khokh from the Rion Valley	297
Source of the Eastern Zenus-Squali	302
Our Camp-fire in the Forest	305
A Native of Jibiani	330
The Tötkmal from above Latal	375
Uschba from above Latal	390
Women of Urusphih	397
Peak in the Tcherek Valley	411
Fort of Dariel	442
Grand-Ducal Villa at Borjom	468
Nicaroflian Wine Jar	479

TRAVELS IN

THE CENTRAL CAUCASUS AND BASHAN.

CHAPTER I.

EGYPT AND PALESTINE.

Introductory—Choosing a Dragoman—Djebel Mokattam—The Nile Steamer—The Mecca Caravan—Sail for Syria—A Poor Traveller—Struck by Lightning—Syrian Sloughs and Storms—The River Kishon—Arrival at Jerusalem—An Idea worked out—'Vive la Mer Morte!'—Jericho—We fall among Thieves—The Jordan Valley—Capture of a Standard-bearer—Ferry of the Jordan.

BEFORE carrying my readers into the primitive wilds of Bashan, and amongst the unknown valleys and ridges of the Caucasus, I must give some explanation of the circumstances which induced me to undertake the journey I am about to describe. In many summer holidays, spent among the Alps, I had acquired a taste for mountain scenery, and when an opportunity of being absent from home for a longer time than usual presented itself, I looked for some country where the zest of novelty would be added to those natural features which chiefly attracted me. For many reasons the Caucasus seemed to be the very region I was seeking. Less distant than the Andes or the Himalayas, its mountains were yet unknown to ordinary travellers, and none of our countrymen had explored the recesses of the finest portions of the chain, although not a few had crossed the great highway of the Dariel, or followed in the footsteps of the Russian armies in Daghestan.

My journey was to begin in January, at which time it was obviously too early to start on a mountaineering excursion, and the ease with which a visit to the Caucasian provinces might be fitted on to an Eastern tour induced me to spend the intervening months in Egypt and the Holy Land.

The plan was definitely settled when my friend Mr. Tucker agreed to join me in the whole of the proposed journey. I had the good fortune to secure a second comrade for our Caucasian explorations in Mr. Moore, who was, however, unable to leave London until the summer, and therefore promised to meet us, at Tiflis, about June 20th. So far our party was complete, but for mountaineering work it was evidently necessary to have the assistance of at least one skilled guide. My old companion, François Devouassoud of Chamouni, was just the man we wanted, and he proved not only willing but anxious to join us. The only question was whether he should meet us at some point in our journey, or should accompany us from its outset. I finally determined to accede to his wishes, and take him as a travelling servant, having full confidence in his intelligence and readiness to accommodate himself to new scenes and unaccustomed modes of life. We had no reason to repent this decision.

After a busy fortnight, spent in getting together the necessaries for our journey—which included a tent, waterproof saddlebags, a portable kitchen, and large quantities of Liebig's soup—we left England on January 4th, 1868. We passed through the South of France in the most intense cold: at Avignon the Rhone was frozen from bank to bank, and the fountains at Marseilles were turned into masses of icicles. On January 8th we sailed for Egypt, on board the Messageries Imperiales' steamer 'Port Said,' with a miscellaneous batch of passengers, including two French officers who were going to Abyssinia, two directors of the Suez Canal, Gerôme the painter, the Viceroy of Egypt's dentist, two missionary ladies bound for Jerusalem, and a party of Algerine Arabs on their way to Mecca, who lay all day and night on deck, huddled in their cloaks. With such variety on board, and a constantly-changing horizon, we found the voyage by no means monotonous.

On the sixth morning the tall lighthouse and low coast of Alexandria came in sight. We landed in a storm of rain, which added to the difficulties of newcomers in an Eastern city. We were at once surrounded by a host of dragomen, and pestered by their persistent attentions, until we at last selected one, whose personal appearance was in his favour, and whose terms and promises were more reasonable than those of most of the men we saw. By the kind assistance of a European resident, a contract was made with him to accompany us during our Syrian tour; his duties were to begin on our landing at Jaffa or Beyrout. The successful candidate was Elias Abbas, a Maronite of the Lebanon.

I have no intention of adding to the already too numerous descriptions of Egypt and the Nile, but I cannot refrain from one hint to all visitors to Cairo. Visit the petrified forest, and make your donkey-boy bring you back by Djebel Mokattam, or you will lose one of the most wonderful views in the East. After riding for miles over the arid African desert, with a narrow horizon, and nothing to attract the attention save a distant train of camels or a troop of gazelles, the edge of an abrupt descent is reached, and the view of Cairo and the valley of the Nile bursts upon the eyes with an almost magical suddenness. The immediate foreground is formed by the quarried heights of Djebel Mokattam, in the centre of the picture rise the taper minarets of the mosque of the citadel, in a valley on its right are the tombs of the Memlooks, a deserted town of the dead, and the vast modern city spreads itself out in the plain below. In the centre of the broad bluish-green ribbon of fertile land, dotted with clusters of pyramids, the Nile itself can be traced to the commencement of the Delta, while beyond, on the west, the yellow sands of the desert mark the limits of its fertilising inundation.

At Cairo we were fortunate in falling in with some pleasant Americans, who were making up a party to hire a steamboat for a trip up the river. They asked us to join them, and although the Nile had not formed part of our programme, the opportunity was too good to be lost, and we gladly accepted the offer.

Although our company consisted of eleven Americans and only four Englishmen, the majority were not at all disposed to abuse their power, and we gave an example of unbroken harmony to the other steamers going up the river at the same time. Indeed, I believe that on this account, as well as from our being so fortunate as to carry with us some pretty and lively Transatlantic cousins, we were an object of envy and heartburning to most of the boats we met.

Time could not hang heavy on the hands of those who, when their admiration was no longer called forth by 'the mysterious type of beauty' peculiar to the broken-nosed sisterhood of Sphinxes, could turn their eyes on the fresher charms which the Far West had sent to compete with the stony loveliness of the East. On February 14th we bade farewell to the little steamer in which we had spent three very pleasant weeks on the Nile, and returned to our old quarters in Shepherd's Hotel at Cairo.

We witnessed the departure of the Haj caravan for Mecca, admired the holy camel, draped with cloth of gold, carrying the annually renewed covering of Mahomet's tomb, and laughed heartily at a sheikh of extraordinary sanctity and obesity, who, stripped to the waist and shining with oil, swayed himself backwards and forwards on his camel with the air of a tipsy Falstaff. A few hours later we bade adieu to Cairo and our Nile friends, and on the next day embarked at Alexandria for Syria.

We had been asked to take out from England a long box, labelled 'Delicate instruments—with care,' for the use of Lieutenant Warren, the officer engaged in superintending the excavations lately undertaken by the 'Palestine Exploration Committee.' On our leaving Alexandria the custom-house officer wanted to examine the box, and it was only by loud protests and threats of official vengeance that we saved the instruments from the risk of being spoiled by the Egyptians. This was the beginning of woes to these 'delicate instruments,' which became celebrated characters with us during the next fortnight.

We spent a day at Port Said, an utterly uninteresting town of third-class villa residences, and wide streets lined with hastily-run-up stores, built upon a sandspit. It is probably destined to future importance as the Mediterranean port of the Suez Canal. We had not time to see much of the works now in progress there, but enjoyed a ramble on the beach, which is entirely formed of lovely little shells of the most delicate shapes and colours. We re-embarked on Tuesday the 17th, and in the evening the sea became very rough. At midnight half the passengers were pitched out of their berths by some terrible rolls;

then the cabin-benches got loose, and tumbled about noisily. At 7 A.M. we were off Jaffa, but landing was out of the question; an hour later the cabin in which I was dressing was filled with a blaze of light, and the ship shook with a report as if she had fired a broadside. Our foremast had been struck by lightning, but, being provided with a conductor, the vessel escaped injury. All that day we ran on through a big tumbling sea, and anchored at night in the roadstead of Beyrout.

On Wednesday morning we disembarked, and went to the 'Hôtel de Damas.' Our original plan, to land at Jaffa and go up direct to Jerusalem, had been thrown out by the storm, and new arrangements were necessary. Mr. Williams, one of our American friends, was in the same position, and now agreed to join us in our Syrian journey, so that we were a party of three. As attendants, besides the dragoman, Elias Abbas, we had a cook and a waiter, with the usual staff of muleteers.

Elias's preparations took him several days, and it was not till Sunday that we succeeded in leaving Beyrout. Meantime we heard complaints from all sides of the extraordinary severity of the season; Damascus was virtually inaccessible, owing to the heavy snowstorms which had blocked up the passes of the Lebanon. The rain fell almost incessantly, and the mock torrents which poured down the streets of Beyrout augured ill for our chance of passing the formidable streams which intersect the road to Jerusalem. At last we set out. We made a long circuit through the hills to Deir-el-Kamr, to find a bridge over the Damur, the first and most formidable of the rivers we had to cross. Along the coast of Tyre and Sidon we journeyed on through rain and mud, until at Acre the tide of our mishaps reached its highest point. We had pitched our tent beneath a ruined villa a mile outside the town; about 9 P.M. the wind rose; an hour later it was blowing a gale, and the ropes began to part; however, by doubling our fastenings, and by dint of constant sallies, we kept a shelter over our heads all night. At 5 A.M. the outside roof of the tent was in rags, the wooden supports of the sides mostly broken, and the wind generally master of the situation. I was too sleepy to stir out of bed, and lay in momentary expectation that we should be caught up and carried away in a whirlwind. My friends, however, did not wish to try this new sensation, so we roused ourselves to action, and with much difficulty succeeded in lowering and fastening up the canvas; then we took refuge, with the horses, in a ruinous cellar.

Next morning, just outside Acre, the passage of a river, which entered the sea by two mouths, rendered necessary a double loading and unloading of the baggage. We crossed in boats, but our animals had to swim. I shall not easily forget the transit of the three donkeys. They were driven into the stream as far as whips would reach them, but just within their depth, and beyond the reach of their persecutors' weapons, the trio unanimously halted. Never was the *vis inertiæ* more strikingly exemplified. In vain their masters hurled on the patient beasts every form of Christian and Moslem imprecation. The donkeys 'were not a penny the worse'; they felt they had the best of the situation, and exhibited a stolid contempt for all the uproar of which they were the cause. At last one of the muleteers stripped, and, entering the water, launched the obstinate little brutes, one by one, by main force. Once committed to the deep, they swam bravely, and emerged on the farther bank dripping and shaking their long ears as if, after all, they were the heroes of the day. An hour's scamper over the sandy beach brought us to the mouth of 'that ancient river, the river Kishon.' It was, of course, flooded, and, considering the combination of difficulties caused by a gale, a sandstorm, helpless ferrymen, and ropes breaking every minute, it was a wonder that we and our mules were not carried out to sea in a body. Altogether our baggage was twelve hours in getting over the nine miles of flat ground between Acre and Caifa.

The weather now changed, and continued fine for our ride down the travel-beaten track that leads through Nablous to the capital of Palestine. Our only remaining difficulty was the mud, which made the Plain of Esdraelon almost impassable: now one mule, now another, stuck in the treacherous quagmire, but the 'delicate instruments' had been confided to an animal equal to his trust, which either kept its legs, or sank in the gentlest and most graceful manner. We reached Jerusalem on February 18th, having been twelve days on the road.

We quartered ourselves in the Damascus Hotel, which is fairly comfortable, and commands a fine view of the Mosque of Omar and Mount Olivet from the windows of the *salle-à-manger*. As soon as possible we enquired for Lieutenant Warren, hoping to deliver to him in person the case of 'delicate instruments,' and to hear how their internal organisation had borne the journey; but he had already left Jerusalem for the trip to the east side of the Dead Sea, which ended so sadly in the death of one of his companions, of Jericho fever.

The English Vice-Consul kindly accompanied me when I went to present a letter of introduction from M. Musurus (the Turkish Ambassador in London) to the local Pasha, who was most courteous, and promised to do anything in his power for us. Thus encouraged, we reflected what boon we should ask. We were all somewhat disappointed with the unadventurous character of a ride through Palestine, so little realising the common idea of Eastern travel, and were eager to seize the first favourable opportunity to escape from the beaten track between Jerusalem and Damascus. When, therefore, the map was produced, and the directness of a route via Jerash, Bozrah, and the 'Giant Cities' of Bashan was pointed out, my proposal to take that course was unanimously adopted. We had read Mr. Tristram's most interesting description of Jerash, and we purchased, at Jerusalem, Mr. Porter's sensational account of the ruins of the Hauran. We knew, therefore, something of the country we proposed to visit, and were aware that to pass from Jerusalem to Damascus by the east side of the Jordan, with all the impedimenta of a dragoman, was not a matter to be lightly undertaken. Travellers who, like Mr. Tristram, have of late years visited Jerash and Amman, have almost invariably paid large sums of money as 'backsheesh' to the Adwan and other Bedouin tribes of the Jordan valley; while those who, like Mr. Porter and Mr. Cyril Graham, have explored the wilds of Bashan have generally been Arabic scholars, and have travelled with little baggage. We could find no record of any traveller since Lord Lindsay, in 1837, who had gone through to Damascus by this route, although several had penetrated eastward from the Jordan valley as far as Bozrah.

Our dragoman, greatly to his credit, at once entered into and heartily furthered our plans, although he warned us of a fact we already knew, that an Arab escort was both an expensive and unsatisfactory luxury. An alternative, however, suggested itself. During the past year (1867) the Pasha of Damascus had made an expedition against the Trans-Jordanic Arabs, had thrashed them soundly, and taken prisoner one of the Adwan Sheikhs, who was now in durance at Nablous. The Arab power was in consequence somewhat broken, and the re-establishment of Turkish garrisons at Es-Salt and Bozrah kept the surrounding districts in more than the nominal subjection they had previously shown to the central authority.

Elias recommended us to have, if possible, nothing to do with the Arabs, but to ask from the Pasha a sufficient escort of Turkish cavalry to insure our safety. Just at the right moment he chanced to meet in the bazaars an old acquaintance, a sergeant of Bashi-Bazouks, Khasim by name. The pair discussed our plans, and Khasim entreated to be allowed to take us in charge. One morning our future guardian was brought, by appointment, to be introduced to us, and first impressions were most favourable. To describe his personal appearance would require the language of an Eastern story-teller; I can only catalogue his beauties like a slave-merchant. Khasim stood at least six feet two inches in height; he had fine features, and was of a fair but sunburnt complexion, with curly brown hair, and long tawny moustaches, which curled behind his ears. We fell in love with him at first sight, and were perfectly ready to promise that we would ask the Pasha to grant him leave to accompany us.

An opportunity of making the request offered before we expected it. We were sitting in the *salle-à-manger*, discussing our plans, when we suddenly observed a commotion in the street below. In another minute the master of the house dashed upstairs, in breathless haste, and announced ' His Excellency the Pasha,' who had come, attended by fifteen soldiers and six attendants, to return our visit. Unprepared for such an honour, we received him as well as we could, but it was not easy to get the coffee and sweets proper for the occasion on the spur of the moment. Nothing could exceed the Pasha's politeness; he accorded us any guard we might choose, and promised us letters to the commanders of the garrisons at Es-Salt and Bozrah.

We now definitely concluded our arrangements, and secured the escort of Khasim, who was to bring with him a second soldier: these two formed our guard to Es-Salt, where the officer in command would, we were told, give us further protection, if necessary.

For the last day or two of our stay at Jerusalem, we were the objects of much misplaced pity and well-meant advice. Certain undeniable facts were thrust down our throats at every public meal. We were reminded that Lieutenant Warren was at that moment paying the Adwan for permission to travel on the east side of the Jordan, we were treated to all the details of the bargain then being made, at the rival hotel, between Goblan, the young Sheikh of the Adwans, and two American gentlemen, who were anxious to visit Jerash, and all the threats which the former had uttered, on being told that some Englishmen meant to pass through his territory without paying blackmail, were repeated for our benefit.

Despite all this, we managed to keep up our spirits, and even to find a companion who was ready to share our luck, in Mr. Cross, an old Oxford acquaintance, who made a most welcome addition to our party. On Thursday, March 12th, we defiled, an imposing train, through the narrow streets of Jerusalem—Cross, Williams, and the dragoman armed with double-barrelled guns, Tucker and I with revolvers, and the two Turkish irregulars bristling with a whole armoury of guns, swords, and pistols. We rode over to Bethlehem, to my mind one of the most satisfactory of the 'holy places' of Palestine, despite the crowds of pert children, who, fearless of another Herod, demand 'backsheesh' with Egyptian pertinacity. As we rode on over the bare hills to the Convent of Marsaba, the beauty of the first view of the Dead Sea so roused François' enthusiasm that, with somewhat Irish brilliancy, he exclaimed, 'Vive la Mer Morte!' We slept in the convent.

Friday was a gloriously fine but very hot day. No one can fail to be struck by the views of the bright blue lake surrounded by red and yellow rocks, and the wastes of sand, every now and then relieved by strips of verdure. Some of the party, of course, bathed in the Dead Sea, and we lunched at the ford of the Jordan, which had as little the appearance of a ford as possible. A turbid stream three feet deep was pouring round the tree, under the shelter of which travellers generally make their midday halt. A hot ride, across a plain covered with brushwood, brought us to the modern Jericho. In the course of the evening a troop of villagers, men and women, came to dance before us; the women exhibited first, then the men, but the performances were very similar—a perpetual swinging of the body and clapping of hands, accompanied by a monotonous chaunt of 'Iwa backsheesh O Howadji!' The people of Jericho bear a very ill name, and we took the precaution to station François at the door of the second tent, to prevent robbery. While he was keeping a look-out in front, some rascal, peering through the opening at the back, where the sides of the tent join, saw Cross's watch lying on the bed close by, put his arm through, and abstracted it. Fortunately, the Sheikh of Jericho, Mahmoud, had been ordered by the Pasha to send two of his men with us from Jerusalem, as a pledge of his protection during our journey through his territory. The Sheikh's brothers were now in the village; their responsibility, therefore, was clearly fixed, and we sent off news of our loss next morning to the English Consul and the Pasha, by the dragoman of a Scotch

CAPTURE OF A STANDARD-BEARER. 13

party who were encamped near us on their way back to Jerusalem. In justice to the Turkish authorities, I must narrate the result of our letters. So effectual a pressure was put on the Sheikh, that he was compelled to disgorge his prey, and on our arrival (five weeks later) at Beyrout, we found the watch awaiting us.

Our ride on Saturday led us off the beaten track of eastern travel. We passed the mounds supposed to mark the site of Herodian Jericho, which a body of Lieutenant Warren's workpeople were employed in excavating. Their labour, as we heard afterwards, was attended with but trifling results.

Our track skirted the face of the hills on the west side of the Jordan valley—now crossing low spurs, now passing through flowery dells. After traversing a wide plain we approached the base of a bold hill, which in form reminded me of Snowdon; its sides were clothed in verdure of the most vivid green. By the roadside were seated a group of twenty Bedouins armed only with clubs. To our intense surprise, Khasim dashed in amongst them, and pounced on one ragged old fellow. The man selected endeavoured, in vain, to kiss his captor's hand and soften his heart; in a minute his 'kefiyeh' was plucked off his head, and his hands were tied with it behind his back. Khasim then galloped off in pursuit of the rest of the party, who had scattered in all directions; he soon returned with a second prisoner, and we rode on, driving the two men before us.

We were naturally anxious for an explanation of the scene, but it was some time before we could come to a clear understanding of the facts of the case. We gathered at last the following particulars. In the war last year the Turks took away their arms from some of the Adwan, and strictly forbad them to appear on the west bank of the Jordan. The party we had come upon were thus on forbidden ground, and were doubtless on the look-out for some defenceless donkey-rider going down to Jericho, whom they might rob. The old gentleman first seized had been the standard-bearer of the tribe during the war, and was a well-known reprobate. After driving our prisoners for several miles, as a warning to them not to be again found on the road, we, reflecting that the men might be an awkward encumbrance on the other side of the river, interceded for them with their captor; the Bedouins were liberated, and, having sufficiently demonstrated their gratitude to us by repeatedly kissing our boots, made off in the direction taken by their companions.

The ford of the Jordan we were now approaching is on the direct road from Nablous to Es-Salt, and is guarded by a few Turkish soldiers, who keep in repair the old ferry-boat, which has been stationed here to maintain the communication of the outlying garrison at Es-Salt with the rest of Palestine. The river flows in a deep trench, a quarter to half a mile broad, and at least 200 feet below the level of the rest of the valley. At the foot of the sharp descent, on a knoll overlooking the turbid stream, we found the tents of the American gentlemen, who, like ourselves, had made up their minds to visit the east side of the Jordan. They had contracted with young Goblan to provide an Arab escort for thirty napoleons—a moderate sum compared with those paid to his father by former travellers. The old Sheikh of the Adwan had, however, failed to appear, according to the contract, to ratify his son's bargain, and our acquaintances naturally hesitated to cross the river without him.

It was still early in the afternoon, and we ordered our baggage forward to cross at once, while we spent a pleasant half-hour in the tent of the Americans. They were most luxuriously provided for by their dragoman, a young and inexperienced man, who seemed somewhat terrified at the prospect of carrying his elaborate *batterie de cuisine* among the Arabs.

The transit was, as usual, a long business, and was made really troublesome by the swollen state of the river, which had lately overflowed its banks and cut off the ferry-boat from the shore by creating between them several yards of mire and water, across which we and all our luggage were carried on men's backs. François was a heavy load; his porter was not up to the work, and the unlucky burden was deposited in the thickest of the mire. He however, as usual, was not at a loss for consolation, and prided himself on being the only one of the party who had fulfilled the duties of a pilgrim by immersion in the Jordan.

CHAPTER II.

BASHAN.

The English Soldier—A Mountain Ride—Es-Salt—Lost on the Hills—The Jabbok—Camp of the Beni-Hassan—Suppressing a Sheikh—The Oak Forests of Gilead—The Tablelands—An Uxorious Sheikh—Derat—The Roman Road—The Robbers repulsed—Ghosm—Bozrah—Honoured Guests—A Ramble in the Ruins—Kurriyeh—Patriarchal Hospitality—Hebran—A Stone House—Kufr—Ascent of El-Kleib—Suweideh—Kunawat—Noble Ruins—Shuhba—Hades on Earth—Visiting Extraordinary—The Lejah—A Lava Flood—Ahireh—Khulab—A Rush to Arms—The Stolen Mule—A Village in Pursuit—Mismiyeh—The 'Giant Cities' are Roman Towns—The Wrath of the Boys—A Friendly Salut—Keeweh—Entrance to Damascus.

OUR tents were pitched, close to the river, in a picturesque situation on the eastern bank. In this our first camp beyond Jordan, we felt, if not all the emotions so eloquently described by the author of 'Eöthen' on finding himself in the Arab territory, at least a pleasant sensation of having escaped from the everyday track of travel, and of being on the edge of a fresh and unspoilt country.

During the evening our dragoman was exposed to the tender solicitations of Goblan junior (or 'young Gobbler,' as Williams preferred to call him), who had, when at Jerusalem, declared that we might cross the Jordan, but that our coming back again was a different matter. He was now perfectly civil, but represented that we were robbing his tribe of their prescriptive dues, by refusing the escort they would be happy to furnish, and that any harm which might happen to us would be on our own heads. Our

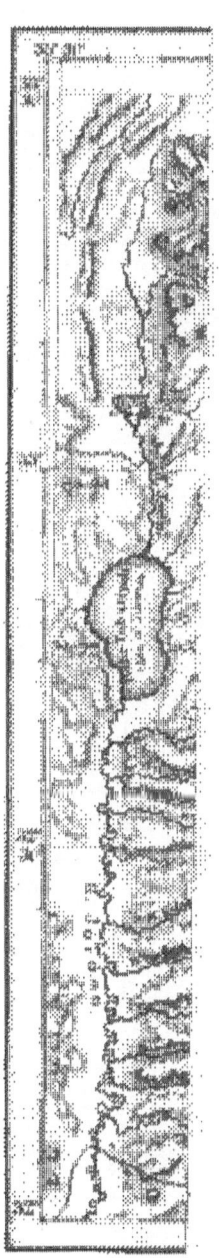

minds, however, were already made up, and we turned a deaf ear to his arguments.

March 15th.—Early in the morning Sergeant Birtles, a bronzed English soldier, the aide-de-camp of Lieutenant Warren, rode into our camp, having travelled all night from Nimrin, where the exploring party was detained by the illness of one of their number. Sergeant Birtles had ridden twice within the week to Jerusalem and back, to procure a nurse and necessaries for the invalid. He was entirely alone, and seemed thoroughly to despise the dangers of the road, as he had proved on the first occasion, by returning with a sister of mercy under his sole charge.

The Sergeant had been up to Jerash with Lieutenant Warren some months previously, and gave us friendly warning against the Sheikh of Suf. and his men, who while with them had begun to show off their pranks, and had required to be checked by a display of revolvers. These were the same villains who plundered Mr. Tristram in 1864. After half-an-hour's talk with us, Sergeant Birtles crossed the river, hoping by hard riding to reach Jerusalem the same night. We also mounted our horses, but our train was scarcely in motion, when two of the mules stuck in a swamp, and had to be unloaded. A well-marked track led from the ferry to the foot of the bluffs which bound the river-bed. Tucker and Williams galloped on ahead, a proceeding which called forth a remonstrance from Khasim, who insisted that, on this side of the river, it was unsafe to divide the party. A short climb through a curious, apparently waterworn, ravine brought us on to the plain, where we met two picturesque wayfarers—a handsomely-dressed Arab with a servant in attendance, both bearing long spears. They returned our greeting with a contemptuous scowl. A little further on, we received an enthusiastic 'bonjour' from a Turkish lieutenant on leave from Es-Salt, who was not a little pleased to display his slight knowledge of French.

Our track turned southward, along the opposite side of the valley to that which we had ridden up the previous day. The ground was bright with scarlet anemones, which tinted the hillsides a mile off; other wild flowers grew in almost equal profusion, although they did not produce such a striking distant effect.

At last we turned sharply to the left, and began to climb, by steep zigzags, the bare hillside. At every corner of the road we extended our horizon; the higher ranges of Central Palestine rose behind the hills which dominate the Jordan valley, and the Dead Sea came into view in the south. After a considerable ascent the path entered a glen, and wound round the hillsides at some height above the dry torrent-bed. During our midday halt our sportsmen went off in pursuit of partridges, but came back empty-handed. On reaching the head of the glen, we faced more zigzags, up which the laden mules climbed laboriously; they led us to a high brow, projecting from the main range, which commanded the finest view we had seen in Syria. We overlooked the whole Jordan valley from the Lake of Tiberias to the Dead Sea. A corner only of the former was visible, but we could see the whole basin of the great salt lake, and trace out the long peninsula of El-Lisan, with the far-off southern shore beyond it, through the heat-haze which always rests on these strange waters.

The hillsides became more broken, and the dwarf oaks which clothed them added to their picturesque features. Circling round to the south of the highest portion of Jebel-Jilad, we crossed two of its spurs with scarcely any fall in the road between them. From the second ridge a short but sharp descent led to Es-Salt, the ancient Ramoth-Gilead. The town is built in a nook among the hills, on steep sun-baked slopes, uglily picturesque, if one may use such a phrase. We found a pleasant camping-ground on a grassy ledge of the slope opposite the town.

Our starting-point in the morning had been more than 1,000 feet below the sea-level, and the ridge we had crossed is 3,676 feet above it, so that we had made an actual ascent of 4,676 feet from the Jordan valley to the pass.

After a short rest we walked over to the opposite hill, to call on the Bimbashi (Colonel) in command of the garrison, who received us in his tent with great politeness. He said there was no difficulty in our going to Bozrah; that if we liked we might have a company of horse, but that he should be quite content to send his own wives under the escort of the two men we had with us. He pressed us to stay at Es-Salt a day or two, which we declined, and withdrew after a somewhat lengthy interview. All our conversation had to pass through two interpreters, being translated by Elias into Arabic, and into Turkish by a servant of the commander. The latter process was doubtless unnecessary; but the Bimbashi's dignity would have suffered, in Oriental estimation, had he not had his own interpreter.

On our return we found the camp in commotion. Cross was the only member of the party who strove to keep up an irreproachable exterior in the wilderness; he generally rode in a somewhat sporting costume, crowned by a white turban, the construction of which cost Elias much time and anxiety every morning. On the present occasion Cross had entrusted his greatcoat to Elias in the morning; the latter ignorant or careless of the respect due to a garment better fitted for Pall Mall than Palestine, had stuffed it into a saddlebag, whence it emerged, naturally sadly creased. We found its owner severely reprimanding the carelessness of the dragoman; but the effect of the rebuke was rather lessened by the untimely mirth excited in Williams by the forlorn appearance of the once shapely coat.

In the course of the evening the Bimbashi sent us a sheep.

March 16th.—We broke up camp early, and rode down the glen under the town, through a ruinous street, passing tombs and some broken columns below the modern houses; the bottom of the glen was well irrigated and cultivated. We soon turned up a lateral ravine, opening on the left, and followed it nearly to its head, through scenery which was quite savage for this part of Syria. Big boulders lay about, and a stream foamed and brawled amongst them. A steep climb led us up to a broad grassy tableland, forming the southern watershed of the Jabbok valley. We came occasionally on circular ponds, exactly like those to be seen on the South Downs, of which the scenery constantly reminded us. The road was continually up or down-hill, till we left on the right the track to Amman, and sweeping round a brow to the north, gradually descended into a broad oval basin, environed by hills—the greater part sown with corn, although no village or inhabitants were visible. The drainage of this curious hollow, which is called by the natives El-Buksah, or the Little Plain, finds its way through a narrow opening to the Jabbok.

Having passed a ruin, apparently that of a small building, surrounded by a courtyard, we traversed the whole length of the basin (about eight miles), and mounted the ridge dividing it from the Jabbok valley. The country now became more wooded, the principal tree being the dwarf evergreen oak. We sat down to lunch, on a spot commanding a lovely view northwards, over a broken and richly-wooded landscape, in the centre of which, in an upland plain, on the further side of the Jabbok valley, we could distinguish the golden-coloured columns of the great temple at Jerash. In the foreground were some picturesque peasants, natives of Es-Salt, engaged in ploughing the ground with the most primitive of agricultural instruments. We rode on in a north-easterly direction, across an undulating country; from the position of the Jabbok valley, I fancied we must be going too far east; Elias, however, turned a deaf ear to my remonstrances, and went off in futile pursuit of some partridges. The baggage mules lingered, and for some time our cavalcade was separated into three detachments, each of which had, more or less, lost its way. A good deal of time, and much temper, was expended before we rejoined our scattered forces, but luckily no foes were at hand to profit by such bad generalship. The glen through which the Jabbok flows is narrow, steep-sided, and not accessible at all points for laden beasts. As we descended the slopes, we passed many prostrate columns and blocks of marble, the relics of some town unknown to fame.

Before making the final descent to the stream, Elias and I trotted off to reconnoitre the country, and look for a good camping-ground. The day was already far spent, and we were not best pleased to discover that the grassy plain we had previously fixed upon as our halting-place, at the junction with the Jabbok of a northern affluent, was occupied by an Arab encampment. We determined, notwithstanding, to put a bold face on the situation, and adhere to our original plan. We were too near to escape the notice of the Arabs, and the best course was to show that we were not afraid of them. We crossed the Jabbok, a clear trout-stream hidden in a dense thicket of oleanders, and rode down its right bank half a mile to the plain. Halting on its verge, we began to unload our mules, while Khasim went off on an embassy to the Sheikh's tent, conspicuous amongst the rest by its size. The fear of European weapons, and the unknown force of government at our back, joined to the Oriental dislike to attack strangers who have assumed the character of guests, overcame even the covetousness of a needy horde of Bedouins, and a deputation soon returned Khasim's visit, bringing as presents some milk and a lamb. For the latter it was afterwards suggested payment would be acceptable.

Our hosts turned out to be of the Beni-Hassan tribe—one formerly of great power, but now down in the world. Their encampment consisted of thirty-five long black tents, each holding about ten men, besides women and children. It was already evening, and the flocks were being gathered in; tall camels strolled listlessly about, cows placidly awaited their milking-time, sheep and goats hustled one another down the slopes, wiry little horses grazed, or were picketed, near the tents, and an odd donkey or two brayed a fussy welcome to his brothers in our train, who were not slow to return the greeting.

While our dinner was preparing, we were surrounded by the most ill-favoured crowd I ever saw. The villanous expression of countenance common to almost all the men reminded me of the Sepoy faces, as they were drawn in the illustrated papers, at the time of the Indian Mutiny. The Bedouin dress, the long burnous, and keflyeh or scarf round the head, though picturesque, did not lessen the savage aspect of the assemblage. All our small belongings were objects of perpetual wonder—in particular,

Williams's carved pipe and Cross's kid gloves. Never before had the Beni-Hassan seen a man with such a peculiar skin, or one so readily put on and off. The revolvers, which appeared to go off for ever, came in for their due share of admiration and awe. We had some difficulty in keeping the tents clear, but it was necessary to draw a line somewhere, and we sternly refused admission to any but the two chiefs. The sheep was cooked entire, and our muleteers, with a select circle of Beni-Hassan, kept up the feast round the camp-fire till a late hour.

March 17th.—The night passed peaceably. At breakfast a stork was brought in which Williams had wounded by a long shot the night before. The poor bird's wing was broken, and he hopped about, pursued by the Arab urchins, in a way that was both ludicrous and painful. Elias had been sharply reprimanded for his wanderings on the previous day, and warned that this kind of thing must not be repeated; he now came with pride to tell us that he had arranged with one of the Beni-Hassan to conduct us to Jerash. He took great credit for his choice, having, as he said, secured the greatest robber in the tribe. There was wisdom in this odd recommendation, as the man who had stolen most sheep was, by implication, he who best knew the roads and bye-paths to the neighbouring villages.

Our guide led us up a dell separated by an intervening ridge from the stream which entered the Jabbok close to our encampment. The country was green and well-wooded, and the soil was free from the detestable crop of stones which Palestine everywhere produces. As we climbed out of the deep valley, the ridges of Jebel Ajlun appeared behind the round top of a lower hill, Neby Hut, crowned by the white tomb of some Moslem saint. Another deep and rugged hollow, the edge of which we skirted, lay beneath us on the left. The landscape reminded me much of some of the finer parts of South Wales, but its beauty was marred by the low clouds which scudded across the sky, and promised us a wetting before long. Near a beautiful fountain, encased in broken masonry, and ornamented with rich evergreen shrubs, the source of the before-mentioned stream, we passed some ruins too dilapidated for our unskilled eyes to make anything out of them. They seemed, for the most part, to be the remains of small houses built against the rocks, with caves at their back, which had served for cellars or storehouses. We rode up a succession of picturesque glades, opening one out of another, till we reached a ridge north-east of the tomb-crowned hill, and suddenly saw beneath us, close at hand, the columns of Jerash.

The scene was very striking : before us were the remains of a noble Roman town, its ruined walls four miles in circumference, not only traceable, but in places almost intact; its public buildings still so perfect that, looking round, one could say, ' Here is the theatre, there the circus, there the baths, there the colonnaded High Street, there the later Christian cathedral '; for it was only after three hundred years of Christian civilisation that the Arabs laid waste the city. The fertile land around is still as capable as ever of cultivation; but a long period of insecurity to life and property has been the ruin of Syria, and now not a single inhabitant is to be found within the circuit of the ancient Gerasa.

We rode through the walls near their south-eastern angle, and, passing the massive ruin of a bath, crossed the oleander-fringed brook which runs through the centre of the deserted city. A very convenient site was selected for our camp, in the vaulted chamber of a second bath, where the tents were sheltered from the thick drizzle which had begun to fall. After an early lunch we set out to explore the ruins, which are fully described by Mr. Tristram, and in Murray's 'Handbook to Syria.' We went first to the magnificent Temple of the Sun, the remaining columns of which, standing on elevated ground facing the east, are conspicuous in all distant views of the city. Near them, in the side of the hill, is the largest theatre. Returning to the great street, we stopped to admire the exquisite carving of a richly-decorated gateway, and then proceeded to the 'Forum,' an oval space surrounded with columns. On the brow above it, near the southern gate of the city, stand another temple and theatre. The latter is wonderfully little injured by time; the stage is almost perfect, and very tastefully decorated. When will some photographer carry his camera across the Jordan, and reap the rich, and as yet almost untouched, field which awaits him amidst the ruins of Ammân, Jerash, and the Hauran?

Outside the town, on the top of the ascent from the Jabbok valley, stands a fine though florid triumphal arch, between which and the city is a circus. We went down to the banks of the brook in search of game, and then, retracing our steps, found a pretty waterfall, and the remains of an ancient mill. Having re-entered the town, we crossed to the eastern quarter by a fine bridge of three arches, and explored its comparatively unimportant ruins.

In the course of the evening we had a visitor. We were engaged in a rubber of whist, when Elias came in and announced that the Sheikh of a neighbouring village requested the honour of an interview. We enquired his name, and simultaneously burst out laughing when told it was the Sheikh of Suf, of whose iniquities we had been reading Mr. Tristram's account not ten minutes before. We declined to see him, but agreed to look over his testimonials. They extend over thirty years, and are probably unique. One Englishman writes thus :—' On no account have anything to do with the bearer of this; he is a thorough villain and awful liar.' The other writers are more guarded in their language, but they all give, in effect, the same advice. After amusing ourselves by a thorough inspection of the documents, we returned them with a message, ' that the Sheikh of Suf's character was already too well known among Englishmen to require to be supported by testimonials; that for ourselves we were travelling under the protection of the Government, as he could see by our being accompanied by a Bashi-Bazouk; and that if he had a commission from the Pasha to escort us, it was well—otherwise we must decline his services.' On the receipt of this message the Sheikh grew angry and violent, and attempted to force an entrance through the archway of the baths to our tent. This was blocked, except in one place, by stones, and Khasim, who had stationed himself in the gap, put a stop to the Sheikh's proceedings by seizing him by the beard, and waggling his head with one hand while he boxed his ears with the other. After this, Yusuf and his followers rode off, not unnaturally in a huff; while Elias shouted after him, that as he had chosen to come over, we should hold him personally responsible if anything was stolen in the night.

To avoid any such misfortune, the mules and horses were picketed in a ruined chamber close by. We could not help speculating on the chance of old Yusuf seeking revenge, and congratulated ourselves on the strength of our position, which was protected on all sides by broken masses of wall. The night passed quietly, except for an alarm caused by my jumping out of bed with the idea that I had been hit over the head. When a light was procured, and it was discovered that one of the wooden poles, used to stiffen the sides of the tent, had suddenly flapped inwards and fallen across my face, I was roundly abused by my companions for so needlessly disturbing their slumbers.

March 18th.—To our great delight the morning was fine, for we dreaded a repetition of the weather from which we had suffered on first landing at Beyrout. We had wished to strike across as directly as possible from Jerash to Bozrah, passing through Um-el-Jemal, where there are extensive ruins, which were visited by Mr. Cyril Graham. Our Beni-Hassan guide, however, seemed unacquainted with this road, and threw obstacles in the way of its adoption, declaring it to be 'desert,' and infested at the present moment by Beni-Sakhr. We agreed, finally, to take a more northerly but somewhat circuitous route, by Er-Remtheh and Derat.

We were on the point of starting, when four horsemen from Suf rode up; the Sheikh was not amongst them. The newcomers told Elias they were going to act as our escort; he, by our instructions, replied, that if they came with us at all, they would come as far as Damascus, and there be handed over to the tender mercies of the Pasha. This prospect was too much for them, and they soon rode off, to be seen no more. We had been careful to leave the men of Suf to suppose that we should follow the route taken by Mr. Tristram to the Lake of Tiberias; and if the Sheikh and his friends were prepared to do us any mischief, they probably lay in wait on that road. We set out in the opposite direction, our course being at first nearly due east. From the slopes behind Jerash we had a most beautiful view of its columns. We noticed the ruins of a small temple situated a mile higher up the valley, and passed several fine sarcophagi, lying overturned and empty amongst the trees.

For some distance we rode through open glades, where

the Roman road was constantly visible, running parallel to the more erratic tracks of modern travellers. When we began to ascend the side of Jebel Kafkafka, the scenery became most picturesque; grey crags jutted out from the hillside, the forest trees grew larger, and the foliage, although evergreen oaks and firs predominated, was more varied. The narrow glade by which we climbed the hill wound every moment between rocks, and branches of trees overarched and shut in the vista. Our long cavalcade lent life to the scene, which was a complete realisation of one of Salvator Rosa's pictures. It was impossible not to remark what an admirable place this would have been for an ambuscade, if the men of Suf had had the pluck to waylay us; but beyond a solitary shepherd and his flock, we saw no living creatures except eagles and partridges. The sportsmen of the party knocked over several of the latter. In the wood the canteen mule, who was subject to fits of obstinacy, charged a tree and upset his load. Five minutes after this incident we reached the top of the pass, the height of which must be considerable, as we overlooked Jebel Osha and the other hills round Es-Salt. The view in this direction was fine; the depths of the Jordan valley were hidden, but we easily recognised the hills round Nablous and Jerusalem. In the east, lower wooded ranges shut out all distant view. Grass rides, at right-angles to the track we were following, branched off constantly through openings in the forest; following one of them, we descended into the Wady Warran, a long and tortuous valley, the sides of which are clothed with park-like timber, in some instances of very large size. The landscape was entirely unlike ordinary Syrian scenery, and we could constantly have fancied ourselves in the wilder part of an English park, but for the absence of running water in the bed of the winter torrent, which, even at this early season, contained only occasional pools. As we advanced, the hills become lower and less wooded, until we at last emerged upon a vast undulating plain, the nearer part green grassland, the more distant, rich brown loam, recently ploughed. The afternoon was hot and hazy, and from time to time fine mirage effects were produced by the state of the atmosphere. Er-Remtheh, the only village in sight, was a long way off, but was conspicuous from its position on a spur projecting from the low range of Ez-Zumleh which bounded the north-eastern horizon. On entering the plain, our Beni-Hassan guide requested to be allowed to return; he had committed some robberies lately in the district we were entering, and was afraid to be caught there. We gave him eleven francs for his services, and let him go.

Soon after he left, a long train of camels met us, laden with black basalt millstones, which seem to be the principal manufacture and, except corn, the only export of the Hauran.

After leaving behind Jebel Kafkafka, the country was quite bare. With our goal full in sight, we pushed on rapidly, and brought a long but fast day's ride to a conclusion in capital time. Mr. Porter says that the dwellers on the Haj road (the route of the yearly caravan from Damascus to Mecca) are remarkable for their fanatical hostility to Europeans; we felt, therefore, some doubts as to what the character of our reception would be. A meadow north of the village, and near the pond which supplied the only water in the neighbourhood,* was selected for our camping ground. After our tents were pitched, two of the party, finding that the people appeared a remarkably mild and

---

* Eastern travellers, who object to swallowing as much mud and insect life as water, should provide themselves before leaving England with portable altars.

inoffensive though inquisitive race, walked off to explore the village. It was of the usual ruinous character, and we found nothing of any interest, except the base of a black basalt column. At a distance of about five miles to the north-west, the ruins of a large village were conspicuous on the plain. In the evening the Sheikh of the place, a good-looking merry old gentleman, visited us in our tent. Strange to say, he refused a pipe, on the ground that his three wives would not let him smoke. With an absence of the usual Oriental reserve, he entered into some amusing details of his domestic arrangements. His wives had each of them, he said, cost him 35,000 piastres; his last acquisition was the dearest of the three, and he was contemplating adding a fourth (the full number allowed to orthodox Mussulmen), still dearer. All these 'dear things,' together, appeared to be somewhat too much for the old gentleman, and he seemed relieved to escape from home and chat with us, even though his fears of being accused of smelling of smoke prevented the enjoyment of a pipe.

We asked when he had last seen European travellers. The Sheikh replied that, three years before, a party, including some ladies, had passed, on the way to Damascus, by the direct road through Mezarib, the residence of the Turkish governor of the Hauran.

During the night our second Turkish soldier, who had been left behind at Es-Salt, and had ridden on to Amman instead of Jerash, came into camp. He was horribly afraid of our anger at his involuntary desertion, and was in an abject state of contrition.

March 10*th*.—Before we started, a caravan of 400 camels, laden with corn for shipment at Acre, passed by. It struck us as curious, that a land described, by the latest authority, as 'utterly desolate,' should be able not only to feed its inhabitants, but to send away such quantities

of grain. What we saw later in the day explained the
mystery, and proved how far preconceived ideas may lead
a writer into misrepresentation. It was interesting to
watch the long train defiling endlessly over the dewy plain.
The camels were attached, head to tail, in batches of about
twenty each, headed by their drivers; and lively little
donkeys, bestridden by boys, trotted alongside the great
solemn beasts. As the sun rose the mists rolled away, and
before we started, the long snowy ridge of Hermon stood
out bright and clear against the blue sky. It looked
very imposing, though at a distance of at least fifty
miles.

Our track led in a north-easterly direction, over a chain
of low hills, where we encountered another train of laden
camels, and in a short two hours from Er-Remtheh, we
approached Derat, a large inhabited village of black basalt
houses, each surrounded by a high wall—the lower six feet
of solid masonry, the upper part built of loosely-piled
stones. Dead animals—dogs, horses, and mules—lay about
the streets in every stage of decomposition, offending equally
the senses of sight and smell.

The principal ruin at Derat is that of a Christian church.
A large quadrangle, surrounded by cloisters, leads into a
low-roofed edifice supported by numerous columns. Architecture there is none, but the building is quite a museum
of capitals stolen from older edifices. Derat is a good
specimen of the modern architecture of the Hauran. The
houses are mere piles of ruins. Having passed, perhaps
by a stone door, through the high outer wall, you find
yourself in a small open space, whence steps lead downwards into sundry burrows, half excavated in the earth,
half built up and roofed in with unhewn stones. Anything
more sombre and unhomelike than these piles of black
basalt boulders it is impossible to conceive. It would be

too ridiculous to imagine the sentiment of 'Home, sweet home!' entering into the head of an inhabitant of Derat, and most of the villages of Bashan are similar to it. The town stands above a deep ravine, at the bottom of which flows the Yarmuk, the ancient Hieromax, which we crossed by a ford close to the river-side, where we found two fine sarcophagi, one ornamented with a human bust, the other with lion's heads. A little above the ford the Hieromax is crossed by a Roman aqueduct, which, spanning the ravine by a very bold bridge, is carried at least eighty feet above the water. Our course now lay almost in a direct line east, across a great plain; on the further side rose the group of Jebel Hauran, with the conical peak of El-Kleib, conspicuous in their midst, and the castled crag of Salkhat on a southern spur. We followed exactly in the line of the old Roman road, the pavement of which was in places intact. After passing the ruins of Gharz, where we failed to discover anything remarkable, we rode on through cultivated land, dotted at intervals by black villages. Presently ten horsemen appeared in the distance; suspecting their character, we at once fell back on the baggage. The Bedouins advanced, then split into two bodies, and wheeled round on either flank of our party; after an interval of apparent indecision, they again united their forces. By this time we were within 300 yards of them, and could judge from their demeanour that they were out on no lawful errand. Khasim rode forward to meet them, when one of the party (apparently the chief) saluted him, and they held a parley together in high tones. As they rode alongside, we admired the skill with which Khasim made his horse dance round the other, constantly keeping his rifle pointed in the face of the Bedouin chief. After a few minutes' conversation, Khasim rode back to us, while the Arabs wheeled round and scampered away. We found out that they were

'Saba,' a branch of the great Anazeh tribe. The gist of their parley with Khasim was, that they requested him and his soldier to stand by while we were plundered—an offer which brought down on them such indignant menaces from our men, that having reckoned our force, and seen that, though superior in numbers, they were far inferior in weapons, they deemed it prudent to let us go our way in peace.

We recrossed to the left bank of the river by a Roman bridge in good preservation. The country was now more undulating, and exceedingly well-cultivated, great pains having been expended in irrigating the soil thoroughly, by means of a complicated system of water-channels. We had some difficulty in persuading our men to push on to Ghusam. Mohammed, the second Bashi-Bazouk, was anxious to turn off the road to some village with which he was acquainted; but we persisted in riding on, and in course of time Ghusam appeared, though not in the position assigned to it on Van de Velde's map. The view from our camping-ground was magnificent. We were in the centre of a vast plain, bounded on the west and north by the mountains of Gilead and Hermon, on the east by the Jebel Hauran, and stretching on the south into the 'vasty wilds' of Arabia. In the distance the black walls of Bozrah glittered in the evening sunshine, like some enchanted city of the 'Arabian Nights.' In our walk round the place, we noticed the remains of a Christian church, with the cross carved on the walls, some reservoirs, and an old stone house, answering better than anything we had yet seen to Mr. Porter's descriptions. A fine pair of folding stone doors, which were thrown wide open at our approach, gave access to the Sheikh's courtyard, into the walls of which were built well-executed carvings of a vine and grapes, the common Christian emblems. On a raised terrace we found carpets and coffee prepared for us; the conversation consisted chiefly of an interchange of the usual laborious compliments, but in the course of it we learnt that the marauding party we had met on the march had stolen two hundred sheep from the next village. The scoundrels must have been sorely grieved when they found that 'very strong man' Khasim and twenty-five barrels interposed between them and the luggage of the Giaour. On our return the Sheikh posted a guard round our tents, and sent us a lamb. It was an uncommonly pretty little animal, and Cross was so overcome by its winning ways that he got up in the night and let it loose; but his kindness was useless, for the lamb was found in the next field in the morning, and was ruthlessly despatched and made into cutlets by Mohammed the cook.

March 20th.—This morning we had a long argument with Elias. He had come to me with several small extra charges since we crossed the Jordan, and now expected us to pay the 'backsheesh' in return for the Sheikh's hospitality. This we declined to do, and produced our contract, which was fortunately precise on this question. Wishing to avoid a prolonged altercation, I told Elias he must make a list of all extra expenses he wished us to pay as they arose, and on arrival at Damascus we would show it to Mr. Rogers, H.B.M.'s Consul, and abide by his decision on the subject. This strategy was entirely successful, and from that moment we never heard anything more of the claim for extras.

A ride of two hours and a quarter over cultivated plains brought us to the walls of Bozrah. Passing, on the left, a fine triumphal arch, we rode straight to the gates of the fortress, over which the Turkish flag was flying. An application to the commandant procured us immediate admission, and a practicable passage having been found for the mules, we pitched our tents on a level plot in the interior of the Roman theatre. The ruins at Bozrah having been lately and well described, I shall make but a short reference to our rambles through the deserted city.

Having first visited a graceful group of columns, we followed one of the principal streets, which was easily traced by a colonnade of Roman date, filled in with the remains of later buildings, to the Great Mosque. This must have been a noble structure, though, like so many other of the Saracenic masterpieces, it is built mainly with the materials of older edifices. The marble monolithic columns, some of which have Roman inscriptions on them, are superb. There is a good general view from the top of the tower attached to the mosque; I nearly got shut up there, owing to the swinging-to of a stone door, which it required all our united strength to reopen. On the opposite side of the street was a bath, the walls of which are decorated tastefully with inlaid squares of Greek pattern. In this quarter are the principal Christian churches. The cathedral must have been a fine building, and is exceedingly interesting, as a specimen of the way in which classical architecture adapted itself to the new religion. The external wall is square, the corners being occupied by four chapels; internally the building is circular, with a lofty dome. Frescoes are still traceable in the aisle. The other two churches are of smaller dimensions, and of that unlovely style of architecture which has been rendered familiar by the London churches of sixty years ago, of which they at once reminded us. In one of these buildings, however, the roof was supported by a pair of noble arches. Passing through the part of the town where the modern population burrows miserably among the ruins of ancient splendour, we came upon a second fine triumphal arch, and the remains of a palace. We found ourselves finally near a large reservoir, in which two of the party bathed, while the others returned to the Castle, now close at hand. To our great amusement the sentinels presented arms whenever we passed in or out. The interior of the theatre, where our tents were pitched, has been more than half filled by a gigantic storehouse, but the upper tiers of benches are in splendid preservation. The stonework is very neatly finished, and the decorations of the stage struck us as being in less florid style than most of the Roman remains in this part of Syria.

Various theories have been broached to account for the strange conjunction of a theatre and a fortress. Should I hazard a guess at the solution of the problem, it would be, that the theatre stood originally outside the walls of the town, and that when the more frequent visits of the Arabs, and the failing strength of the legions, rendered it liable to injury, the still-existing fortifications were built round it. At night a sentry was posted outside our tents, but the quiet of the dark hours was undisturbed except by a furious storm of thunder, lightning, and rain. Cross, always ready with sympathy for man or beast, pitied the sorrows of the poor man on guard, and having hunted up some piastres from his coat-pocket, he suddenly appeared in his night-shirt at the tent-door, and presented them to the astonished soldier.

March 21st.—Early in the morning, we climbed up to the battlements to enjoy the view of the plain. Large flocks were being driven out to pasture; as we watched them, and gazed over the wide expanse of cultivated land we had ridden through from Derat, we were naturally led to contrast the facts under our eyes with the desolation described by Mr. Porter, and to indulge in a hope —which even the most ardent enthusiast for the fulfilment of prophecy might share—that better times may be in store for Bashan.

Before leaving Bozrah, we sent to thank the commandant, whom we had not yet seen, for his courtesy. We were at once invited to visit him in his quarters, an airy little room on the house-roof. The commandant expressed great disappointment at our short stay, said that he had meant to offer us an entertainment, and excused himself for not having called on the previous day, on the ground that he thought we should be tired after our journey, and prefer to repose. He prepared sundry documents for the villages of the Jebel, which were handed over to Khasim. I do not fancy they did us much service, for the Druses pride themselves on maintaining a practical independence of the Turkish Pacha at Damascus, and are little disposed to obey the orders of his lieutenants.

We were now within the borders of a district which has acquired great celebrity from the extent and peculiar character of its ruins, and has been recently brought into prominent notice by the well-known Syrian traveller, Mr. Porter. A perusal of his pages had set before us the exciting prospect of seeing whole towns, deserted indeed, but so little ruined, that they might be inhabited again at a moment's notice, although said to be of an age compared to which Pompeii may be considered a modern city. We naturally laid our plans so as to include the places considered most noteworthy by our predecessors, and arranged a zigzag route by which we might in three days reach the northern extremity of the Jebel Hauran, visiting Kureiyeh, Suweideh, and Kunawat on the road. From Shuhba we intended, if possible, to follow the plan which Mr. Porter found impracticable, and to ride through the centre of the strange volcanic district known as the Lejah, and celebrated equally for the bad character of its roads and its inhabitants. Our first day's journey was a short one, for we did not mean to push beyond Hebran. As we rode out of Bozrah, we passed several small reservoirs. On the way, Elias told us an amusing story of native manners. The head-man of the village had the previous night given a feast to our muleteers. One of them, a Christian of the Lebanon, ate with a fork. 'Mashalla!' exclaimed his Bozran host, 'what a brute; he has not yet learned to eat with his fingers!'

We rode across a plain strewn with volcanic boulders, with patches of cultivation between them. Here, for the first time, we saw the Druse women, with their extraordinary horns and long white veils. The latter only cover one-half the face, the division being made vertically, so as to show one eye and cheek, instead of both eyes and nothing else, in the Egyptian fashion. The Druse style leaves room for a good deal of coquetry, and the girls with any pretensions to good looks are at no pains to conceal them; but beauties are rare in the Hauran, and the ugly women are, much to the traveller's relief, uniformly bashful.

After fording a clear Welsh-like stream, one of the feeders of the Hieromax, we rode over a perfect wilderness of stones into Kureiyeh. It was a marvel how our animals kept their legs on such ground, but it takes a great deal to puzzle a Syrian horse.

Under its ancient name of Kerioth, Kureiyeh is one of the places distinguished by having had a special judgment pronounced on it by Jeremiah. We explored its ruins on foot, and found an old tank, beside which is a curious edifice, supported by stumpy columns. We saw no stone doors equal to those at Ghusam, and the houses were all more or less dilapidated. On the whole, though we strove to repress our feelings, we were decidedly disappointed with the first of the 'Giant Cities.' I thus recorded, on the evening of the same day, the impression made on us by the famous stone-houses attributed by some recent writers to the Rephaim mentioned in Deuteronomy:—'Among many houses, the comparatively recent date of which is evidenced by fragments of Roman sculpture built up into the interior walls, a few of earlier times probably exist. These may be of the time of Og, or they may not; there is nothing to show they were built by giants.'

The Sheikh of the village, a powerful Druse chieftain, was away, but his steward pressed us hard to stay to partake of a sheep, saying, 'What will my lord say, when he returns, and finds travellers have passed his door without tasting food? He will be angry with me, and I shall have nothing to answer him.' Throughout the Jebel Hauran we were almost oppressed by the hospitality of the inhabitants, and did not find any inconvenience from being unattended by a Druse escort.

Hebran is in sight from Kureiyeh, and the track to it is a gradual ascent. The first half of the way is dreary and monotonous, but, as we neared our resting-place, dwarf oaks clothed the hillsides; and though, being unprovided with magnifying glasses or poetical imaginations, we failed to discover the 'dizzy crags' and 'deep ravines' described by a previous traveller, the landscape redeemed itself from the charge of actual ugliness. We met many parties of villagers on the road — some returning from labour in the fields, others driving laden donkeys. The position of Hebran itself is really fine. It stands out boldly on a spur of the mountains, if a range rising less than 3,000 feet from the tablelands at their base may be dignified by the name. From the ruined temple which crowns the crest, a wonderful panoramic view is obtained of the plain-country of Bashan. The temple itself is not very remarkable, compared with the ruins we afterwards saw. In the village we found no antiquities worthy of notice, although the inhabitants were very civil in pointing out any old carvings or inscriptions likely to interest us. The younger portion of the population, never having seen so many Europeans before, thought us a capital joke, and enjoyed themselves immensely at our expense.

An isolated building, about a quarter of a mile to the south-east of the village, attracted our attention, and we were well repaid for visiting it. As it was the first specimen of stone architecture we saw which was at all perfect, I will describe it in detail. On the ground-floor a stone door led into a long room, the ceiling of which, made of huge blocks of stone, was supported by two circular arches. In this instance I did not notice any staircase, but in other buildings it was outside the walls; above were several small rooms, where a giant must often have knocked his head, and one curiously small door, about four feet high. The windows were closed by stone shutters, which still swung more or less easily in their sockets.

Our tents were pitched in a field below the village, whence the view of the conical peak of El-Kleib, 'the Little Heart,' now temptingly near at hand, rising behind low wooded hills, suggested an ascent on the following morning. Elias and Khasim went off in the evening to take coffee at the Sheikh's, where, if the account Elias afterwards gave was true, guests and host must have required all their Oriental politeness to get through the evening pleasantly.

In the course of conversation, it came out that Khasim, in some affray with the Druses, had killed a brother of the Sheikh; but matters were squared by the discovery that the Sheikh had killed a near relation of Khasim, and they agreed to postpone all discussion on the subject. I fancy that during the next few days, Elias's diplomatic talent, which was very great, found constant employment in keeping things smooth between our escort and the Druses, who, living in a normal state of resistance to the Government, naturally look with dislike on its officers. Our dragoman must have burthened his conscience with no slight weight if he really uttered half the untruths for which he afterwards took credit. We were represented by him as princes, Williams being specially distinguished as the American prince, while Elias modestly described himself as a commissioner sent by the Government to secure us proper attention, and to report where it was found wanting. Doubtless he played the part well, for he was a great dandy in his dress, always wearing a splendid gold-shot kefiyeh, and, to use his own words, he had a 'certaine politesse, tout à fait particulière,' in the Arab tongue.

March 22nd.—This morning the ground was hard with white frost. This was not extraordinary, for we were encamped at a height of 4,000 feet, and snow lay in patches not far above us. We despatched our baggage-mules, under Mohammed's charge, straight over the hills to Kunawat, as we intended to take a considerable circuit in order to visit Suweideh and Atil.

A sea of mist covered the plain below us, and before long the fleecy billows broke against and rolled up the hillsides, enwrapping us in their chilly folds. A short ride across a brow, covered with stunted oaks, and watered by numerous springs, brought us to Kufr. The place must have been very large, to judge by its ruins, but its inhabitants do not now appear to be numerous. In a long ramble in search of the gates, 10 feet high, mentioned by Porter, I came across relics of a Roman temple, many old houses, and a curious window, consisting of two square apertures, with a circular one in the middle, all sheltered by a projecting eave. The doors turned out to be 7 feet high; they were folding, and each half was of a single block; they did not fully fill the gateway, which was a foot loftier. The ruins formed a perfect labyrinth, and we separated, each taking a quarter in order to examine them as thoroughly as time would permit.

It was but a short distance to the foot of El-Kleib, 5,725 feet, the finest in form and second in height of the summits of Jebel Hauran. Tucker and I started on foot, with François, to ascend it, leaving the rest of the party with the horses. The mountain presents the appearance of a symmetrical cone clothed with a dense forest of evergreen oak, except on the south, where the lava and scoriæ are entirely bare. Just under the highest point is a small but very perfect crater. On the summit, layers of squared stones, the foundation of some ancient building, are visible. The view must be very striking; unluckily a dense mist hid it from us, and we waited in vain in hopes of its clearing off. In descending, we contrived to miss our way, and wandered about for some time, discharging revolvers into the fog, till we fell in with some peasants ploughing, and managed to understand their directions. Rejoining our friends, we remounted, and rode nearly due west to Suweideh, one of the chief seats of the Druse power. When we emerged from the cold mist that had enveloped us, the country was by no means interesting, and we were glad when Suweideh appeared in the distance. On arriving there, we avoided halting near the house of the chief, not wishing to waste time in receiving his tedious hospitality, and therefore cast but a passing glance at the ten columns of a ruined temple close to his door. Riding on through the town, we dismounted to visit the interesting remains of an ancient house; the masonry was extraordinarily massive, like that of a Cyclopean wall, and the building impressed us with the appearance of greater antiquity than any other we saw in the Hauran. We then continued our journey, and crossing the stream, which flows in a deep bed on the north side of the town, rode up to a fine Roman tomb, erected by a husband to his wife.* A broad track led us over the bare spurs, through which the hills sink down into the plain, to Atil. Here there are the ruins of two diminutive but exquisitely decorated temples. The immense size of the stones employed in the construction of these small buildings was very remarkable. Close to one of the temples we found some good pieces of carving—a winged figure of Victory, a horse, and a fine head of Apollo. Nothing could exceed the intelligence and courtesy of the inhabitants in pointing out the Roman fragments which lay about everywhere. We were led, through an exceptionally heavy stone door, into a house of very ancient-looking and massive masonry, to inspect a bit of the frieze of one of the temples, which was built into the interior wall. This must be taken for what it is worth, but it aided to shake our belief in the extreme antiquity of the greater number of the Hauran ruins. The columns of Kunawat were already in sight over intervening woods, and half-an-hour's scramble up a rocky path brought us to our tents, which had been pitched in a charming situation between the town and the beautiful temple, which stands apart from the other ruins on the west.

March 23rd.—Early in the morning, a dense mist again veiled the plain; Hermon alone stood up above it, flushed with a sunrise glow, and as the bells of the outgoing flocks rang through the air, we could almost fancy ourselves

* It is well described in 'Giant Cities,' p. 55.

in the Alps. Kunawat, the ancient Kenath, celebrated for the worship of Astarte, whose image now lies prostrate before her ruined temple, is built on the edge of an upland plateau, on which its principal group of buildings stands; the remainder of the place, enclosed by walls, runs down the slope, and overlooks the great plain. In the background rises a circle of wooded hills, and on one side a strong green torrent forces its way in numerous cascades through a narrow ravine. On a brow visible from the plain, and forming a landmark for the traveller approaching Kunawat, stands a temple raised on a high artificial platform. The wall and many of the columns have been overthrown, but enough is left standing to form one of the most picturesque ruins in Syria.

A little higher up the hillside seems to have been the fashionable cemetery, and we found numerous mausoleums scattered amongst the thick underwood. Though varying in size and architecture, they agree in their general form, which is that of a small square tower, with a chamber inside containing shelves for coffins. Entering the old city walls, which are still well preserved, we were led to a most remarkable group of ruins, now called the Serai, or palace. Here we found the remains of a temple, and of an extensive building, or rather several buildings, which must have been used either as a palace or for some public purpose. The richness of the architectural ornaments, and the picturesque irregularity of the whole mass, would make these ruins beautiful without their additional attractions. The space round them is paved, and (as is seen where the pavement has fallen in) supported on the arches of large subterranean reservoirs. The streams, which formerly filled them, now burst out everywhere among the ruins, and cause the growth of a mass of vegetation which conceals the surrounding desolation with

a mantle of greenery, such as is seldom seen in the East.
Some of the water flows down a conduit, to work an ancient
mill, still perfect, and in use. It is an old one-storied
chamber; the windows have the usual stone shutters, and
nothing, except perhaps the millstone, has been changed
since it was built. On the steep path leading down into
the gorge, we remarked several fine stone doors, one
of which was ornamented with vine-leaves, and another
with bosses. In the centre of the town is an old Christian
church, now used as a storehouse; inside another building we saw a very flat arch of great span. Kunawat
would repay a much more careful inspection than we
were able to give it. Descending over a beautiful piece
of Roman pavement, to the bridge, we turned aside to
inspect the remains on the right bank. Here there is a
quaint little theatre, and a temple with a fountain in
its centre, both on a very small scale. Steps cut in
the rock lead up to the brow, on which stood a tower-
tomb. Kunawat is the religious centre of the Hauran
Druses, and a great Sheikh lives here; but, anxious to
spend all our time in the ruins, we did not make any
advances, and our intercourse, although friendly, was slight.
To-day we again sent our luggage on by a short cut. The
ride down to Suleim is very pretty. Although there are no
large trees, the environs of Kunawat are clothed in luxuriant vegetation, and the ground was painted with anemones,
varying between bright scarlet and pure white, through
numerous shades of pink and purple.

In the middle of a wood we turned off the path, to visit
an isolated ruin placed on a slight eminence. There was
a large quadrangle, enclosed by walls, against one of which
stood a small building, with some good carving on the
doorway. Suleim is in the bare country; its principal
attraction is a small temple with an exquisite frieze: we

also noticed a fine doorway, over which was a long inscription in Greek hexameters. Close to the temple is a subterranean reservoir similar to those at Kunawat. We were now approaching the boundaries of that curious tract of country which has been successively known as Argob, Trachonitis, and the Lejah. In point of fact, it is nothing but a huge lava-glacier, if one may be pardoned the expression. The northern summits of Jebel Hauran are all volcanic cones; from these the lava-streams have issued forth, and flowing northwards have spread fanwise over a vast extent of country.

The limit of the inundation is in general sharply defined, and those who have ridden along the borders of the Lejah, and wondered at its broken crags and forbidding aspect of desolation, have not unnaturally taken a part for the whole, and described the entire tract as absolutely unproductive and desert. The wild and rapacious character of the inhabitants has added to its reputation for inaccessibility, and to the vague feeling of terror with which it has been often associated. We were now in full view of its south-west border, and the towns of Nejran and Edrei, the latter celebrated as the capital of Og, were plainly distinguishable. Our track bore away to the east, along the lower slopes of the hills. Leaving on our right the hamlet of Murduk, the inhabitants of which were ploughing the neighbouring fields, we soon entered a barren volcanic tract of country. Our road gradually ascended, in the direction of a depression under the most southern of the three conspicuous cones which are the northern outposts of the Jebel Hauran. The surface over which we were riding was rocky and broken, and lava-crags protruded on all sides, with little beds of withered grass lying amongst them. The desolation of the scenery increased as we advanced, and its effect was rather heightened than diminished by the gaunt arches of a long Roman aqueduct which had supplied the ancient Shuhba with water drawn from Abu Tumeis, the fourth in height of the summits of Jebel Hauran. As we rounded the south-east corner of the volcano, we came suddenly upon the old gateway of the Roman town, the very name of which has been lost. Inside the walls, all was as desolate as without; the old roadway was torn up, and the modern track zigzagged in and out, and up and down, among the lava-crags, which were contorted into the most extraordinary forms. What motive can have led to a large town being built on a site so gloomy, and so little adapted to human habitation, it is hard to divine. Its inhabitants had at least one advantage; the journey to the shades must have been robbed of half its terrors to men who, on their arrival in Hades, found the scenery just like home, and Pluto's palace not quite so sombre-looking as their own theatre. In its prosperity Shuhba must have been a strange city; in its desolation it is the weirdest spot imaginable.

The ruins of the ancient town and the modern village do not together occupy quite a quarter of the extent of ground included within the ancient walls. We found our tents pitched in an enclosure near the inhabited houses. Our muleteers had managed to get into a dispute with the people about the camping-ground, and there had been some disturbance; the question had, however, been referred to the Sheikh, who, on hearing they were the servants of Englishmen, at once ordered all civility to be shown them, and on our arrival peace was quite restored.

The ruins at Shuhba are not so ornate as others in this country, but are peculiarly interesting. There are two temples, similar in character to many we had seen elsewhere, and a mysterious building which looks as if it might have formed the apse of a 'basilica,' and which seems to have puzzled most travellers. The four main streets are still easily traceable, and, at their point of junction, the pedestals, once probably crowned with groups of statuary, are still entire. From this point to the southern gate of the city, a distance of full one-third of a mile, the pavement is as perfect as the day it was laid down. The baths were the best preserved we had yet seen; they contained several large and handsome chambers, and the stucco still adhered to the interior walls. Many of the pipes remain in their places, and the great aqueduct which supplied the water still exists, and terminates beside the building. We found a staircase which led us on to the top of its arches, whence we had an excellent general view of the bare northern slopes of Jebel Hauran, and the plain, dotted with conical mounds, which spreads to the east of it. South of the walls are some large open reservoirs. On our way back we visited the theatre, which looks spacious externally, but the building is so exceedingly solid that the size of the interior is disappointing. The massiveness of the masonry and the hardness of the material (black basalt) have been the causes of its preservation. The stage, the rooms and passages behind it, are uninjured, and very slight repairs would be wanted to make the building again serviceable.

Sheikh Fares, who received Mr. Porter so hospitably, was dead, and had been succeeded by his son. The Sheikh of Shuhba is one of the most powerful of the Druse chieftains, and although of late we had made it a rule to avoid invitations and visits of ceremony, we felt it would be wrong not to call on him, especially as we wanted some advice and information as to the best route through the ill-reputed Lejah. Having sent notice of our intention, we went in a body to pay our respects. A large gateway led into a courtyard surrounded by buildings. As we entered, seven dignified white-turbaned Druses bowed to the ground; then the Sheikh—a fine-looking man about thirty-five years of age—came forth to welcome us, and ushered us into his abode. We were received in the winter residence, a large ill-lighted room, with a fireplace in the centre, and divans round it. The Sheikh took his place on one side of the fire, attended by a younger brother, and several white-bearded elders. We, with our guard and François in the background, sat opposite to him. The room was quite filled with villagers. The conversation opened with the usual compliments and enquiries as to the success of our journey, but after we had requested the young brother to sit down, which he at first declined to do, it diverged into more general topics. We enquired as to the possibility of traversing the interior of the Lejah, and received satisfactory replies. We were told that in two days we might easily reach Khubab, on the north-western border, but that Ahireh, half a day's journey distant, was the only stopping-place on the road, as Damet-el-Alya was now uninhabited.

Meanwhile coffee was prepared. Among those Easterns who maintain their primitive customs, this is a very important ceremony, and must always be performed in public. The coffee-maker is an old servant, well practised in the art, and any failure on his part would be considered a disgrace to the whole household. In the present case the beans were first roasted over the fire in an open pan, which the Sheikh himself took occasionally into his hand for a moment, in order to keep up the appearance of serving his guests in person. Then followed the pounding. This is done in a finely-carved wooden mortar, and must require considerable dexterity, as the operator is expected to beat a lively march, like the rat-a-tat of a French drummer. The coffee was next boiled in a small tin-pot, and when ready was tasted by the maker, to show that it had not been poisoned. The Sheikh himself sweetened the fragrant beverage, which was handed round to us in the usual Eastern cups. The taste and aroma were delicious, but an unfortunate detail of etiquette prevented our indulging in such deep potations as we should have liked. In the filling of cups, as in greater things, the Eastern rule is exactly the reverse of the Western. Instead of filling a friend's glass to the brim, as a mark of goodwill, you give him a mere spoonful at the bottom of his cup; to pour out a full cup is a declaration of enmity to the man to whom it is presented. After the select circle round the fire had been served twice, the coffee was sent round to the crowd who filled up the background. As soon as the ceremony was concluded we rose to go, but the Sheikh came out with us, and showed us his summer residence, the façade of which was rather striking. A flight of steps led up to a portico, built to catch the cool northern breezes from Hermon, and supported by two pillars crowned by magnificent capitals, stolen from some ancient building. The interior was gaudily painted, in the usual Eastern style, with quaint representations of birds and beasts; built into the walls of the courtyard we noticed two pieces of sculpture, one representing a seated figure, the other a winged wind; the latter struck us as good.

On the way back to our tents we were taken into a camel-stable, above the door of which was a beautifully-cut Greek inscription. Later in the evening the Sheikh and his son, a sleepy boy of about twelve years old, returned our visit. The Sheikh talked a great deal of the constant friendship which had existed between the English and the Druses, and of his pleasure at seeing any members of our nation in the Hauran. Our visitors stayed so long that our stock of conversation became completely exhausted, and we were immensely relieved when they departed.

We hired a Druse to guide us through the Lejah, as Khasim was unacquainted with the paths in its interior, and set out on the morning of the 24th, which was dull and showery. We rode out of Shuhba, by a gap near the north-west corner of the walls, and skirting the north base of the same cone we had passed on the previous day, descended a long slope covered with the most extraordinary lava-streams, cracked in places exactly like the broken portion of a glacier. Mohammed, Khasim's subordinate, managed to be left behind for the second time, and did not come up with us till we had been nearly an hour on the road. Khasim meantime was alarmed lest the Druses should have done him some mischief. When the truant appeared, he told us that he had been purposely misdirected. If there was any truth in his statement, which I very much doubt, it was the solitary unfriendly act we met with among the Druses of the Jebel Hauran, whom we found (as Mr. Porter well describes them) 'a people of patriarchal manners and genuine patriarchal hospitality.'

The ground after a time became rather less rugged, and some traces of cultivation appeared before we passed the hamlet of Selakhid, a quarter of a mile to the right. Its Sheikh rode out to invite us to turn aside and rest in his house. He was well-mounted, and was a most picturesque figure, as he caracoled by our side, accoutred in jackboots, and clad in loose-flowing garments, which rivalled the rainbow in their varied colours. Finding we were not to be persuaded, he rode with us for some distance, and then, wishing us a prosperous journey, turned back to his home. Crossing the Roman highroad from Bozrah to Damascus, which ran through the centre of the Lejah, and leaving another village behind us on the left, we came to a tract bristling with lava-crags, and scantily covered with gaunt deciduous trees. This kind of country continued till we came in sight of Ahireh, which is situated in a sort of oasis with a good deal of corn-land to the westward. The village is built at the foot of Tell Ahmar, a green mound about the size of Primrose Hill, which is crowned by a Mahommetan 'wely,' or tomb. It is the highest eminence in the Lejah, of which it commands a complete view. The proportion of green grass and brown rock seemed pretty equal: here and there a black spot showed the position of a village. We spent the afternoon in rambling about among the houses of Ahireh, and came upon four Greek inscriptions, some heavy stone doors, and fragments of carving from a small temple. The people, far from showing any jealousy of our copying the inscriptions, took pains to point them out. The only other curiosity we lighted upon was a great cave, probably used as a tank. The Sheikh of the village came down to our tents, but his looks were not prepossessing, and we did not cultivate his acquaintance. He was, however, reputed to be a gallant soldier, and a deep sabre-cut across his face confirmed his reputation, although it increased the ugliness of an otherwise ruffianly countenance.

March 25th.—The night was disturbed by rain-storms and howling dogs; the latter we quieted by firing off a revolver at one of the noisiest. No more formidable animals made their appearance, and we had not the luck of Mr. Porter and his friends, who, during the night they spent within the Lejah, were surrounded by jackals, wolves, and hyænas, and afterwards somewhat naïvely congratulated themselves on the fulfilment of Isaiah's remarkable prediction: 'The wild beasts of the desert shall also meet with the wild beasts of the island, and the satyr shall cry to his fellow.' The village Sheikh had, as usual, entertained our guard in the evening, and had bragged to Khasim of his power, saying that he might tell the Pasha of Damascus to come with 5,000 men at his back, and he would beat him. In the morning this warlike hero condescended to cheat our muleteers out of six francs, in settling for some provender they had bought; but before riding off we sent for him, and told him very plainly our opinion of his conduct.

The morning was damp and lowering, and we made up our minds for a wet day, or (as Williams put it) 'guessed it was going to flop.' We were not deceived, and when we reached Damet-el-Alya, were glad to stable our horses and shelter ourselves in a mosque with stone doors and windows, while we ate our lunch. The pelting rain interfered somewhat with our explorations, but I found one interesting old house. The folding-gates of the entrance arch, still in their places, led into a courtyard, from which several doors opened into rooms of various sizes. The basement, ground and first floors were all perfect; the staircase was, as usual, external. Our baggage-mules had gone on while we rested, and we therefore shortened our halt, and rode on in pursuit, as fast as the nature of the ground would permit, for the character of the district was so bad that we were uneasy at leaving them long unprotected. The interior of the Lejah is not such a desert as it has been represented, and the path was decidedly better than the highway from Jerusalem to Nablous. In the wilder parts little green paddocks are interspersed between the banks of lava, and we several times during the day came upon considerable tracts of corn-land. The outside rim answers better to the description given of the whole in Murray's 'Syria'; but even here the language must be modified, and 'mound' must be read everywhere for 'hill,' and 'crack,' or 'depression,' for 'ravine.'

We met with no dangerous characters during the day to justify the bad name the district has acquired—perhaps the rain kept them all at home; but so easy and unadventurous was our progress that we had some difficulty in realising the fact that, with the exception of Burckhardt and Mr. Cyril Graham, we were the only European travellers who had succeeded in penetrating into the interior of the Lejah. As we neared the edge of the 'black country' the scenery became wildly picturesque; several villages occupied the knolls before us, their dark towers at a distance reminding us forcibly of feudal castles. Tucker and I turned off, under Khasim's escort, to visit Zebireh, the nearest of these villages, and found it entirely deserted. There were plenty of old stone houses, and in one was an upstairs-room, with a fireplace, and stone window-shutters. The roofs were in some instances supported by quaint pillars, primitively constructed of stones of unequal sizes, piled, like cheeses, one on another. The manner of building the interior walls in these strange dwellings is very curious. A framework is first constructed of large stones, with square pigeon-holes left between them; these are generally filled up, but sometimes left open, when they look not unlike wine-bins. We found here a stable with a stone manger, and also saw an inscription recording the erection of some monument, 'on account of the safety of the Lord Autocrator Severus Antoninus Cæsar, Britannicus.' Having finished our explorations we remounted, and, leaving behind us several mounds covered with towers, rode through a sort of pass or gap in the high bank of lava which runs along this side of the Lejah. In about an hour we reached Khubab, situated on the edge of the plain, where our companions had arrived before us. They amused us with the account of a row, of which they had been the cause, and which had only just terminated. The history of the dispute was rather complicated. The Sheikh, it seems, indicated to Elias a plot of ground where our tents might be pitched, and his son, a lad of seventeen, came down to superintend the proceedings. The owner of the ground and his child, a mere boy, then appeared on the scene, and objected to the arrangement which had been made; the boy, by some remark, angered the Sheikh's son, and got his ears boxed; whereupon the outraged father knocked the Sheikh's son down, and made his nose bleed. There was an immediate melée; swords were drawn, the whole village ran together—the men with their weapons, and the women screaming: but the culprit very discreetly put an end to the disturbance by running away, and leaving the Sheikh to swear revenge at his leisure.

The people of Khubab call themselves Christians, but they are dirtier and less well-to-do than their Druse neighbours. The women, who go about unveiled, are peculiarly hideous.

March 26th.—When we awoke in the morning, we found ourselves wrapt in a dense fog, but the mist soon cleared off, and the day became brilliantly fine. Our encampment was surrounded by most peculiar-looking hillocks of lava, which, together with house-walls of huge stones, formed a characteristic specimen of Lejah scenery. A few minutes' ride brought us out on the level plain. On one side the snowy mass of Hermon rose grandly over nearer green ranges; on the other was the rugged coast of the 'black country,' which here juts out in a promontory, and there recedes, leaving room for a grassy bay. At the cross-roads opposite Shoarah, we met a man riding furiously from the plain. He shouted to us, in passing, that his mule had just been stolen by a party of Arabs, and galloped on into the village. We halted on a rocky projecting brow, and then waited to lunch, and see what would happen next. While Tucker demolished the sardines, Williams, who even in ordinary life was a martial personage, and Elias, who was anxious to acquire the reputation of a fire-eater, exchanged the shot in their guns for bullets, with a military air which must have struck terror into the heart of the boldest Bedouin. No opportunity, however, occurred for the display of valour, and we had to content ourselves with the amusement of watching the shepherds driving in their flocks hurriedly from the plain, and the villagers issuing forth in twos and threes—some mounted, some on foot—in quest of the marauders. Of course nothing was seen of them, and we continued our march in peace.

The track to Mismiyeh, which is well within the borders of the Lejah, leads over very rugged ground. Our horses scrambled over great solidified waves of the lava-flood, divided by little grass-grown hollows, in one of which Elias surprised and slew a partridge. We determined to pitch our tents just outside the Roman temple, which is the most striking ruin of Mismiyeh. There was not sufficient depth of soil to drive in the pegs, but we tied the ropes to big stones, which answered all the purpose. Our ride had not been long, and we had plenty of daylight left to explore the place. The little temple was the most perfect we had yet seen; part of the portico was destroyed, and the central dome had fallen in, otherwise it was in good preservation. On either side of the doorway were niches for statues, under which were carved the words 'Pax' and 'Eisis' (*sic*). The building was square externally, but a sort of chancel was formed in the interior by shutting off a small vestry for the priests, and a staircase which leads on to the roof. The dome was supported on four columns, which are all standing, and the walls were decorated with

statues. The most striking feature of the interior was,
however, a beautiful fanshell apse, in very good preservation. From the roof there is a wonderful panorama, more
extensive, but resembling the view already described: on
one side the green plain and hills, backed by snowy
Hermon; on the other, the black Lejah, the most desolate portion of which is here visible, with the summits of
Jebel Hauran rising in the distance behind Tell-Ahmar,
which was easily distinguished by its white 'wely.'

A further ramble was rewarded by several discoveries, the
most important being a large house in the Bashan style of
architecture, but evidently of Roman date. An arched
gateway led into a courtyard, from which staircases gave
access to the first-floor, which contained one noble room—
the ceiling decorated with a fine cornice, and supported
by an arch eighteen feet in height, from the floor to the
keystone. The fact of all the roofs being constructed of
stone renders some such support necessary in every room
of too large size, to admit of the heavy blocks stretching
from wall to wall. We noticed curious recesses in the
walls, which may perhaps have been intended for the Pe-
nates. This fine building may have been the residence of
the Roman governors of Trachonitis, as Mismiyeh was, we
know, the capital town of that province. We saw some
well-executed stone-carving, such as twisted snakes, and
a double Greek pattern, and encountered numerous stone
doors. We found one pair eight feet high, and saw six *in
situ* in one courtyard. All that we observed confirmed our
opinion that the stone houses—which, from their peculiar
construction, and especially from the rude massiveness of
their stone doors, window-shutters, and rafters, have been
represented as of extreme antiquity—are of comparatively
modern date. Surely no one without a preconceived
theory to support, will maintain that where every public

building—whether temple, theatre, triumphal arch, tomb, or church—is of Roman or later date, the private dwellings are, as a rule, 1,800 years older.

In the larger buildings, the frequent use of the arch, and the introduction of classical ornamentation, are of themselves proofs of a late origin, and our wish to recognise in the smaller and ruder houses the dwelling-places of a prehistoric race, was frequently frustrated by the discovery of friezes and classical inscriptions built into their interior walls. The stone doors and shutters, which attract the attention of all travellers, are characteristic of the country, not of any period in its history, and we found them alike in the Roman temple, the Christian church, and the Saracenic mosque. The finest specimens, notably that of which a picture is given in Mr. Porter's book, are covered with Roman ornaments.

The Pentateuch tells us that Bashan was once inhabited by giants, and it has been argued that the size of the stone houses shows that they were built by a race of abnormal stature, and proves the date of their construction. In reality, however, the private dwellings are the reverse of gigantic, and the rooms they contain are to modern ideas small. If gates are sometimes found eight feet in height, they are (as far as we saw) always in positions where animals as well as men had occasion to pass under them, and those found at the present day in similar situations are of the same dimensions. The stone doors guarding the entrances to the vineyards around Tabreez are larger and more massive than any we saw in Bashan.

The extent and number of the ruined towns are used as an argument that they are the remains of the sixty fenced cities conquered and destroyed by Moses. Travellers are too apt to forget that Syria formed a portion of the Christian Empire of Constantinople, and that in the fifth century there were thirty-three Christian bishops in the Hauran alone. The population which built the churches and the theatres was quite numerous enough to have filled the ruined houses which now remain. If any buildings older than our era still exist in the Hauran, they are, I believe, exceptions, and do not disprove our conclusion that a false impression is given by describing the ruins of Bozrah, Kunawat, Suweideh, and Shuhba—in fact, those of Roman provincial towns—as 'Giant Cities.' It is not of Og but of the Antonines, not of the Israelitish but of the Saraccnic conquest, that most modern travellers in the Hauran will be reminded.

Mismiyeh is inhabited, at present, by a few families of beggarly Sulut Arabs, who have so far abandoned the traditions of their race as to condescend to live within walls. They are great rascals, and much addicted to petty thieving. Our muleteers got into a dispute with some of them during the afternoon, whereupon Elias, on his own responsibility, ordered the arrest of the leading villager, and proclaimed that 'the Beys willed he should be carried to Damascus.' The elders came down to represent the youth of the culprit, and to beg Elias to deprecate the wrath of the Beys. No reference was in reality made to us, but the prisoner was released, with an admonition to the natives in general, that they had better be careful for the future, as a word from us to the Pasha would ensure their ruin.

March 27th.—We now finally turned our backs on the Lejah, and prepared to cross the strip of 'debateable ground' which lay between us and Deir Ali, the frontier village of the Damascus district. The plain across which we rode was for some miles covered with scrub, bright yellow flowers, and green herbage, on which immense flocks of sheep and camels were feeding. We passed close to the tents, seventeen in number, of their owners. A tall

spear stuck into the ground before the door marked the
abode of the Sheikh. He came out and entreated us to
alight and partake of coffee, and when we excused ourselves,
brought out a huge bowl of milk. It was rather a relief
to meet with so pleasant a reception, as the Sulut tribe, to
which these Arabs belonged, bears anything but a good
character.

The ground grew more stony and barren as we approached the foot of Jebel Munia; we noticed curious rows
of artificial pools, made to catch and retain the waters
of the rainy season, but now dry and fallen into decay.
The first building we came to was an isolated farmhouse,
built like a fortress, with strong iron gates to resist the
marauders of the neighbouring desert. Another hour's
ride over a bleak plateau, during which Hermon, now
comparatively close at hand, towered grandly before our
eyes, brought us to Deir Ali, a large and prosperous Druse
village. The neighbourhood is rendered fertile by abundant springs, and for the first time since leaving Bethlehem,
we saw the fig, the vine, the olive, and the poplar growing
luxuriantly.

We lunched under the shade of some gnarled old olives,
finer specimens of the tree than are usually seen in Syria.
The further ride to Kesweh was round the bare flanks of
Jebel Munia, and had nothing but the distant view of
Hermon to make it interesting. We saw, away to our left,
the great caravanserai called the Khan Denun, where the
Mecca caravan rests on the first night after its departure
from Damascus. As we neared the Nahr-el-Awaj (the
ancient Pharpar), the white clean-looking houses and
minarets of Kesweh appeared on its further bank; the
stream itself was hidden in the thicket of fruit-trees which
lines its course. The river was crossed by a stone bridge;
but so swollen were the waters, owing to the recent rains,

and the melting of the snows on Hermon, that they touched the keystones of the arches, and looked as if they would soon carry away the whole fabric. Stalls for the sale of provisions and saddlery showed that we had entered a district where there was some security for property and attempt at trade; a paved piece of road bore witness to the fact of our being on an old highway of commerce; while the recently-erected telegraph-wire between Damascus and El-Mezarib, the capital of the Hauran, proved that the Turks are not altogether blind to the advantages to be reaped from the adoption of the discoveries of Western science. Kesweh is a neat but unremarkable Syrian village; our tents were pitched on a grassy brow before the place, near some turban-capped tombstones. Tucker and Williams went in pursuit of birds down the banks of the Pharpar, but came back empty-handed. One of Khasim's pistols was stolen in the night, but he got it back next morning by paying a small 'backsheesh' to the thief, or (as he preferred to call himself) finder, of the missing weapon.

March 28th.—A broad beaten track runs over the hills which separate the basins of the Pharpar and Abana. Winding through a gap in the range, we came in sight of Daroiya, a town some miles west of Damascus, surrounded by orchards. After turning to the right we crossed a low spur of Jebel-el-Aswad, and caught our first view of the city, spread out across the plain, backed by a mass of verdure, and the tawny slopes of Anti-Lebanon. The scene, though very striking, did not impress us so much as similar views of Cairo. Damascus is singularly poor in the minarets which lend such a charm to its Egyptian rival. While cantering carelessly over the flat expanse between us and the gates, the sudden failure of a stirrup-leather gave me a tumble upon the hard ground. Luckily

I did not hurt myself seriously, but neuralgia, a stiff back, and barked knuckles served to moderate any feelings of triumph at the successful conclusion of our novel journey from Jerusalem, and the opening of what may prove a new route for Eastern travellers, and I needed François' prompt consolation, 'Ah, monsieur, vous faites bien de suivre l'exemple de St. Paul,' to reconcile me to the complication of bodily ills. A review of Turkish troops was going on outside the city; the cavalry were remarkably well-mounted.

After passing the gates we rode along the shabby boulevard which traverses the suburbs, and forms an entrance to Damascus. Leaving the bazaars on our right, we at last reached Demetri's Hotel, a pleasant house built, in the usual Damascene style, round a courtyard full of lemon-trees.

We remained a week in these comfortable quarters. Here our connection with Khasim ended as satisfactorily as it had begun, for he was more than contented with the 'backsheesh' we gave him. Our Trans-Jordanic trip added only 5*l*. a-head to the usual dragomanic expenses, which, considering where we had been, and what we had seen, was a very small sum.

CHAPTER III.

LEBANON AND THE LEVANT.

Damascus—Bazaars and Gardens—An Enthusiastic Freemason—Snow-storm on Anti-Lebanon—Baalbec—An Alpine Walk—The Cedars—Return to Beyrout—Cyprus and Rhodes—Smyrna—The Valley of the Mæander—Excavations at Ephesus—Constantinople—The Persian Khan—May-Day at the Sweet Waters—Preparations for the Caucasus.

THERE are very few sights in Damascus, unless one considers as such the window from which St. Paul was let down, and the tomb of the legendary porter who aided his escape. The Great Mosque is fine, but not so interesting as that at Jerusalem. The commercial aspect of the place is the most striking; the bazaars, the rough wooden roofs of which rather spoil their otherwise rich effect, are very extensive; and though Manchester goods meet you at every turn, the ways and manners of the people are purely Eastern. It is a very seductive place to go shopping in; Williams once spent a whole day in a silk-mercer's den in the Great Khan, and came home in the evening followed by a man laden with gorgeous scarves. Our friend, despite the time and bargaining his purchases had cost him, was troubled with an uneasy suspicion that, to use his own expression, 'the old fellow had regularly waggled him.' The gardens round the town are rather orchards than gardens in our sense of the word; but at this season, with the fruit-trees in full blossom, they were very beautiful, and it was amusing on Sunday to stroll amongst the numerous companies of citizens, sitting in circles, chatting and telling stories under the shade. The fashionable ladies' dress is a white sheet, and a coloured handkerchief over the head; but the infantine population swell about in scarlet and gold tunics, and all manner of 'pomps and vanities.'

At the *table-d'hôte* there was much discussion about the expedition which the Pasha was said to be about to make to Palmyra; he proposed to spend a month in the trip, and to take with him a small army, including artillery. Some of the travellers at our hotel were staying on in order to avail themselves of the opportunity of going in his suite; among them was an elderly American, a professor at one of the universities in the Western States, who in his quality of a Freemason had already called on the Pasha and Abd-el-Kader, both of whom are brothers of the craft, and now announced his purpose of 'planting the banner of Freemasonry on the ruins of Palmyra.' He was unfortunately prevented from fulfilling his mission by the Pasha abandoning his design. Before leaving Damascus we had the pleasure of seeing Mr. Rogers, the English Consul-General, to whom we handed over our letter to the Pasha, which we had not found an opportunity of presenting, with a request that he would use it in securing for Khasim promotion to a higher grade, in which we felt sure he would not dishonour our recommendation. Mr. Rogers kindly showed me some of the most valuable trays of his fine collection of Eastern coins, and also a coat-of-mail taken by the Pasha from an Adwan Sheikh in the previous year, and a noble sword, which, as was recorded by an inscription in gold letters on its blade, had belonged to a son of the famous Saladin.

The last days of our stay at Damascus were so cold that

we had fires in our room. We left on April 3rd, and travelled in three days along the ordinary track to Baalbec. The general character of the Anti-Lebanon scenery is poor, but there is one charming spot, Ashrafiyeh, perhaps the most picturesque in Syria, and the glen of the Abana is pretty for some way above it. The weather was cold and misty; the night we slept at Surghaya, it first blew and then snowed, and when we woke in the morning we found the ground frozen hard outside the tent. In such weather we were glad to take shelter in a clean room at Baalbec, instead of tenting, as is the custom, in the temple enclosure. About midday the snowstorm, through which we had ridden all the morning, passed over, and we had a fine afternoon to visit the ruins. Magnificent as is the scale and superb as are the architectural details of the great temples, we agreed in thinking the general effect less impressive than that of Karnac.

The morning of April 6th was bright and frosty, and the chain of the Lebanon shone out clear on the further side of the Plain of Cœle-Syria. Its summits are rounded and lack character, but the effect of the long snowy range against the blue sky was very grand. Tucker and I felt it would never do to let a little snow prevent our reaching the Cedars, and we therefore arranged to divide our party into three sections. Williams and Cross started, with the boy who owned their horses, to ride down the valley to Shetawâra (pronounced 'Stora'), the halfway station on the Damascus-Beyrout road; our baggage-train was ordered to Shelfa, a village at the foot of Lebanon; while Tucker and I, with Elias and François, set out for Ain-Aat, the highest hamlet (5,317 feet) on the eastern side of the Cedars' Pass. After crossing the plain we rode up a steep ascent, clothed with dwarf oaks. Even below Ain-Aat the snow lay deeply in the hollows, and gave our horses some trouble. We slept in a cottage, inhabited by a family of about a dozen peasants, and an unknown but very appreciable quantity of insects.

Ain-Aat is situated on a shelf immediately under the backbone of the Lebanon. We started for the Cedars at 5 A.M., with François and a villager, leaving Elias behind. It took one hour and forty minutes' sharp climbing, up a steep but perfectly easy snow-gully, to reach the ridge (7,624 feet), whence we looked down on the Mediterranean. So far the snow had been in excellent order, but on the western side of the pass, the horseshoe of mountains, within the hollow of which the grove of Cedars stands, had shut out the sun, and prevented the surface from ever melting sufficiently to form a hard crust by regelation. Getting down to and up again from the grove was one of the heaviest three hours' work I ever did. We sank at every step up to our knees. The trees are in very flourishing condition, and well repay a visit, especially when seen, as we saw them, with the snow resting on their broad-spreading branches, the only green things visible on the great white slopes. The little chapel was almost buried in snow, and it was only just possible to get in at the door. On our return we met several parties of villagers, who seemed equally surprised and pleased to see travellers capable of walking over a mountain-pass. We were back again at Ain-Aat at 12.15 P.M., and in the afternoon rode down to Shelfa, a prettily-situated hamlet at the foot of the mountains. There we found our tents pitched, and a good dinner cooking. I have described our visit to the Cedars, in order to show that there is often no difficulty, to men of active habits, in making the excursion when the dragomanic world of Damascus pronounces it quite impossible.

A day's ride through Cœle-Syria brought us to Muallakah, in the neighbourhood of Zahleh, the most flourishing

Maronite town in the Lebanon; there we slept, and on the following day cantered along a fine road, constructed by a French company, which crosses Lebanon at a height of 5,175 feet. The scenery reminded me, at times, of the lower parts of the Italian Tyrol. Beyrout was hot and hazy; we never saw the summits of Lebanon clear till the last day of our stay, but then the bay was really beautiful.

Elias during the last three days, when Tucker and I were alone with him, grew more confidential than was his wont, and treated us to the story of his early life. A native of a village in the Lebanon, he had been left an orphan at an early age. His father had been a man of some property, and the riches Elias inherited enabled him to indulge to the full his boyish taste for smart dress. To this he soon added a passion for donkeys, and gave large sums for animals of the best breed and most showy appearance. A fall, caused by the stumbling of one of his favourites, disgusted him with donkeys, and he took to horseflesh. The pursuit of this last fancy had brought him almost to the end of his inheritance, when he was aroused to a sense of his position by the sneers of his former friends. Elias sold his stud, and started afresh, until, having amassed sufficient capital to set up as a dragoman, his love of horses and out-of-door life led him into that profession. He had now, he told us, succeeded in buying back most of the property he had sold in his youth, and was a well-to-do man.

Having paid off Elias, and arranged for the despatch of our Damascus purchases, to which we added some specimens of the work of the Lebanon, we embarked on board an Austrian steamer, and finally bade adieu to Syria, on the afternoon of Easter Sunday, April 12th.

Next morning we landed at Larnaca, the chief port of Cyprus—a dull ugly town, where we failed in our search

for good wine or pretty faces. Many of these classical
places have nothing left but their associations. The west
end of the island and the Bay of Baffa (the ancient Paphos)
are well seen from the sea. On Wednesday morning we
had two hours in which to run over Rhodes, a most interesting old town, full of monuments of the Knights Templars.
Sailing on all day under the lee of the Isles of Greece, we
found ourselves at sunrise on Thursday steaming up the
Gulf of Smyrna; the shores looked fresh and beautiful, but
the water was sadly discoloured by the recent floods of the
Hermus.

Smyrna, like Alexandria, brings into vivid contrast the
East and West; Paris fashions and bearded camels come
into constant collision in its narrow streets. At the theatre
a French company was performing 'La Belle Hélène.'
Homer's ghost can scarcely view with pleasure his heroine
in the hands of Offenbach. Our stay at Smyrna—where,
owing to the kindness of friends, we enjoyed most agreeable society, and the comforts of an English home—was a
very pleasant interlude between the mild roughing of
Syria, and the real hardships of travel in the Caucasian
provinces of Russia.

Ionia, into the interior of which we made two short
excursions, is as far superior to Syria in scenery as Kent is
to the Pays-de-Calais. Our first expedition was to Aidin,
a large and flourishing town, charmingly situated under the
hills on the north side of the valley of the Mæander, over
which there is a lovely view from the neighbouring heights.

Next day we returned by rail to Dalachik, and rode
thence to the site of Magnesia ad Mæandrum: the broken
columns of a temple are the principal remains, and there
was nothing to compare with what we had recently seen in
Bashan; but the ride was delightful, amongst tall olives
and fig gardens. Our classical recollections were aroused

by meeting a boy playing the primitive Pan-pipe, and by seeing a pretty fountain at which a bevy of nymphs were bathing. We got back to Aiasalook (the station nearest Ephesus) in the evening, and were kindly housed and entertained by Mr. Wood, who has spent some time in excavating the ruins, with a view to the discovery of the site of the famous Temple of Diana. Before his excavations, the ruins of Ephesus left above-ground had suffered too severely, from time and violence, to be of great interest to anyone but an antiquarian; much, however, has now been brought to light. The theatre, the scene of the goldsmiths' riot, is the most striking sight; the stage has been laid bare, and many inscriptions have been found. Some of the recently excavated marbles are as white as on the day they were cut. The city was built mostly of brick, encased in various marbles, of which fragments strew the ground in every direction. Mr. Wood has also discovered a small building, which, on the strength of some Christian symbols, he rather boldly calls the Tomb of St. Luke; a marble basin of noble dimensions, and a sort of 'Via Sacra,' outside the walls, lined with sarcophagi and funeral inscriptions. When we were there he believed himself to have settled, within a square mile, the position of the Temple of Diana, and seemed quite confident of turning it up sooner or later.

Our second excursion was to Manissa (Magnesia ad Sipylum), a fine Turkish town built on a steep slope at the base of the splendid crags of Mount Sipylus. We drove on several miles, in a Turkish cart, to see the statue called Niobe, a rude figure, probably of Egyptian origin, carved on the face of a cliff. On the way we had a distant view of a fine snowy mountain, Boz-Dagh, far away in the interior, beyond Sardis. We returned to Smyrna the same evening.

On Saturday, April 25th, we left Smyrna on board an Austrian steamer for Constantinople. The boat was crowded with Russian pilgrims returning from Jerusalem, who occupied themselves alternately by eating salt-fish and fighting; hideous females, perpetually smoking cigarettes, were strewn all over the deck, and from time to time neglected infants raised dismal howls. Happily the sea was calm; what the state of things must have been during the run from Alexandria to Rhodes, when the vessel encountered a severe gale, it was easy but not pleasant to imagine. The poor pilgrims had been terribly frightened during the storm, but were now rather elated, as they attributed their safety to the prompt piety of a man who threw into the waves a taper lit from a candle kindled in its turn from the sacred fire in the Church of the Holy Sepulchre on Easter Sunday.

We entered Constantinople at sunrise on Monday, and admired, as everyone must, the enchanting aspect of the city from the water. We spent six days at the Hôtel de Byzance, during which we were fully occupied in sight-seeing, and making the necessary arrangements for our journey in the Caucasus. Of course we did the 'lions'; were first hurried round the mosques, perhaps the most tiring day's sightseeing in the world; and afterwards paid a quiet visit, by means of 'backsheesh,' to Santa Sophia, which more than realised our expectations. While admiring the effect of the vast unbroken area under the dome, even when merely dotted with the bright dresses of Turkish worshippers, we could form some faint idea of what must have been the splendour of a state ceremonial of the Byzantine Court in this noble basilica. The old walls, the seven towers, the burial-ground at Scutari, all had to be visited.

One evening we were recommended by a gentleman, staying at our hotel, to visit the Persian Khan, to hear the wailing for Hassan. We found a long room decorated with buffets covered with ornamental glass and candlesticks. On the floor squatted at least 700 high-capped Persians; in the centre of the room was a low pulpit, from which a Mollah recited the piteous tale of Hassan's death. When he came to an exciting point in the story, the audience wept and beat their breasts, or 'oh-oh'd' their indignation against the murderers, like an election mob hooting an unpopular candidate. Excellent coffee and sherbet was handed round to everyone, including our own party, who had been given seats in a recess commanding a full view of all the proceedings, and were treated in every way with great civility.

Our row in a caique to the Sweet Waters was well-timed. May-day, the date when the picnics at the Sweet Waters usually begin, fell on Friday, the Mahommedan day of rest, so that the concourse was greater than usual. Our caique jostled a crowd of boats filled with Turkish ladies, plump little dolls who make themselves fair to look upon by adding artificial brightness to their eyes, and wearing transparent veils over the lower part of their faces. Their balloon-shaped dresses, mostly of the brightest colours, present a charming *coup-d'œil* when massed in groups. The Sultan has a villa up at the Sweet Waters, which consist of a stream (about the size of the Cherwell at Oxford) with a drive on one side, and gardens on the other. The place is just pretty enough to make it an excuse for a promenade, whether by road or water. There were many European carriages and Parisian costumes on the drive, the latter far more extravagant than anything the East can produce.

We were lucky in meeting at Constantinople Mr. Gifford Palgrave, H.B.M.'s Consul at Trebizonde, who was on his way home. When consul at Soukhoum-Kalé, Mr. Palgrave made several journeys into the interior, and had been twice to the foot of Elbruz; he was consequently able to give us much valuable information as to the character of the country. But the most important aid we received was the recommendation of a Mingrelian servant, who would act as our interpreter. The need of some such attendant, and the difficulty of finding one who would fall in with our plans, had long been a weight on our minds. The man Mr. Palgrave suggested to us was a native of Sugdidi, between Kutais and Sonkhoum-Kalé, and had been employed as cook in the consular household at Trebizonde. He spoke French, Russian, Turkish, and Georgian.

I presented, at the Russian Embassy, the letters of introduction which had been forwarded to me from England, and received from General Ignatieff much politeness. He gave me letters to Count Leverschoff, the Governor of Mingrelia, and to a gentleman attached to the Grand Ducal court at Tiflis. We were warned that the country was still in an undeveloped state, and that we should find rough roads and meagre fare, but were also told that the worst danger to which we should be exposed from the mountaineers was having a horse stolen.

On May 1st we parted from our friends Cross and Williams, who sailed for Italy; and on the following afternoon embarked, with all our traps, on board a Russian screw-steamer, which looked very small in contrast to the large boats in which we had voyaged of late. She was named the 'Gounib,' after the scene of Schamyl's last resistance and capture in Daghestan. The boat was built more for freight than passengers, and the accommodation was very scanty. Tucker and I were lucky, however, in getting a comfortable cabin to ourselves, owing to the courtesy of a Russian officer, who exchanged his berth with one of us. The deck was littered with all sorts of odd passengers, bound for the ports of Asia Minor. The way in which an Eastern, immediately he gets on board-ship, spreads his rug, wraps himself round in his cloak, and resigns himself to destiny and seasickness, is worthy of all praise. The vessel was delayed so long before the mails came on board, that it was dark before we got under weigh, and we saw but little of the beauties of the Bosphorus. Passing the lighthouse which marks the entrance to the Black Sea, we watched the steamer's head swing sharply round to the eastward, and felt that we had abandoned the ordinary track of travellers, and that a new stage in our wanderings had indeed been entered upon.

CHAPTER IV.

TRANSCAUCASIA.*

On the Black Sea—Trebizonde—Rival Interpreters—Paul—Running a Muck—Batoum—The Caucasus in Sight—Landing at Poti—The Rion Steamer—A Drive in the Dark—Kutais—Count Levershoff—Splendid Costumes—Mingrelian Princesses—Amiras—The Valley of the Quirili—A Post Station—The Georgian Plains—Underground Villages—Gori—First View of Kazbek—Tiflis—The Hôtel d'Europe—The Streets—Silver and Fur Bazaars — Maps — German Sarants — The Botanical Garden—The Opera—Officialism Rampant—A False Frenchwoman—A Paraclednsia—The Postal System in Russia.

THE weather on the Black Sea was cold and rainy, but the water was never really rough. Half our fellow-passengers were English—an engineer with his wife, and two young men going out to aid in the construction of the Poti-Tiflis railroad. Our other companions were a young Russian colonel, a little man who talked familiarly and affectionately of 'votre John Stuart Mill,' on the strength of his having read his 'Utilitarianism,' and some Armenian merchants, more or less uninteresting.

On Monday we called at Samsoun, and on Tuesday afternoon arrived at Trebizonde, where the boat remains twenty-six hours to take in cargo. The weather was vile, the rain falling like a waterspout, and we were glad to escape from the rather rough and monotonous Russian fare, and the uneasy roll of the steamer, to a nice little hotel on shore, kept by an Italian, who had served in the Sardinian

* The political division of the Russian empire ruled by the Viceroy of the Caucasus, extends from the Manytch, on the north, to the Araxes on the south. The provinces on the north of the great Caucasian chain are called Cis-Caucasia, those on the south Trans-Caucasia. Russians and natives of the country never restrict the name Caucasus to the mountain-range.

army during the campaign in the Crimea. No sooner had we acquainted the landlord with our plan of travel, than a candidate for the post of interpreter appeared in a good-looking man, showily dressed in Caucasian costume. His acquirements, by his own account, were marvellous; he spoke perfectly at least seven languages, including English. We thought he was too much of a dandy to appreciate such rough work as we meant to undertake, and were moreover unpleasantly reminded of dragomanic tyranny by his way of saying, 'I am sure you cannot get on without me; you will be very sorry if you do not take me.' We fortunately had an easy answer to his importunities in our previous understanding with Mr. Palgrave, and we started off through the rain to find the dragoman of the English Consulate, for whom we had letters. He at once sent for the Mingrelian whom Mr. Palgrave had recommended to us. He was a handy-looking fellow, young and active, dressed in ordinary European clothes, and he was quite ready to accept such an engagement as we offered him; so a bargain was at once struck with him, and he promised to be ready to start on the following day.

On Wednesday we had a pleasant walk to an old Byzantine church, mutilated and whitewashed by the Turks, outside the town. Trebizonde itself is a picturesque place. Its houses rise in terraces above the water, on the lower slope of a bold green hill, backed by finely-shaped, well-wooded mountains. The modern town has spread along the coast on either side of the old fortress, the walls of which are still perfect; two ravines, which cut it off from the adjacent slopes, make it a very fine and strong position. The great article of manufacture seems to be wooden cradles, very gorgeously decorated; we saw store after store full of them. The bazaars are well stocked with game, among which we noticed some woodcocks, from the neighbouring hills. When we were there the place was very quiet, but shortly afterwards its peace was disturbed by a tragic incident. A Mussulman fanatic, either mad or drunk, took it into his head to run-a-muck through the bazaar, and so far succeeded in his horrid purpose, as to stab no less than seventeen people before he was himself waylaid and despatched with a poleaxe by a discreet butcher. Eleven of his victims died of their wounds. These outbursts of fanaticism sometimes occur among a Mahommedan population, but they are quite exceptional phenomena, and as a rule your person and pocket are far safer in an Eastern city than they are in London.

In the afternoon we climbed by a steep path to the Flagstaff Hill, behind the town, which commands a very good view of the coast and the mountains of the interior. The brow was covered with the most wonderfully smooth turf, like an English lawn. On the way down we turned aside to visit a very curious rock-hewn church, decorated with frescoes, some apparently of great antiquity. We left at 6 P.M., and at daybreak next morning were in the harbour of Batoum. The weather had cleared during the night, and, to our great surprise and delight, we found ourselves for the first time in the presence of the 'mystic mountain range' of which we had talked and thought so much, but of which we as yet practically knew so little.

As we looked from the deck of the steamer, our eyes followed a long line of snowy peaks, the most western of which rose directly above the waters, like a ship at sea when only its white sails are visible. Next to these came a cluster of fine rocky peaks, which reminded me of the Dolomites as seen from Venice; in the centre the outlines were tamer, but on the east was a very massive group, probably Koschtantau and its neighbours, which stand midway between Elbruz and Kazbek. The harbour of Batoum is the only safe one at this end of the Black Sea; it is formed by a long spit of sand, which runs out in a northerly direction, and the bay faces the north-west. The town stands on low ground, and is poorly built; it is only some twelve miles distant from the frontier fort of St. Nicholas, and it seems curious that, in some of their accessions of territory on this side, the Russians have not managed to obtain possession of the harbour, which would be of great value to them. Poti, at present the port of Trans-Caucasia, is a most miserable place, and the bar of the Rion is so shallow that no vessel of any size can cross it. All the Black Sea steamers, consequently, either stop at Batoum or Soukhoum-Kalé, and transfer their cargoes into smaller boats. The steamer which ought to have met us had not arrived, and we were compelled to spend the whole day at Batoum. At a brook in the outskirts of the town we found several men engaged in capturing frogs: no sooner were the victims secured, than they were beheaded and skinned; a revolting spectacle from which we quickly fled. Crossing, by a ruinous wooden causeway, the swamp which intervenes between Batoum and the hills, we climbed up a projecting knoll covered with rhododendrons in blossom, and crowned with beech-trees. The vistas of sea and coast through the trees were exquisite. A hamlet built on the hillside reminded me of the pictures of South Sea island habitations; it consisted of huts built of rough interlaced wood plastered with mud, surrounded by quaint little square boxes raised upon poles, and looking like young châlets starting for a stilt-race. I believe they are used for storing corn. We made provisional arrangements with the Russian consul at Batoum to remedy our Mingrelian servant's want of a passport, and were much amused by his name, which proved to be Bakoua Pipia. Pipia was the family title—Bakoua a term of endearment which he had acquired as a boy. We preferred to call him by the more familiar appellation of 'Paul,' by which he had gone when in European service. In the afternoon time hung heavy on our hands, and, having exhausted our last 'Saturday Review,' we had recourse to a café, kept by a Frenchman, which offered some bad beer and an atrocious billiard-table.

Towards evening the little steamer arrived from Poti. It had been detained to aid a vessel, laden with the ironwork for the railroad bridges, which had stuck on the bar. No certain intelligence could be obtained as to when we should start, and we were finally allowed to turn in with the impression that we were not to be disturbed till the morning. The captain, however, changed his mind, and at midnight we were awoke, and told to go on board the small boat. Meantime, François and Paul, in preference to sleeping on the deck, had gone ashore to seek quarters in the town, and no one knew where they were to be found. We sent off men to go the round of the lodging-houses, and promenaded the quay ourselves, shouting their names and 'jödelling' at the top of our voices. All was in vain—no trace or sign of the truants was to be had. The little vessel got its steam up, and we were obliged to go on board. Of course we complained loudly to the officers of their mismanagement in first giving notice that we should not leave till morning, and then routing everybody out of bed at midnight. While we were venting our indignation, the ropes were cast loose, and the paddlewheels began to revolve; we had actually gone a hundred yards when a movement took place on the shore, and the burly outline of François was seen standing like Lord Ullen, when left lamenting on the waterside by his heartless daughter. I made a last energetic appeal; the engines were stopped, and the lost ones were brought off rapidly in a boat. François nearly tumbled into the water in his hurry to get on deck, and both men looked, as they well might, very sheepish and ashamed of themselves.

It was about 2 A.M. when we got off from Batoum. The cold soon drove us below, but we came on deck again at sunrise, so as to lose nothing of our approach to the Caucasian shores. The steamer was running quickly across the fine bay which forms the eastern end of the Black Sea; behind us lay the ranges on the Turkish frontier, grand masses rising to 8,000 or 10,000 feet in height, carrying at this early season, and after an unusually inclement winter, a great quantity of snow, but still clearly mountains of the second class; before us rose ridge behind ridge, until behind and above them all towered the peaks of the central chain of the Caucasus, scarcely telling their height to the eye uninitiated in mountain mysteries, but showing us plainly enough that we were in the presence of an array of giants, armed in like panoply of cliff and ice to those we had so often encountered in the Alps. One great dome of snow, which conspicuously overtopped all its neighbours, we hailed at the time as Elbruz, and I do not doubt we were right in our recognition of the monarch. On our right lay a low wooded coast, the basin of the Rion; a group of twelve vessels anchored about a mile off shore, and a tall lighthouse marked the mouth of the river and the position of Poti. The meeting of the fresh water and the salt was most curious; the Rion is at all times a muddy stream, and the line between the brown and blue water was marked sharply enough to be visible from a considerable distance. The town lies about half a mile up the Rion, on the southern bank; we ran up alongside a wharf, close to the custom-house—a long log-building, where our luggage was very leniently examined, and our passports were taken away; we were told they should be sent after us to Tiflis. There is no restriction against bringing

arms into the country. We had expected, on landing in
Russia, to be struck, after the universal untidiness of the
East, by the appearance of a well-dressed European soldiery, but, to our surprise, the men we saw were clad in
worn-out grey suits, and were physically of the most
wretched appearance. Partly owing to the exertions of
our companion the Colonel, the departure of the river-boat
up the Rion was delayed until passengers from the
Black Sea steamer could get on board, and after a stay
of only an hour and a half, we left Poti behind us; I shall
therefore postpone its description till our return, and at
once carry my readers up the country.

The voyage up the Rion from Poti to Orpiri occupies
eight hours, and on a clear day, such as we were favoured
with, is most beautiful. The stream, a short distance above
its mouth, makes several bends, each of which discloses a
charming vista. Thick forests clothe the banks; and over
the trees glitter the peaks of 'the frosty Caucasus.' One
summit, exactly at the end of a long reach of the river,
strikingly resembled in form the snowy side of the Grivola.
On the right we had always the Turkish ranges, which
sink in beautifully-shaped hills into the basin of the Rion.
For the first four hours of our voyage, both shores were
covered with primeval forests, and the country was low
and swampy, the only signs of life being a few log-huts,
or a Mingrelian horseman riding past. One man raced
the boat for some way, and we had time to remark his
costume. The most striking part was the long frock-coat,
the breast of which was decorated with a row of cartridge-
pouches; and the 'baschlik,' or Caucasian hood, with two
long tails, used to wind round the neck in case of wet;
this, with the big sheepskin cloak common to the country,
forms a most efficient protection even against an Eastern
deluge. The stream, averaging from 200 to 300 yards in

width, now bent to the south; the forest became thinner, and the country more inhabited, while orchards, fields of Indian-corn, and clusters of cottages, appeared on either shore. The lowest outposts of the southern hills here advance close to the Rion, above which they rise in steep banks covered with fine timber. Our little steamer contained a good saloon, where an excellent dinner was served to a very mixed company. The most marked characters at table were a French baron, absent from home for political reasons, who had been down to Poti to fetch two of Ransome's ploughs for his farm near Kutais, a fat roaring Mynheer-van-Dunk of an official, connected with the post-service, and the captain of the steamboat, a little scrap of a man, who did his best to be polite to a very rough English engineer, incapable of speaking any language but his own, and labouring under the suspicion, for which he probably had sufficient grounds in Russia, that everyone was taking advantage of him in consequence. The amiable captain paid severely for his politeness to our countryman, in being compelled to swallow a tumbler of porter, as a proof of the sincerity of his sentiments.

Before reaching Orpiri, we noticed, on the hillsides, the road which runs to Fort St. Nicholas through the district of the Guriel, celebrated for the personal beauty and picturesque costumes of its inhabitants. The stream was exceedingly rapid, and the steamer had some difficulty in cutting her way past the mouth of the Zenes Squali or Horse River, the largest affluent of the Rion. The villages of Orpiri and Meran stand on either bank. The former—a cluster of wooden cottages, at one of which food and even beds may be obtained—is on the right or northern banks. At Nakolakevi, in this neighbourhood, some antiquaries believe that they have discovered the site of the ancient Aea, whence Jason carried off the golden fleece. Those who wish to read a rationalised view of the early legends of Colchis, and whose feelings can support the intelligence that Jason was, in fact, only the first man who made a rush to the diggings, and that Circe was Medea's niece, a very discreet young lady, who put Ulysses' companions into the police-station because they got tipsy and riotous, but let them out on the entreaty of their insinuating and polished commander—will find all this, and much really valuable information besides, in Dubois de Montpereux's 'Caucase.' Meran is the place of banishment of the Scoptsi, a religious sect whose tenets enjoin self-mutilation. There is no mistaking the appearance of one of these men, who have all the look of loutish old boys; their faces resemble one another, and change little with years. They are said to make honest and intelligent servants, a rare article in Mingrelia, if one may believe the universal report of European residents.

Our voyage ended at Orpiri, whence a diligence starts in correspondence with the steamer to carry on the passengers; but all the places had been secured by telegraph, and not being provided with a 'podorojna,' or order for post-horses, we were obliged to seek some other mode of getting on to Kutais. A peasant's waggon was the only resource; in this we packed ourselves and luggage, and at 6.30 P.M. started to gain our first experience of road-travelling in Caucasia. Our vehicle was a long and narrow trough covered with a tilt, and had no springs or seat. It was drawn by three horses, which however did not drag it at any great pace. The road was level and straight, and as we jolted slowly along, bumping over every stone, we all in turn felt aweary, and wished we were in bed. Sometimes we passed a village where the lights showed that the people were still awake, and we often met waggons, similar to our own, journeying in the opposite direction; between times there was nothing to divert our minds from the perpetual croaking of the frogs, till, like Dionysus in the play, we wished they and their 'quack' might perish together, by a fate similar to that we had seen inflicted on their brethren at Batoum.

Halfway, one of our horses had to be shod; this caused further delay, and we only reached Kutais at 1.30 A.M. A slight descent leads into the town. Passing a barrier, and crossing the Rion by a fine bridge, beneath which its waters gleamed in the moonlight, we drove up to the Hôtel de France, where, after our week's confinement on board steamers, we were glad to install ourselves in a large and comfortable bedroom.

We spent the next two days at Kutais in roaming about, and making arrangements for the drive to Tiflis, where we were anxious to arrive as soon as possible, in order to catch some of the officials and residents to whom we had letters, before they all dispersed for the summer to the numerous retreats in the hills, whither they fly from the heat of the Caucasian capital.

The situation of Kutais, which stands at the point where the Rion emerges from the hills into the plain, is extremely pretty, although the low wooded eminences which surround the place shut out entirely the snowy chain. The view looking southwards, across the Rion basin to the ranges on the Turkish frontier is, in a favourable light, very beautiful. The main part of the town, including the bazaar and the public gardens, is on a level space on the left bank of the river. The houses are all new within the last twenty, and most of them within the last ten, years; the streets are straight, and the shops, fitted up with glass windows in the European style, are under arcades. The principal native articles of manufacture seemed to be silver-work (of which, however, the display is inferior to that at

Tiflis), jet, and quaint-coloured chests. A hatter's shop at Kutais is wonderfully brilliant, owing to the variety and gorgeous character of the headpieces worn by the inhabitants. The shops seemed well stored with European goods, from saddles and flasks, to opera-glasses, goloshes, and cosmetics. We failed, however, to discover any of the famous Circassian cream, of which Western ladies have been known to request friends starting for the Caucasus to bring back a store, in the belief (I need scarcely say unfounded) that it is really a product of this country.

We walked out to a botanical garden which has been established on the opposite bank of the river: here there are shady walks and a greenhouse; and although its present attractions are limited, it will no doubt develope into a very interesting collection of the trees and plants of the country. In the afternoon I called on Count Leverschoff, the Governor of Mingrelia, who was living in a prettily-situated villa outside the town, on the Tiflis road. He was most polite, promised a 'crown-podorojno,' and advised us to send to the postmaster and order a carriage. The postmaster kept Paul for two hours, and then sent him away with a message that he had no carriages at home, and that we must wait for the next diligence, which did not leave for four days. Unaccustomed as yet to the difficulties attendant on all negotiations with post-officials, and deluding ourselves with the belief that a 'crown-podorojno' was treated with some respect in Russia, we were both surprised and indignant at the reply, and I returned to the Governor, to inform him of the result of my inquiries. I was shown into a room most gorgeously decorated in Eastern style; the windows, still unfinished, were draped with Persian carpets, hung as tapestry; others were spread over divans, and one of the walls was decorated with a trophy of Caucasian arms, from amongst which a chamois-head looked down on us. I was promised that the hitch at the post should, if possible, be got over, and received some useful information as to the mountain districts. The Count told me that we should find the Ossetes (a tribe living on the north side of the chain, in the valleys round Kazbek) the 'gentlemen' of the Caucasus; and that Suanetia, the name given to the upper valley of the Ingur, was the most primitive and, in some ways, most interesting district in his government. He spoke in the most glowing terms of the scenery of those parts of the country which he had visited, and of the defile of the Dariel, the beauties and horrors of which the Russians are all fond of descanting on. From a postal map, which he kindly got out, I discovered that seven passes were laid down over the main chain between Kazbek and Elbruz. One of these, the Mamisson—running up the valley of the Rion to its eastern source, and thence descending along the Ardon to Ardonsk, near Vladikafkaz—is a well-known route, and a carriage-road has been traced, though never completed, over it. Of the other six, some at least, as we found afterwards, are mere glacier-passes, used only by the people of the country. All this information was quite new to us, for, during the short time at our disposal before leaving England, we had not succeeded in finding any account of the country between Kazbek and Elbruz, and our programme of ascending those two mountains, and following out the main chain between them, was based only on the German maps of the Caucasus which we could obtain in London.

On Sunday morning the postmaster came to call on us, to say that he had one carriage at home, which should be prepared if we liked it. We went to inspect the proposed vehicle—a long-bodied trap, something like a Swiss 'bergwagen,' which had been disused for some time, and left out in the rain; consequently a small hay-crop was growing inside. The framework, however, seemed solid, and the necessary repairs being promised, we settled to start next morning. This was the first illustration we had of the extraordinary mismanagement of the post-yards in the Caucasian provinces. Every carriage, as soon as it gets out of order or often before, instead of being repaired or kept under a shed, is left to rot and to fall pieces in the open air.

We amused ourselves during the day by strolling about the outskirts of the place, which consist of detached dwellings surrounded by little gardens, and entered a Russian church, where the singing was remarkably good. On the hill on the western bank of the Rion, behind the great hospital which overlooks the town, are the ruins of a very fine Byzantine cathedral. Four lofty pillars, still remaining, once supported a central dome. The porch, now fitted up as a chapel, is very curious, and we remarked the ram's head introduced into its sculpture, as though the legend of the Golden Fleece had been known and appreciated by its builders. In the graveyard near is a very pretty monument, a small bronze angel raised on a pedestal.

On our return we found Count Simonivitch, the police-master of the district, looking out for us: he proposed to make arrangements for horses for our use on the morrow, if we wished to visit the old monastery of Gelathi, some five miles distant; but we were anxious to arrive at Tiflis, and declined his kind offer. The Count proved an exceedingly pleasant acquaintance, and amused us much by his account of journeys in which he had accompanied Sir Henry Rawlinson, for whose knowledge of languages he seemed to entertain a great respect. Amongst other anecdotes, he told us of a curious superstition still prevalent in Armenia. In that country (I have forgotten the locality) is a well named after St. John, which is venerated even by the Kurds. When the locusts eat up the land, a child—too young to have committed any deadly sin—is let down into the well, and brings up a cup of water. The holy water thus procured is scattered over the fields, and in a few hours a miraculous flight of birds arrives, and eats up the locusts.

The public garden at Kutais is a plot of ground the size of a large London square, with walks down the middle, and a few trees, but no flowers; it is, in fact, like an unkempt piece of the Regent's Park. On Sunday afternoons, when a military band plays, it becomes a most amusing promenade, owing to the immense variety of costumes which meet the eye. In this part of the world fashion runs wild in head-dresses. There is first the hideous Russian military cap, white, bulging at the top, and much like a baker's, which some of the inhabitants have the bad taste to adopt: then there is the tall sheepskin hat, like a lady's muff set on end, with a round cloth cap, generally scarlet, to crown the edifice; this has a smaller and humbler relative of the pork-pie order, of the same family is the Tartar cap, conical in form, like a sugarloaf. Besides these the poorer peasants are to be seen in every variety of felt wideawake, from a bell-shaped fancy article, with gilt braid and a button on the top, which looks as if it had been stolen from the great Panjandrum himself, to an almost shapeless piece of battered material. But the two most characteristic headpieces have yet to be mentioned—the 'baschlik,' and Mingrelian cap. The first is a cloth hood with long flappers attached, and is used by both sexes. The men wear them plain, but for the ladies they can be made as gorgeous, with gold embroidery, as the fair owner pleases. When worn with the hood over the head, and the flappers allowed to fall loosely down the back, they give a man the appearance of Touchstone in the play, but the native oftener binds the ends up into a happy combination of a fool's-cap and turban. The Mingrelian cap is a small oval-shaped piece of cloth, or with the higher classes of embroidered velvet, stuck on the back of the head, and fastened by strings under the chin. It is about the size of a fashionable lady's bonnet, and I am disposed to think that some Parisian milliner must have been thus far, and carried home the idea for future use.

A curious legend, illustrating the thievish character of the race, even in the first century, is recounted at Kutais, as an explanation of the origin of this peculiar headpiece. The story runs thus:—St. Peter, who is said to have visited the Black Sea shores, and first preached the Gospel there, was one day travelling through the Mingrelian forest. The saint was on foot, the heat was great, and the road long; he threw off his hat and shoes, and, lying down under the shade of a spreading beech-tree, fell fast asleep. Before long two natives, a Mingrelian and an Imeritian, rode by. They observed the sleeping saint, and the first idea which suggested itself to their profane minds, was to see what they could get out of him. He had no silver belt, not even a dagger, but the discarded hat and shoes offered an obvious booty. The Mingrelian secured the hat, the Imeritian the shoes, and the pair hurried off. Some time afterwards St. Peter awoke, and discovered the robbery of which he had been the victim. Finding his property irretrievably lost, he had recourse to the natural consolation of cursing the thieves, which he did in the following form: 'May the posterity of him who has taken my shoes go for ever barefoot! May no son of the man who has got my hat ever wear one on his head!' From that time no Imeritian peasant has ever had a pair of shoes on his feet, no Mingrelian a sufficient covering for his head.

All the townspeople, except the Russian officials, wear the long cloth frock-coats, reaching considerably below the knees, and confined at the waist by handsomely-worked silver belts, to which are suspended silver-sheathed daggers. The row of cartridge-pouches on the breast, which is *de rigueur* even for small children, is made a vehicle for much tasty ornament, and the binding of the coat and silk undershirt is often of silver or gold braid. This costume gives an air of immense height to the really tall and fine men, whom we often met promenading in twos and threes. The poorer folk cover their shabby garments in great sheepskin cloaks, and struck us as a sleepy inoffensive-looking people.

The women show their half-civilisation by the harsh mixture of colours in their dress. They are distinctly a handsome race, with fine eyes and good complexions; but after the bloom of youth has passed, their features sharpen, and assume a shrewish air, which bodes ill for the peace of their husbands. We saw many faces which might have served as models for Medea, who, as some of my readers may recollect, is described by Propertius as a native of Kutais. The hideous fashion of wearing a great plait of hair, or two corkscrew ringlets, over the cheeks, detracts much from the charms of the modern Mingrelian belles, and the unfortunate spread of civilisation has led them into imitations of Parisian costumes which, as they are out of date by at least three years, are likely to find but little favour in a Western eye. Large crinolines, of the stiffest make, were in full vogue, and a devoted husband—surely in his honeymoon!—was seen on one occasion riding home, with his dagger and sword at his side, brandishing proudly in his hand an iron framework, destined to support the heavy skirts of his spouse. The only trace of local costume worn by the ladies, besides the 'baschlik,' is a Greek cap fastened on the back of the head by a lace veil or handkerchief.

Count Leverschoff has the reputation of being a man of progress, and, with the assistance of his wife, has done much to promote the welfare and gaiety of Kutais, by encouraging balls and theatricals, making all officials wear their uniforms in the streets, and instituting a military band in the gardens. It was curious enough, among such a company, and after an impromptu burst of wild harmony, or (to speak the truth) discord, from a party of country-folk, to hear the band strike up the familiar Mabel waltzes.

The hotel at Kutais is fairly comfortable, and English tastes are well understood, owing to the number of engineers who have been out here for the last few years, to direct the works of the railroad now in course of construction between Poti and Tiflis, which it is proposed to continue, at some future date, as far as Baku, on the Caspian. The mistress is an untidy voluble Frenchwoman, and, as we afterwards learnt to our cost, her promises are in no way to be depended upon.

On Monday morning our trap arrived at the door, soon after the appointed hour. For the first time in our wanderings, we assisted in making our own seats, by twisting a piece of rope in and out of holes left for the purpose in the framework of the carriage, and spreading our plaids on the top. François and Paul had a wooden bench slung forward, and the driver perched where he could. It was soon evident, despite François' determined endeavours, that all our luggage could not be carried with us; and we reluctantly confided our tent, with one of the portmanteaus, to the charge of the mistress of the hotel, who promised faithfully to forward them that evening, by a German carrier. The weather was most lovely, and we set out in high spirits, for our vehicle had springs enough to save us from any painful jolting, and the road, for the first two stages, is excellent. Passing the Governor's house, we emerged on to a common, golden with wild azaleas in full blossom, the perfume of which was delicious. Sharp zigzags led down the opposite side of the hill to a narrow stream, over which a new bridge was being constructed. The road now ran over low wooded hills, the last spurs of the Caucasus, and offered a succession of charming views towards the Turkish or (to use a convenient name suggested by Mr. Palgrave) Anti-Caucasian chain *—large rounded mountains, not unlike the Tuscan Apennines. The whole scenery was delightful, and the country vividly green and spring-like—a striking contrast to the bare brown regions, too common in the East. The azaleas, however, formed the distinguishing feature of the day's drive; the commons were bright with them, the oak-woods sheltered a dense undergrowth of them, and higher in the hills their golden blossoms mingled with the purple masses of the rhododendron, the white flower of the laurel and the hawthorn, pale yellow brooms, and beds of the bluest forget-me-nots. We dipped into a pretty wooded glen, and then came suddenly on the first station— a low white building, which overlooks the basin of the Quirili, a great tributary of the Rion, believed by geographers to be the ancient Phasis. The postmaster was a surly and impudent little monkey, and refused to give us horses, on the ground that our 'podorojno' was made out for two, instead of four, persons. We declined

---

* German geographers seem to have adopted the epithets Great and Little Caucasus, to distinguish the ranges south and north of the basins of the Rion and the Kur.—See Peterman's 'Geographische Mittheilungen.'

to bribe him, and eventually, by the threat of returning to Kutais, to lay a complaint before the Governor, brought him to his senses. His object was to make us take an extra cart with three horses for the servants, by which manœuvre he would have been paid for six horses instead of three.

The second stage was along the right bank of the Quirili, which now flows through undulating country. Snowy peaks, bold in form but of no great height (perhaps 11,000 feet), rose in the distance on our left. Near Simonethi, the second station, we saw numerous clusters of the clean white tents of the Russian soldiery. Up to this point the line from Poti will probably soon be opened; but unless the works are pushed with greater vigour than is now shown, it will be long before the iron road pierces the Suram chain, and reaches Tiflis. The earthworks are being constructed by the soldiers, who, besides being, as a rule, weak physically, are sufficiently enlightened to appreciate the principle of a fair day's work for a fair day's pay, and naturally hold that two copecks a day is amply repaid by a very little work and a great deal of shuffling.

The road now enters the Suram chain of hills, which separate the basins of the Rion and the Kur, and form the watershed between the Black Sea and the Caspian, and the connecting link between the Caucasus and the mountains of Armenia. The Georgian highway, which is very rough and bad for several stages, follows to its head one of the main sources of the Quirili, which has found itself a way through a long and tortuous valley. The scenery consequently changes every minute, and is additionally varied by frequent glimpses up lateral glens. An old castle guards the entrance of the valley; higher up the vegetation becomes richer; box, laurel, and bays clothe the banks, and the beech grows to a great size. A steep hill leads up to a picturesque ivied tower, and a solitary house stands on the opposite bank of the stream, which is suddenly confined between bold precipices of limestone crag, beneath which the road passes. The defile soon opens out, and the third station comes in sight. Here there were no horses to be had, and after an inspection of the stable, to ascertain that we had been told the truth, Tucker and I set off up the nearest and steepest hillside, to while away the two hours we were obliged to wait. A climb of nearly 1,000 feet up a sledge-track brought us to meadows where the hay had just been cut; we now overlooked the lower hills, and had a good view of the finely-shaped peaks which stand in a semicircle round the head-waters of the Ardon, and of an icy mass to the west which we could not then recognise. We returned to the station, to find the expected horses arrived and resting. At last we got them put to, and started. The valley was much narrower; castles peered at one another, like the cat and the mouse on the Rhine, from wooded knolls; and the road was driven into close companionship with the foaming torrent by steep banks clothed in deciduous forest trees. We gained frequent glimpses up lateral glens to the higher snow-streaked ranges on the south.

Our horses were tired, and it was dark before we reached the fourth station, fifty miles from Kutais. It was a wretched place, but there was no alternative; so we stopped, and were ushered into a small room, clean, but furnished only with a long bench. Ham was the only food we could procure; the posthouse itself supplied neither tea, coffee, nor wine, but we got some very strange effervescing drink, said to be made from grapes, at the village store. Although Caucasian posthouses differ too much in their size and internal fittings to admit of any very accurate general description, they have all one feature in common—an absence of comfort paralleled in England only in second-class railway refreshment-rooms.

May 12th.—In the morning our rug-straps were missing, a warning that honesty was not a common virtue here, and at 5 A.M. we were obliged to depart without them. The road, which from this place to Suram is very good, continues to follow the narrow valley, although, leaving the stream, it winds along the northern slopes, making from time to time a long circuit to cross the ravine of a lateral torrent. The hills were covered with timber, resembling that of an English copse, and the azaleas perfumed the morning air. Clusters of untidily-built wooden cottages crowned the knolls on the opposite side of the valley. We met long files of camels carrying merchandise down to the seacoast; many of the young animals were frisking about by the side of their dams, others, too young to walk, travelled strapped on to their mothers' backs, where they seemed more comfortable than might have been expected.

After changing horses at a village close to the top of the pass, we drove over the green ridge, and looked for the first time into Georgia. The day was misty, but I doubt if the view is ever very fine, as higher hills must shut out the great chain on the north. The road, which had been well engineered for the last stage, made itself supremely ridiculous in the descent to Suram, by wandering aimlessly backwards and forwards on the hillside, in enormous and ill-constructed zigzags, by means of which the bottom, with great waste of time and trouble, is reached at last. Suram was in view from the top—a small town, gathering round a castle perched on a bold rock, which stands in the middle of the valley. The station is beyond the town, at the junction of the branch-road from Borjom and Achaltzich, which, in company with the Kur, here emerges from the southern hills through a narrow glen. The aspect of the country had now entirely changed for the worse. Instead of the varied landscape and rich vegetation of Mingrelia, we had before us a rolling plain bounded by distant ranges, so brown and bare as almost to make us fancy ourselves back in Syria again. The 'chaussée,' as the Russians invariably

A Georgian Church.

call a regularly-made road, had come to an end, and we wandered over the fields at our driver's will, selecting the least rough and muddy line of country there might be within a quarter of a mile of the telegraph-posts, which marked our general direction. The novel sight of village churches was, however, a source of interest. In Turkey they are of course unknown, and, except at Kutais, we had hitherto seen little external evidence of Christian

worship in Mingrelia. We also noticed villages of odd
underground houses, or rather burrows, marked only by a
brown dome of earth, and approached by steps descending to a sunken doorway, somewhat like that of an ice-
house; a hole, lined with basket-work, serves as the
chimney to these dreary abodes, and, as François remarked,
one of the little pigs which swarm hereabouts might
easily tumble down and be boiling in the pot before he
well knew where he was. Gargarepi is a large village
buried in fruit-trees, with a handsome church. The
drive into Gori was hot and dusty; the road crosses the
Kur, halfway, by a long wooden bridge.

The station at Gori is on the right bank of the river, but
the town lies about half a mile distant, on the opposite side;
it is picturesque, at a distance, owing to the bold outline
of the castle-hill, and the contrast of colours between the
cool grey of the houses, the bright-green church-towers,
and some red-roofed buildings in the foreground. We
walked into the bazaar in search of novelties, but discovered nothing specially worthy of notice, except a glass
paper-weight with the word 'Balaklava,' and a picture of
our Light Brigade 'sabring the gunners there,' which one
would scarcely have expected to find in this part of the
world. We visited a small chapel, built of ruddy stone,
the front decorated with a large carved cross. In the interior we were shown a finely-illuminated missal, and a
silver reliquary with figures of the Four Evangelists.

At Achalchalaki we forded a stream, which now covered
only a portion of its wide stony bed, and the track then
took for a time to the hills on the southern side of the Kur.
The sky was clear, and, to our great delight, our constant
search of the northern horizon was rewarded at last by
the first appearance of Kazbek. The mountain towers
far above all its neighbours, and, seen from the south,

shows two summits, of which the eastern is evidently the higher. We fancied it looked loftier than any Alpine peak from a similar point of view, and made ourselves happy with the belief that it was too large to be inaccessible on all sides. We descended to a pretty village, surrounded by vines trailed in the Italian fashion, and enlivened by a large encampment of railway workmen, a motley and picturesque crowd of Persians, Georgians, Kurds, and Russians—each nationality easily distinguishable by its peculiar dress.

We now entered a fine defile; the Kur, a smooth swift stream, flowed beneath us in a deep bed, with cliffs on either side, perforated by numerous rock-tombs, for which the most inaccessible positions had been chosen. Where the Dariel road comes in from the north, over a lofty bridge, stands the posthouse of Macheti, the first out of Tiflis. The large building, with its extensive stabling, looked so imposing in the dusk, that François fancied he must be at home again, and wanted Paul to ascertain the hour of the *table-d'hôte*. We had already driven eighty-eight miles, and, wishing to make our entry into Tiflis by daylight, determined to sleep here, as we found we could get some dinner, and hire mattrasses.

May 14th.—We had a drive of twenty versts (or nearly fourteen miles) between us and Tiflis; the first part was exceedingly rough, as the new road and the railway were both in course of construction, and the space between the river and the hill being limited, carriages had for the time some difficulty to get along anywhere. Macheti, surrounded, after the fashion of the country, by battlemented walls, stands on the left bank of the Kur, in a fine situation above the junction of the stream which comes down from the Krestowaja Gora. Once a large and flourishing town, it is now decayed, but contains a curious church, in which many of the kings of Georgia are buried. A castle on an opposite height commands the pass. When the hills retire, and the Kur bends southwards, Tiflis comes into sight for the first time. A bare dull-coloured basin opened out before us, at the end of which, about eight miles off, we could see the buildings of the city, apparently crowded into a narrow space beneath the steep ridges which bounded the view. A more unlovely spot at first sight it is impossible to imagine. The road was nearly finished, but, with the usual Russian habit of leaving difficulties till the last, several steep-sided gullies remained unbridged. Just at the entrance to the town we passed a monument which records the upset of a Czar, caused by one of these perilous descents. Rain began to fall heavily as we drove down the long wide German-looking boulevard. A sharp turn to the left brought us up to the door of the Hôtel d'Europe, which stands in an open square at the back of the opera-house, nearly in the centre of the town.

We had always looked on Tiflis as our depôt and base of operations during the summer months, and we were naturally anxious to ascertain what sort of quarters we should meet with, as the hotels in Russian towns are not always pleasant resting-places for those unaccustomed to the ways of the country. We were therefore delighted to find that our host and his wife were French, and that the house was fitted up in European style. The bedrooms were large and amply furnished, and the beds had good spring mattresses, instead of being (as usual in Russia) mere sofas with hard leathern cushions, and a sheet spread over them. Moreover, the master of the hotel was also the head-cook, and many of our dinners would have done credit to a restaurateur of the Palais Royal; while 'Madame,' besides constantly attending to our comforts, was always ready to help us in our final struggle with some greedy Georgian or Armenian, whose wares had previously taken our fancy in the bazaars.

On this our first visit we spent a week at Tiflis; but after our return from Persia, and again ere setting out on our homeward journey, we made short halts in the same comfortable quarters. I must now endeavour to throw together the impressions which were the result of our several visits. Our first feeling was, undoubtedly, one of disappointment. We had heard one way and another, while in the East, a good deal of the attractions of Tiflis, and now we found a town, which consists of a Russian quarter roughly handsome, and ostentatiously European, and two strangely incongruous suburbs, Persian and German. The covered bazaars of the one are small and, after Damascus and Constantinople, comparatively commonplace; the other is neat and snug, with its 'biergarten' and band, where the German mechanic and 'mädchen' promenade together, fondly and dully, as if in their native archduchy. The environs of the town are certainly not commonplace, but no one can call them beautiful. Bare green downs lie on the left bank of the Kur, and over the town on the right rise steep cliffs of clay, dried and parched up by the suns of many summers.

A better acquaintance, gained by many drives and rambles through the town, greatly modified these first impressions. We found that the Russian quarter contained many well-built private houses and excellent shops, and if the bazaars did not make the outward show of Damascus or Cairo, there was no lack of temptation to spend money within.

The first thing which struck us in the business quarter of the town was the eagerness to sell shown by the occupants of the various stalls. In the East you may generally stop, and turn over one piece of goods after another, and their owner will not deign to interrupt the enjoyment of his pipe until you take the first step by enquiring the price of some article. No such notions of etiquette restrain the hungry-faced Georgian or Armenian artificers. After our first visit to the Silver Row, our appearance was hailed by a crowd of eager merchants, and we were exhorted, and beckoned to on all sides, by rivals for our custom. The shop-fronts are about the size of a small cupboard, and in dark recesses behind, the workmen may be seen hammering out objects similar to those which the master offers you for sale. These are of a varied and attractive character; there are silver-belts—some for men, consisting of handsome links of solid silver—others for women, of lighter and more delicate workmanship. A common conceit is to hang from them a model of a Caucasian dagger, the size of a penknife, neatly cased in its sheath. The Georgian family drinking-cups are both quaint and handsome: some consist of a cocoanut, mounted in silver-work, and furnished with a long straight spout; others have a bowl entirely of silver, and a curiously-twisted mouthpiece, with three funnels, which must, one would fancy, be very awkward to drink out of. They are used chiefly as loving-cups at the family picnic parties, to which the Georgians are much addicted. The big ladles and bowls, hammered out into quaint designs of birds, beasts, and flowers, are also exceedingly handsome. The stalls belonging to one trade are mostly in the same row; close to the silversmiths, the armourers and furriers display their respective wares. Here we saw tiger-skins from Lenkoran, on the Caspian, hung side by side with lamb-skins from Bokhara, and bear-skins from the neighbouring mountains; and had offered for our inspection a choice of every size and quality of dagger and sword, and every variety of flint and steel pistol and gun. One of the traders boasted a medal obtained the previous year, at Paris, for the excellence of his workmanship. The number of wine-skins exposed for sale is another curious feature of the bazaars. The skins are of all sorts and sizes, from that of an ox to that of a sucking-pig. The wine kept in them is generally Kakhetie, the produce of the grapes of the Telaw district, which is very cheap, and is said to have the peculiar properties of curing gout and never causing headaches. Despite these recommendations, the flavour imparted by the skins will prevent most travellers from partaking largely of the commoner sorts. The best quality, after being kept some time in bottles, is a full-flavoured wine much resembling Burgundy.

We were struck with the entire absence of any Turkish element in the crowd, and the consequent want of the bright fezzes which give such colour to the streets of many Eastern cities. Their place is poorly supplied by the tall cloth-caps of the Persians, or the conical sheepskins of the Georgian and Armenian merchants. On the whole, we did not see such variety of costumes here as at Kutais. In a town full of government offices the Russian uniform of course predominates, and the number of unhappy creatures doomed to walk the streets with a sword always dangling between their legs is very great. An occasional turbaned mountaineer from Daghestan, or a handsomely-accoutred Ossete, may of course be met, but here, as at home, the domestic servants are pre-eminent for gorgeous apparel. A Tiflis major-domo is got up regardless of expense; his belt and dagger-sheath are massively wrought in silver, and his cartridge-pouches and fur hat are of the most elegant and expensive kind. Such a costume costs from 25*l.* to 40*l.* Winter is the Tiflis season, and then, I am told, the variety of dresses is really marvellous. As it was, Georgian, Armenian, Persian, Russian, and German make up a fair list of nationalities, and a member of each would probably be met with in a five-minutes' stroll.

One of our first business visits was to the Topographical Department, which was in the same square as our hotel. From the officials there we met with unvarying courtesy, and no difficulty was made in allowing us to purchase any sheets we liked of the Ordnance or (as it is generally called, from being on the scale of five versts to the inch) the Five Verst Map. We inspected with great interest a beautiful relief model of the whole Caucasus, constructed on a large scale, a copy of which has lately been presented, by the Czar, to the Geographical Society of St. Petersburg; and we were also glad to add to our collection a panoramic outline of the chain, giving the heights of the principal summits. These works are most creditable to the officers engaged in them, especially when the scanty time and means at their disposal are taken into consideration.

We had brought with us several letters of introduction, explaining the object of our journey, to the officials attached to the Court at Tiflis, and we were disappointed to find that the Grand Duke had already left for Borjom, his summer residence, taking of course half Tiflis society with him. We were fortunate, however, in meeting Monsieur Barthelemi, Attaché for special missions to the Grand Ducal Court, who was living at the time in our hotel; and through his kindness, we were introduced to the Russian and German gentlemen then resident in Tiflis, who were best acquainted with the natural features of the country. We had the good fortune to catch Herr Abich, on the eve of his departure for Germany, from whom, in the course of a half-hour's conversation, I obtained two hints, which were both afterwards of the greatest service to us—namely, that there was a very lofty névé-plateau, at the northern base of the summit of Kazbek, and that Elbruz might be attacked, with good prospect of success, from the glaciers at the head of the Baksan valley. Herr Radde, chiefly known in England by his Siberian travels, although now settled in Tiflis, as curator of the Museum of Natural History, which has been lately founded, has not given up his roaming habits. He kindly presented us each with a copy of his work, 'Die drei Langhochthäler Imeritiens, Rion, Ingur, and Tskenis-Squali,' the fruit of his wanderings in the southern Caucasian valleys. It is the first German book which has been printed at Tiflis. The museum has not been long formed, but the collection is already most interesting. Specimens of the geology, natural history, the costumes, and household articles of the inhabitants of the neighbouring regions, are grouped together as effectively as the limited space will allow. The most striking object is a magnificent 'auruch' from the mountains west of Elbruz, a region which, now it has been depopulated by the expulsion of the Tcherkessian tribes, will perhaps offer a safe asylum for some years to come to this rare and noble beast. Two very well-stuffed tigers from Lenkoran occupy the middle of a room, round which are grouped bears, chamois, and bouquetins from the Caucasus.

One of the pleasantest of our Russian acquaintances at Tiflis was General Chodzko, under whose superintendence the Government Survey and the 'Five Verst Map' of Trans-Caucasia have been executed. During the progress of the survey he ascended Ararat, and remained camped for nearly a week a short distance below the summit, engaged in scientific observations. The General had also made attempts on Kazbek and Elbruz, but he laughingly admitted that mountaineering had been with him rather a necessity than a pleasure; and he strongly dissuaded us from wasting our time in attempting the higher summits, which, from

his experience, he thought would certainly each take us a
month to vanquish. So anxious was he to put us in the
right way, that he drew for us an itinerary, the fatal objection to which was that no one but a Russian could
expect to survive sixty miles a day, for five weeks, of post-
travelling over the steppes, swamps, and boulders which
are called roads in the Caucasus. The General's kindness
did not end here, for he constituted himself our 'cicerone,'

The Georgian Castle, Tiflis.

and took us a round of all the sights in Tiflis. From the
Persian quarter we climbed, by a very steep road, to the
Botanical Garden, which is 'sown in a wrinkle of the
monstrous hill' overhanging the town. The southward-
facing slope of a narrow glen has been cut and built up
into terraces, planted with rare trees and shrubs, and connected by vine-trellised paths and flights of steps. Over-

head are the ruined towers of the old Georgian castle; below, a stream, scanty in summer, has worn a deep ravine, the bare hillside on the further bank of which lends a charm, by contrast, to the fresh vegetation and shade of the garden. There are several shallow caves in the rocks which support the castle ruins, where the townspeople used frequently to resort for family picnics, a kind of entertainment beginning with a light meal and frequent passage of the loving-cup, and carried on by story-telling, music, and dancing, until late in the evening.* The return home was a service of some danger, since the road is exceedingly steep, and the drivers were apt to refresh themselves at a wine-shop near the gates of the garden. Upsets and accidents used to be of frequent occurrence, and perhaps this, in conjunction with the making of the new gardens near the Grand Ducal palace, has served to render this pleasant retreat no longer fashionable. A zigzag path leads up to the ridge, on the outmost crags of which the castle stands. A wall runs along the top, and when the door in it was unlocked, a grand general view of the city burst upon us. Directly below were the straight streets and gaily-coloured houses of the Russian quarter, the bright roofs of which formed a pleasant contrast to the cool grey of unbaked bricks in the Persian town. In the distance we looked straight up the valley of the Kur, to the wooded hills behind Mscheti, and (had it been clear) to the snows of the great chain. Whenever this is free from clouds the double head of Kazbek is a conspicuous object from Tiflis. It is seen, together with several lower snow-peaks on its left, from the boulevard, and from many of the houses in the town. General Chodzko pointed out the watercourses, made partially for irrigation, but also to check the floods, to

* The Czar has a very pretty watercolour drawing of one of these parties in his study at Livadia, the Empress's Crimean villa.

which Tiflis is subject from the hills above. After any
heavy rain, torrents pour down every street, and we were
told it was no unusual thing for children to be drowned in
the middle of the city. We were next taken to the principal covered bazaar (an arcade about the size of the
Burlington), where the goods exposed for sale are mostly
European. Below the castle, and between it and a commanding spur of the opposite hills, also fortified, the Kur
is so closely confined between high banks as to be crossed
by two bridges of a single span. Tall houses, with
balconies over the water, are built on either shore, and the
river, beaten back and turned at a sharp angle by the
right-hand bank, rushes away with a fine swirl of water,
which must put a stop to all navigation. Wood is brought
down as far as Tiflis in large timber rafts like those of the
Rhine. We often admired the adroitness of the steerers,
but the Kur is not easy to navigate, and accidents sometimes happen. Near Gori we had seen a crowd on the
river-bank, and been told that a raft had capsized, and two
men were drowned. On the left bank of the river we
visited an interesting old church inside the fort I have
mentioned, and a large building—partly used as a warehouse, partly as offices by Persian traders. We returned
by the new bridge, a handsome stone structure of several
arches, at one end of which stands a statue of the Prince
Woronzoff, who did so much for Tiflis, Odessa, and the
Crimea.

We went twice to the Opera, a pretty house in Moresque
style, and heard 'La Traviata' and 'Faust' very fairly
performed by an Italian company, but what amused us
most was a farce, in which an English tourist played the
principal part. He was drawn, not after real life, but
after the caricatures of the boulevards, immensely tall,
enveloped in a plaid, and with long sandy whiskers.

When he refused to fight a duel, because the pistols provided for him had not been made at Birmingham, the mirth of the audience reached its climax. The rest of the time, not occupied in making arrangements for our further journey, we spent in some tempting shops of the European style near our hotel, where a greater choice was to be found than in the bazaars, though the prices were somewhat higher. Our window furnished an amusing lounge at spare moments, for the wood-market was held in the square below, and from an early hour in the morning it was filled with carts, drawn by scraggy buffaloes, and a constant jabber of bargain and sale went on all day, wet or dry. Once we saw a funeral pass: the coffin-lid was carried first, then the open coffin; the body was gaily dressed, and covered with flowers, and the priests who accompanied it raised a fine chant as the procession moved onwards.

We arrived at Tiflis on May 13th; but as Mr. Moore was not to join us till June 20th, we had five weeks at our disposal, and determined to employ them in a run into Persia, combined with a visit to and, if possible, an ascent of Ararat.

Before we could feel ourselves in order for this journey, much had to be done. Only those who have been in Russia can understand how officialism may be brought to bear on every detail of travel, and a man must go to the Caucasus to appreciate how a great system like the Russian post, which would be admirable if carried out properly, can become, by imperfect organisation, and gross incapacity and dishonesty on the part of those employed, a positive hindrance to travellers.

Our most important needs were to have our passports properly signed for leaving and re-entering the Russian dominions, and to obtain an order for horses, without which it was useless to make enquiries at the post. Mons. Barthelemi kindly came to our aid, and offered to introduce us to the Governor of Tiflis, who received us most cordially, and promised that all necessary documents should be prepared forthwith. In due time, we received special passports for leaving and re-entering Russia, a 'crown-podorojno,' good for all Trans-Caucasia, and an order for a Cossack escort wherever it might be needful.

On Monday morning I sent Paul to the post, to order horses for the next day, and to ask some questions about the different kinds of carriages and their cost. He returned with a message, that we might perhaps have horses in two days, and no answer to my enquiries about carriages. I immediately drove to the post; the postmaster was for once at home, but was excessively offhand in his manner, and tried to walk away while I was talking to him, a manœuvre only prevented by my placing myself between him and the door. Fortunately, I had occasion to see the Governor that evening, and I took the opportunity of calling his attention to the way in which a crown-order was treated by minor officials. My remonstrance at head-quarters had its effect, and no further difficulty was made in providing us with horses; but as to a carriage, our Russian acquaintances agreed in advising us to reconcile ourselves to the carts of the country, and their reason afterwards appeared in the fact, that the road, halfway to Erivan, was for the present impassable for spring-vehicles, and that we should, therefore, according to the rules of the post, have paid for our carriage without having the use of it. Meantime day after day had passed, and still our luggage did not arrive from Kutais. We were consequently obliged to make up our minds to start without our tent or our mountaineering boots, which we had hoped to make use of on Ararat.

Our ice-axes, however, of which we had brought out only the heads and spikes from England, had, under François' supervision, been mounted by a French workman, and were now ready for use. The weather was not brilliant, we were annoyed at the non-arrival of our luggage, and the appearance of the trap provided for us by the postmaster of the Transcaucasian capital was admirably adapted to render still more surly the 'winter of our discontent.'

A 'paraclodnaia' (so great a name does the country cart bear in the Caucasus; in Russia proper it is oftener called a 'telega') is the ordinary conveyance of the Russian posts, and the only one to be obtained at any but the largest towns; even at so considerable a place as Erivan, nothing else was procurable. This hateful vehicle is so bad as to be almost beyond description. The body of the cart is sometimes flat-bottomed, like a punt — sometimes rounded, like a tub boat; the boards of which it is composed are ordinarily rotten, and nails stick out wherever they have a chance of injuring the clothes or flesh of the occupant. The driver sits on a plank in front, while the travellers, if they have any experience, carefully draw and tighten a piece of rope, through holes left for the purpose, until a sort of cat's-cradle is contrived at the back of the cart, on which they spread their rugs and seat themselves. None but a native could bear to lie on a quantity of hay at the bottom, and allow himself to be jolted like a pea in a rattle. The body of the cart rests on two blocks of wood, which are in their turn directly supported, without any intervening springs, by four wheels of the rudest construction. There are, however, degrees of badness even in the framework of a 'paraclodnaia'; if the framework and the road are ordinarily bad, the jolting is painful; if either is very bad, it is maddening. The luggage, which is generally stuffed under the rope-seat, has, as well as the seat, to be rearranged six or eight times a day, as the conveyance is changed at every station. Nothing of glass can be carried without breakage, and if the road be muddy, clothes and face are covered in five minutes with a thick layer of dirt. Three horses draw these traps; the two trace-horses are quickly fastened on either side; the centre animal goes between the shafts, and over its neck is fastened the 'duga,' or wooden arch, to which one or more bells are attached—probably intended, by their incessant clang, to drown the groans of the suffering travellers. Scarcely a stage passed without our having to stop in the middle of it, to rearrange this clumsy structure, in the beauty and fitness of which the native drivers seem to have implicit belief, and to which they attach a sort of mystic importance.

Such are the carts which the Imperial Government provides for its couriers! Its traditional policy seems to have been to develope towns, and supply every luxury and amusement for a swarm of official drones; while commerce and industry were discouraged by the neglect of the communications of the country, for which the Government, by keeping in its own hands the entire management of the roads and postal system, had made itself responsible. What provision it does make I have endeavoured partially to show; but it would fill a volume to narrate all our own experiences, and the stories we heard from other, and partly Russian sources, both of the badness of the roads, and of the insolence, ignorance of truth, and rapacity of the postal officials. Imagine a place like Tiflis, the residence of a brother of the Czar, a town of 80,000 inhabitants, with a large European society, and an opera-house, unconnected by any pretence of road with the Black Sea coast, the

Caspian, or the Persian frontier! A more enlightened spirit now happily prevails in high quarters, and the Lieutenant of the Caucasus has ordered the construction of roads to Kutais, and to Erivan, while the highway of the Dariel is almost completed; but the dawn of intelligence is late, and light spreads but slowly through the dense mists of jobbery and peculation which impede, if they cannot stifle, the coming of a better day for this as yet undeveloped region.

In no country has the transition from utter want of the means of transport to the facilities of a large railway system been so sudden as in Russia, and the extent of the change likely to be produced there during the next few years can scarcely be exaggerated. Amongst its smaller results will doubtless be the sweeping away of those petty but troublesome safeguards with which police and post officials combine to hinder and render disagreeable all travel in the interior of the country.

CHAPTER V.

THE PERSIAN POST-ROAD.

The Banks of the Kur—Troops on the March—A Romantic Valley—Delidschan—A Desolate Pass—The Gokcha Lake—Ararat—Erivan—The Kurds—The Valley of the Araxes—A Steppe Storm—A Dangerous Ford Nakhitchevan—A Money Question—Djulfa—Charon's Ferry and a Modern Cerberus—A Friend in Need—A Persian Khan—Marand—Entrance to Tabreez—Chez Lazarus.

We left Tiflis on May 20th. The road, once clear of the rough pavement of the capital, follows the valley of the Kur. The suburbs of Tiflis stretch far in this direction, and the views of the town and castle from this side are often striking. At the first station we got a better 'paraclodnaia,' which, by bribery and argument alternately, we contrived to keep for several stations, and we further improved our condition by taking a second cart for the men, which was a great boon, both to them and to us, as we had been sitting previously in a terribly cramped position. The new high-road from Tiflis to Erivan, which will run along the Kur valley, is yet unfinished, and we had to make a détour of four stages, over low hills and high plains, before we rejoined the river. The steppe was fortunately, for once, in fair driving order, and we made good progress. A curious circular hollow, containing a lake at its lower end, is crossed before reaching Kody, the second station, distant twenty-seven versts* from Tiflis by the road, but only twelve by the short cut over the hills, which we made a

---

* Three Russian versts equal two miles.

great zigzag to avoid. At the third posthouse we crossed a small stream, the Algeth, and stopped half an hour to lunch on cold turkey; for we had profited by experience, and started well supplied with provisions. We now followed, for some twenty-six versts, the valley of the Khram, a large tributary of the Kur, which is principally fed by the streams from the Mokruja Gora, and other chains which lie between Tiflis and Alexandrapol; their wooded summits, broken here and there by castellated crags, formed the southern horizon. At the Red Bridge we were delayed for want of horses. This brick structure, the centre arch of which is of considerable span, is a relic of Persian rule, and is probably the oldest in the country. It was repaired by Rostom, king of Georgia, in 1647, and still remains perfect, although it is being slightly widened for the new road. The Russian engineers must have been relieved to find so substantial a structure ready-made, as bridge-building is not one of their strong points, and they too often neglect either to make proper approaches, or to direct the course of the stream with dykes. In consequence, their arches are often left high-and-dry, with the water sweeping over the road a hundred yards on one side of them. We had now returned to the Kur, which has here entirely quitted the mountains, and entered on the dull green steppes, through which it winds a weary way to the Caspian. Its bed, a belt of swamp and forest, is considerably below the general level of the plain, which breaks suddenly into it. This part of Georgia is exceedingly wild and thinly populated, and forms the borderland of the steppe country, inhabited only by wandering tribes of Turcomans and Kurds. We passed several underground villages, the existence of which is indicated, at a distance, only by a brown blotch on the surface of the plain; on nearer approach a low mound of earth, with perhaps a thin column of smoke issuing from it, shows the position of each house.

The men wear the great sheepskin coat and the conical fur hat, the women dresses of crimson-lake hue, which lit up wonderfully the dull green landscape. Every half-hour we came to a weird group of ruddy tombstones, averaging six feet high, and often delicately carved; they resemble upright sarcophaghi in shape. These strange graveyards make much more show than the villages. Companies of camels, their day's work done, and their heavy cotton-bales ranged in a circle, sauntered lazily about in search of herbage. Gaily-feathered birds perched on the telegraph-wires, which were our constant companions and guides, scarcely cared to fly away as we passed. It was hard to realise that it was scarcely twelve hours since we had left a town supplied with every European luxury.

The seventh station was a mere Tartar hut with a large underground stable. The post-horses had first to be driven in from the steppe, and then harnessed; a party of very merry-looking natives did both in less time than an ordinary postmaster would have taken in examining a 'podorojno,' and reflecting what he should write on the back of it. Night was now coming on, and we quickened our pace. Lighted by a rising moon we cantered over the plain, passed a large stream, flowing towards the Kur, by a crazy bridge, and five minutes afterwards alighted, stiff and weary, at the door of the large posthouse of Akstn-flnsk, situated at the junction of the Erivan and Elizavet-pol roads, and 110 versts from the capital. Unluckily for us, a large detachment of troops, on a roadmaking expedition, had halted here for the night, and the resources of the house were employed in providing for the comfort of the officers, who occupied all the accommodation. Under such circumstances we, who wore neither official caps nor decorations, could not expect, and did not meet with, even the commonest civility and thought ourselves lucky when the surly postmaster, with a very bad grace, accorded us permission to roll ourselves up in our rugs on the floor of his room—quarters which we shared with a huge dog conscious of fleas, and consequently provokingly restless during the dark hours.

May 21st.—We were awoke from sleep, if the uneasy rest we obtained deserved the name, by a most horrible discord, some idea of which might be obtained by hiring itinerant performers on the bagpipe and barrel-organ to play different tunes simultaneously. The soldiery were starting, and their drum-and-fife band was cheering them on the road. I imagine that the dregs of the Russian army are kept in Trans-Caucasia; anything more wretched and slovenly than the uniforms, marching, and general appearance of these men we had never seen even in Turkey, but the work in which they had been employed might account partially for their unsoldierly aspect, as some of the Russian troops we saw afterwards were very different.

We changed our direction to-day from east-south-east to nearly due south, and entered a valley among the hills which separate Georgia from Armenia, the basin of the Kur from that of the Araxes. Half an hour was spent in passing the troops and their long trains of baggage-waggons. The road was narrow and bad, and our driver timid, but at last we left even the vanguard behind. Our course lay along the banks of the stream we had crossed overnight, which were ornamented by magnificent forest-trees. The morning was lovely, soft clouds were clearing off the hills on the south, while the snowy crest of the Eastern Caucasus ran along the northern horizon, rising beyond the Karaja steppes. The next time we saw it was from the slopes of Ararat. The depression in which the river

flowed gradually narrowed and deepened, and after leaving
the station where we first changed horses, we fairly
entered the hills. The track was rough, the pace slow,
and the jolting incessant. The second station was in a
considerable village; the houses, one-storied and flat-
roofed, were built up the hillside in an angle of the
beautiful valley, the windings of which we were now
following. Up to this point the scenery had reminded us
of some of the more richly-wooded parts of Wales on a
larger scale—henceforth it grew bolder, castellated crags
alternated with forest-clad slopes, snow-streaked summits
appeared in the background, the stream danced and
sparkled at our side; every prospect was pleasant, and the
road alone was vile. The mountain-sides were abrupt and
picturesque, and the richness of the vegetation suggested
a comparison with the neighbourhood of the Italian lakes.
We were struck by a very curious rock-formation, which
at a distance gave the face of the cliffs and the porphyry
fragments strewed at their base the appearance of masonry.

At our third halting-place, a solitary house, no horses
were to be had for two hours. During our compulsory halt
here we were amused by a struggle between a horse and its
master. The animal bolted into the stream to escape
capture, and was carried down for some distance by the
force of the current; on regaining its feet it came wisely
to the conclusion that captivity was better than a watery
grave, and quietly surrendered to its owner. We walked
on for five versts, leaving the men to look after our traps.
This stage was very heavy and hilly, and even after our
conveyances caught us up, we often preferred to walk, except when for a few versts we had the advantage of a
finished piece of the new road. The valley now opened
out, and everything showed we were approaching a more
elevated region. We passed a hamlet on the side of a

pretty wooded basin, out of which we climbed by a long
ascent, and then wound along, or rather up and down,
the slopes into Delidschan, a large village situated at
the foot of the pass into Armenia, looking up a wide
and somewhat bare upland valley, along which runs the
road to Alexandrapol. We had only travelled sixty-
eight versts during the day, but it was too late to cross
the pass. The horrible jolting of our carts had given us
all headaches, and made us feel generally out of sorts, and
we had made up our minds to give them up and take to
riding, which now indeed became a necessity, as the stage
over the mountain was impassable for carriages, owing to
the destruction of the old track by the works for the new
road. Delidschan turned out to be the destination of the
troops we had seen in the morning, and there was already
a considerable force collected in the white tents picturesquely grouped in the valley below. The evening
was cold, for we were at a height of 4,230 feet, and we
were glad to solace ourselves after our fatigues with a
brew of mulled wine.

May 22nd.—Heavy rain was falling when first we looked
out of window, and we set about our preparations in a
gloomy frame of mind. Our luggage was soon packed
on horseback, and, mounted on animals more used to
draw than to carry, we formed a very queer cavalcade
when we started for the ascent of the pass. Paul, like
most of his race, was a good but rough rider, and
bullied his beast, until the animal plunging, and the
saddle turning, gave our friend a tumble and a lesson in
moderation which was not unneeded. The road led up a
lateral glen of the valley we had left, through forests
carpeted with cowslips, and past several villages, untidily
built of wood. On the bare slopes, near the top, we found
the soldiers at work; they were blasting a terrace for the

new road, above the line of the old, which was almost covered with the fragments sent down, amongst which we had to pick our way under a desultory fire of small stones from above. The watershed between the Kur and the Araxes is here a broad grassy ridge, on which the snow still lay in patches; the rich herbage has given the name of the 'Eshak Meidan,' or 'donkey's pasturage,' to the pass, from the custom of wayfarers to reward their beasts, after the labour of the ascent, by turning them out to graze. It does not seem to command much view. For some distance we bore to the right, with but little descent, until presently as much of the big Gokcha Lake as the mists did not enshroud came into sight. Size seemed to be its chief merit; there was not a tree to be seen, the ground had a dull and sodden appearance after the heavy rains, and the surrounding mountains, though many of them are 10,000 feet in height, produced but little effect from being seen over the broad surface of a lake, itself 6,000 feet above the sea-level. The day was very unfavourable, and the low clouds, which swept rapidly across the landscape, added to its grim and desolate character. Here, as elsewhere, we noticed the great difference between the northern and southern slopes of the Anti-Caucasian chain. All the valleys facing northwards, towards the Kur, are full of luxuriant vegetation; while the southern slopes, falling to the Araxes, are always bare, burnt, and arid in summer, and swamps in the rainy season. Tucker's horse evidently had a dissipated owner; it made a dead halt at every drinking-shop on the road, and its misdemeanours culminated at a village just below the pass, where no persuasion could get the brute past a well-known halting-place. It was not a pleasant spot to dismount, for the mud was deep; and finally, I had to return, to withdraw ignominiously my friend and his misguided beast from the scene of temptation. Unavailing as the hunting-whip he always carried had just proved, it effectually took the bark out of two dogs that ventured to set up derisive howls at the discomfited horseman.

Having reached a better piece of road, we trotted briskly on in pursuit of Paul, who had been sent ahead to procure horses at the next station, leaving François, a safe but not brilliant rider, to keep an eye on the baggage. The post-people refused to give us riding-horses, on the ground of having no saddles or bridles. We had learnt by this time to be too glad to get on anyhow to argue the question, and thankfully took possession of the miserable 'paraclodnaia' provided us. The posthouse was close to the lake, along which our course lay for the next stage. The shores are steep, and the road consequently climbs up, down, and round the promontories, occasionally venturing on a pitch about the steepness of an ordinary house-roof. A rocky island, about a quarter of a mile from the shore, and a village built on a bold peninsula, are the only objects which seem worthy of notice. Elenovka, situated at the west end of the lake, where the shore is low, and the waters find an outlet, stands in a wide sea of mud, with hovels, arranged more or less in the form of a street, scattered amongst it. A strong odour of dried fish revealed at once the staple of industry; the salmon-trout of the Gokcha Lake are famous, and are sent both to Tiflis and Erivan. The postmaster was a Jew, and talked a little German; but he had neither horses nor 'telegas,' and did not seem to know when he was likely to have them. After a delay of three hours we got two horses, and, with Paul as my companion, I pushed forward, to make, if possible, all ready at the next station; as we were unwilling to give up the hope, faint though it was, of reaching Erivan that night. Fancy the wildest, ugliest part of Wales in bad weather, with mountains, swamps, and rainstorms all on an enlarged scale, and some feeble idea may be formed of this part of Armenia, as we saw it. The track—a broad belt of mud streching across the swampy downs—was not difficult to find, despite the driving mists; the carcase of a camel, or a dying horse, by the wayside, and the telegraph-wires singing a quiet tune of their own overhead, sufficiently revealed the whereabouts of what the Russians naïvely call a road, and illustrated the happy definition we afterwards heard, 'Une route de poste en Caucase, c'est où il y a ni route ni chevaux.'

A short descent brought us to Achta, after a ride of sixteen versts, during which my steed and I, to my great surprise, did not once part company; as trotting with a saddle about the size of a lady's bonnet, and stirrups which exalt your knees to the level of your face, is an exercise more sensational than safe. Tucker, François, and the baggage arrived in due time, and we continued our journey in carts. The stream from the lake, reinforced by contributions from the western range, flowed in a deep depression on our right, while we continued to traverse the swampy downs which spread round the base of Ak-Dagh. So deep was the mire that we could seldom get beyond a foot's pace, and it was dark ere we reached the next station, a lone house standing in a hollow, near the base of a conical hill of apparently volcanic origin. Soon after our arrival a 'tarantasse' (a carriage with a hood and rough springs) drove up. Its occupants were a gentleman, a member of a Greek firm at Teheran and Tabreez, who spoke a little English, his wife, and a lady's-maid. He told us that Mr. Abbott, the English Consul-General, was at present at Urmia, but that we should find a hospitable welcome at Tabreez from Dr. Cormick, the English physician in charge of the heir-apparent of Persia, who holds his court at Tabreez. He also gave us a note to an old servant of his firm residing at Djulfa, on the Russo-Persian frontier, which proved invaluable. In return he asked for information as to the road we had travelled. We were sorry not to be able to give a better report, as he was considerably perplexed how he should get his wife and heavy baggage over the pass. He started, however, next morning, to try his luck, and we afterwards heard of his safe arrival at Tiflis.

May 23rd.—The station supplied a comfortless room, where I was privileged to enjoy repose on a sofa which had once been stuffed; but as the middle had, by some incomprehensible means, risen two feet higher than either end, my position was somewhat constrained, and I envied Tucker his level, if hard, boards. The morning was tolerably fine, and on starting we traversed uplands of the same description as yesterday's, only that we could now distinguish the snowy summits of Ak-Dagh on the left, and Alagoz, with two summits about the size of the cone of Piz Languard, perched on an enormously bulky base, on the right. In front a rise of the ground shut out all beyond. As we overtopped the brow, Ararat burst suddenly into view—a huge but gracefully-shaped mass, rising to a height of 16,910 feet, from a base of about 3,000 feet. It stands perfectly isolated from all the other ranges, with the still more perfect cone of Little Ararat (12,840 feet) at its side. Seen thus early in the season, with at least 9,000 feet of snow on its slopes, from a distance and height well calculated to permit the eye to take in its true proportions, we agreed that no single mountain we knew presented such a magnificent and impressive appearance as the Armenian giant. I can only compare it to the popular idea of Atlas— a huge head and shoulders supporting the sky. One is ready immediately to admit that the Ark must have grounded there, if it grounded anywhere in these parts.

François went off at once into somewhat Colenso-like speculations as to the mode of the elephant's descent, and how many years the tortoise must have taken to reach the bottom; and he was scarcely satisfied with the suggestion that the tortoise turned on his back and made one long glissade of it, while a 'special' avalanche was engaged to transport the more unwieldy animals. As we thus examined and talked over the mountain, we were glad to observe that, though steep, it did not appear to offer any serious difficulties; still the quantity of snow was so great that we decided to postpone our attack on it till after our return from Persia, when we might hope to find the weather more settled, and the slopes in better order for an ascent. Already we fancied we discovered a 'Grand Mulets' in a rocky tooth which projected from the eastern side of the greater peak, and anathematised anew the faithless Frenchwoman at Kutais, by whose carelessness in failing to forward our mountain-tent we were likely to be prevented from sleeping at so great a height. After we had gazed our fill at Ararat, we noticed that away to the west spread the broad upper basin of the Araxea, bounded by the snowy mountains of Armenia and Kurdistan. The last station before Erivan, perhaps an old caravanserai, stands in a most picturesque gap, the rocky sides of which serve as a frame to the stupendous snowy mass of the two Ararats. We were required to write our names and nationality, by an officer of unusually courteous manner. A few versts more, over bare downs, where herds of camels were picking up the scanty herbage, or kneeling with lugubrious grunts to receive their loads, brought us to a large village on the verge of the last steep descent into the valley. Thence we got our first view of Erivan; a grey flat-roofed town, nestling under the shelter of the hill we were about to descend, and embowered in groves of lime and acacia—a pleasing contrast to the bare plain around. We were driven through its wide streets, which have the unkempt air of most provincial Russian towns, to the post-station; but not liking the quarters there, we insisted on returning to the 'Gostenitza Ararat,' where we found a tolerable room, clean beds, and excellent food, including even such luxuries as coffee-ices. Erivan is a place which belongs to no one nationality, but shows in its buildings, and still more in the crowd in its streets, the traces of several. Two-storied stone houses, wide streets, an abundance of town-carriages, and an untidy public garden, where a military band performs every evening, mark the presence of Russian rulers. The bazaars are thoroughly Eastern, and a stroll through them will be sure to afford some amusement. The large open space between the public garden and the fortress was always crowded with camels and bales of merchandise. The principal mosque, standing on one side of a quadrangle, is covered externally with blue tiles, which give its minaret a very bright appearance. There is a certain Persian element about the place, which manifests itself most prominently in the paintings with which any blank space of wall was decorated: here of a company of high-capped horsemen, there of strange wild beasts, amongst which the Persian lion—a near relation of our red lions at home, with a sword in his paw, and the sun rising out of his back—took the first place. Russian, Persian, Armenian, Kurd, and Tartar jostle one another between the stalls, and it is strange to reflect on the different pasts, and probable futures, of the races they represent. Now you pass an Armenian priest or merchant, distinct in type from the Russian—like the Greek clever and successful as a man of business, and renowned throughout the East for his sharp practice, and yet, also like the Greek, incapable of

combining to form a wise polity, and insensibly yielding to
his destiny, soon to be merged with the slower but more
steadfast Russian. By his side may be seen the Kurd
chieftain, from the slopes of Alagoz, or Bingol Dagh,
armed with a round leather shield and dagger, Turk or
Russian, as suits his convenience—in reality paying
neither allegiance nor dues to any man, and looking what
he is, the free Arab of the mountains. The strong
pressure from without is producing some sort of union
amongst the Kurdish tribes; they are constantly reinforced by emigrants from the Russian side of the
Araxes, and, if we may believe one who has had long experience of Oriental races, it is to them we must look for
the continuation of the struggle against the northern
flood, which has now finally swept over the Caucasus, and
is breaking round the slopes of Ararat.

In the afternoon we called on the governor of the
province, who spoke excellent French, and showed a
desire to render us all the assistance in his power. We
learned that the Araxes was impassable, which of course
confirmed our intention of postponing our visit to Ararat.
The governor gave us letters to the commander of the
district of Nakhitchevan, and the Colonel of Cossacks
stationed at Aralykh, the frontier-post on the further side
of the Araxes, at the foot of Ararat. This letter we hoped
to use on our return from Tabreez. We sent Paul to
endeavour to procure some kind of spring-vehicle, but no
such thing was to be had in Erivan.

May 24th.—We were off at 4.30 A.M., in the usual
'paraclodnaia'; both the carts and the road were horrible
for the first four stages. The ruts in this part of the
world run across, instead of parallel with, the track, and
in consequence inflict a series of short sharp jolts on the
unlucky traveller. During the second stage we passed

between a succession of orchards and vineyards, nourished by a careful system of irrigation, which has made this part of the Araxes valley like a great garden. These vines are locally reputed to be descendants of those planted by Noah after the Deluge, and some support is given to this tradition by the fact that the juice is still famed among the Russian officers for retaining the peculiarly intoxicating quality it possessed in the days of the Patriarch. From Kamirlu, a large village where the track to Aralykh turns off, the view of the two Ararats, now close at hand, is superb. The Little Ararat, on the left, is a perfect cone, looking a volcano all over; the higher mountain rises from the gap between the two summits, in a long slope, broken, about 3,500 feet below the top, by a huge rock-tooth. The snow-dome falls away gently towards the north for some distance, and supports a large névé-plateau, below which the mountain breaks down steeply for several thousand feet.

We now left cultivation behind, and drove at will over the grassy steppe, passing every now and then a group of Turcoman tents, large and comfortable erections. A sheet of dingy canvas forms the roof, while the sides are constructed of wicker-framework, hung with gay-coloured carpets, woven by the women. During the fifth stage, we passed through a curious gap in a range of hills, which ran out from the barren chain on our left, and came in view of a broad lagoon formed by the flooded Araxes. Clouds had been for some time gathering round Ararat, and now slowly detached themselves, in black masses, from its sides. We noticed, first, apparent puffs of smoke on the further side of the valley; then a pillar, as of cloud, rose into the air, and swept towards us across the Araxes; with it came the howling wind, lashing up the waters of the lake; and a minute afterwards, we were overtaken by a storm of rain, which laid the dust-cloud which had been its forerunner. The storm did not last long, but it made a great impression on us at the time, though it was a trifle to one we afterwards encountered. In the worst of the wind and rain, our driver pulled up short, and jumped down to secure two bales of serge which had fallen from the back of some overburdened camel, and lay on the steppe the prize of the fortunate first-comer. The driver of the second cart claimed his share of the booty, and we were obliged to insist on their postponing the argument of the case until our arrival at the next station. There we received the unwelcome intelligence that it was 140 instead of 80 versts, as we had previously believed, from Erivan to Nakhitchevan, and that immediately before us lay a large river, now dangerously swollen, which it was doubtful whether we should be able to cross.

Reflecting that, in the present state of the weather, the stream was more likely to be larger than less next morning, we determined to take our chance, and declined to follow the rustic policy of waiting till the river should sink, recommended by François, among whose strong points fording of rivers was not included. We very soon came to the brink of a formidable-looking stream, but our driver was plucky, our horses faced the water bravely, and, piloted by a native horseman, we emerged safely on the further bank. In the same way we crossed a second branch of the river, and were just congratulating ourselves on the ease with which we had vanquished the enemy, when the third and last branch came into sight, as big as the other two put together. Before venturing on the passage we confided the saddlebags, now our only luggage, to a horseman who was to precede us, and took off our boots and socks, in case the water should come into the cart. The stream before us was 100 yards wide, and was coming down from the hills in lumps of brown water. Fortunately, we had plenty of assistance in our difficulties, from a band of natives who were on the look-out to earn some honest copecks. One rode alongside, to direct and cheer the horses; two others, half-naked, hung on to the side of the 'paraclodnaia,' which met the force of the stream, and helped to prevent our being borne away.

When all was ready we plunged in. The stream, four feet deep, poured through the cart, but the horses fought gamely, and we soon found ourselves in shallow water, in the centre of the flood; another plunge, another sharp but short struggle, and we were landed in safety on the further bank. As we looked back on the cart containing Paul and François, still surrounded by the water, the burly form of the latter, standing erect, to escape wetting, reminded us ludicrously of Pharaoh in the Red Sea, as represented in children's Bible-pictures. After distributing a well-earned 'backsheesh' among the men who had aided us, we pursued our journey over the wide dull plain. Our driver was, in more ways than one, a cool hand; and having done his business so well in the passage of the river, seemed to consider he was now entitled to take his pleasure, which he did by deliberately pulling up at a halfway house, and keeping us waiting while he took a glass of 'vodka' and smoked a pipe. As rain was again beginning to fall heavily, our patience did not endure long, and having captured our sybarite, we induced him to hurry on to the next station, which was in sight, on a low hill at the further end of the plain. On arrival we were met with the dismal intelligence that there was nothing to eat, but, like most official assertions, this turned out not to be strictly true; and Paul succeeded in unearthing new milk, tea, and eggs—so that, with the chickens and cheese we had brought from Erivan, we made no bad supper. If the sleeping accommodation had been better, we should have had no reason to complain. In this, however, lies the cardinal defect of the Russian post-stations; in no single one of those between Tiflis and Djulfa, a distance of 300 miles, can a mattress or a blanket be procured for love or money. The eye, in its despairing search for creature-comforts, is met by a wooden framework with a sloping board at its head, representing the pillow of civilised life. This is the couch awaiting the traveller, weary with 100 versts of road, which have given him a horrible headache and a pain across the chest, and made every joint in his body stiff and sore. Stretched on one of these barbarous contrivances, he shifts himself restlessly from side to side; and if he is so lucky as to snatch a short slumber, still urges in his dreams the inevitable 'paraclodnaia' over the interminable steppe. Unwieldy as such an article is to carry, a mattress of some sort is a necessity in travelling in the Caucasian provinces.

May 25th.—The morning was clear, and the two peaks of Ararat, now well behind us, and brought almost in a line with one another, looked very imposing. Two stages separated us from Nakhitchevan; the first of fourteen versts, we accomplished in fifty minutes. The ground was soft, and the disagreeables of jolting were exchanged for the doubtful pleasure of being plastered with mud. The track led up and down over bare hills; every few miles we came to a Cossack station, one of a chain extending all along the Persian frontier. Nakhitchevan is a small and decayed town, built on a high brow which overlooks the basin of the Araxes; it boasts of a large but now ruined mosque, a governor, a passport-bureau, and a custom-house. We first sought the untidy but not ill-supplied military restaurant, where we got a good breakfast, and a bottle of Allsopp's beer. Wherever there are

Russian officers quartered, there is at least a semblance of European cookery, and European cookery is an excellent thing after a course of post-station fare. We next presented our letter to the governor, a stout man, whose final cause, as far as we could make out, was to serve as a receptacle for Russian decorations, the rage for which, in this country, reminded us constantly of the South Sea Islander's passion for a coat with brass buttons and a cocked hat; the same instinct prompts both. He told off two Cossacks to aid us in crossing the streams which still intervened between us and Djulfa.

Officialism is very rampant at Nakhitchevan. Our passports were long retained and anxiously studied by clerks of every degree. After being led from custom-house to police-office, from police-office to custom-house, for nearly two hours, till our patience was wellnigh exhausted, we were curtly asked by an official, 'how much money we had got?' Being by this time fairly irritated, I answered, as curtly, that he might find out. Our inquisitor persisted, and it was not till the officials found we could be as obstinate as they were, that they condescended to explain that a convention existed between Russia and Persia, by which all money that had been registered before leaving the Russian territory could, if lost by robbery on Persian soil, be recovered from the Shah's Government. As the custom-house is twenty-five miles distant from the actual frontier, the money has to be registered at Nakhitchevan, where it is tied up (in the present instance in my pocket-handkerchief) and sealed with the double eagle, not to be broken until the moment before the ferry-boat leaves the Russian soil. This is the official story; the real object I believe to be, to prevent the exportation of Russian silver, which is scarce enough already. When treated with civility, we made no further difficulty, and handed over our cash. Our Napoleons puzzled the officials greatly, their value being utterly unknown to them; indeed, I believe they had never seen gold before, and the discovery of some 'Victor Emmanuels' in the roll added greatly to their perplexity. At last the formalities were completed, the Cossack escort arrived, and we started for a forty versts' drive to the frontier-post. At the foot of the hill on which Nakhitchevan stands we encountered a stream, which, though troublesome, was nothing after our sensational feat of the day before. It had been more than enough, however, for some Armenians, whom we met just before we reached it, and who by exhibiting their soaked state tried to dissuade us from attempting the passage. One of our Cossacks was conducting a prisoner, whom he drove at the trot—probably some Persian who had committed a theft, or come across without a passport, and was being relegated to his own country. The poor wretch was allowed to mount in one of our 'paraclodnaias,' for which he was very grateful. We waded for fifteen versts across a plain more than half under water, and then passed over a low chain of hills, beyond which we came to the second river, comparatively a small one, on the further bank of which stood the solitary posthouse. The whole scenery of the Araxes valley is wild, not to say dreary; but it is so utterly unlike anything we are accustomed to in Europe, that it has at least the charm of novelty. The landscape now grew more and more savage. Ararat, which had long served as a kind of familiar landmark in this, to us, unknown region, was lost to view behind lower hills. In front a wild confusion of mountains gathered round us, amongst which towered one huge and, as one would have said before the fall of the Matterhorn, inaccessible rocky mass, tower-like in form, and rising at least 3,000 feet above its base. The Araxes is here obliged to force its way through a gorge in the hills, and the tributary we had just crossed has a similar task. The track follows its channel, between two walls of sandstone rock, and it is necessary twice to ford the stream. We had barely effected the first passage when the off-fore-wheel of our crazy vehicle fell to pieces. Had it done so a minute earlier, both we and our luggage would have been drenched, and, as the current was deep and strong, probably seriously injured by the accident. Fortunately, we were only some six miles from our destination, and the road was smooth; so we crowded into the remaining cart, the harness of which instantly gave way under the extra strain, but was promptly repaired.

As we trotted quickly down the long slope which leads to Djulfa, we amused ourselves by contrasting our wretched trap and magnificent escort. Had our carriage been a little better, we might have fancied ourselves royal personages, with an outrider cantering in front, and two Cossacks trotting on either side. The Cossacks here have no fixed uniform; their dress is a conical Tartar hat, a cloth coat, generally blue or brown, a silk shirt coming down to the knee, and long leather boots with turn-up toes. They carry a more or less extensive armoury, but none have less than a sword, gun, and pistol. On the road they amused themselves and us by feats of horsemanship. One of them was particularly clever in picking up his gun from the ground without dismounting, and with apparent ease. These men neither rise in their stirrups English-fashion, nor sit close like the Arabs, but trot for hours, alongside the 'telega,' in a standing position. We afterwards made trial of this mode of riding, and Tucker asserts that he found it a grateful change from the ordinary style, to which a Tartar saddle, short, narrow, and hard, is certainly very ill-adapted.

Djulfa, the most southern outpost of the Russian Empire

(at least on this side of Bokhara or Samarcand), is situated at the climax of the wild scenery of the Araxes; it might be the frontier-post of the habitable world, and the entrance to some other region, such as the ancients imagined Hades. A treeless plain, of a dreary brownish-grey, slopes down to the Araxes, which flows out from behind rugged hills, through which it has forced a way by some hidden cleft. Behind us were the low red hills, and the gap through which we had come; before us a bold mass projected from the higher chain on the Persian side of the river, which was reft by a gap exactly opposite, and corresponding to, that by which we had entered. To the north-east, where one might have expected to look down the lower valley, the view was suddenly barred by a grand snowclad range, the summits of which towered 10,000 feet above our heads; their lower slopes were as arid and desolate as those above the Dead Sea. The only signs of life were the two custom-houses on the opposite banks, and a few miserable buildings clustered round each. When we reached the river's edge, and gazed on the ferry-boat, now rendered useless by the flood, and the frayed and worn-out rope, which scarcely saved it from being borne away towards the Caspian, the similitude of Hades, the Stygian flood, and Charon's tub, was yet more forcibly recalled to our minds. On our explaining our thoughts to François, he carried out the idea with his usual readiness, and replied 'Oui, monsieur, et je pense que le voyage en enfer se fait en para-clodnaia'—an allusion to our late sufferings which gave us a hearty laugh.

Our vehicle pulled up before the door of the one good house, where travellers are generally received by the officer in charge of the station. We were about to enter, and ask for a room and beds, when we were met in a most chilling manner by a man in uniform, who informed us that this was not the place for us, and in answer to our enquiries, ordered a soldier to point out where we might sleep. We were accordingly conducted to a newly-built and as yet unfinished mud hovel, scarcely approachable for mire, without door, window, or any vestige of furniture, except a chimney. There was a big puddle outside the door, and the mud floor was so damp that we could stick our heels into it for some inches. Naturally imagining there must be some mistake, we returned to the big house, where I succeeded in speaking to the head-officer, who informed us we must put up with what was offered, or shift for ourselves. Paul simply expressed it, 'Pardon, monsieur, mais il vous dit en Russe, que si vous n'êtes pas content, vous pouvez aller au diable.' We began to think this really was Hades, and felt at a loss with what sop to appease the modern Cerberus.

The Russian ferry-boat being disabled, and the Persian not crossing till morning, we were compelled to remain where we were, and make the best of a bad job. We were indebted to the good-nature of some soldiers (the Russian private is almost invariably a kindly fellow, ready to lend a hand to anyone in difficulty), for a table, some boards with which we closed the door, firewood, rough rugs to lie on, and some new milk. The sound of merriment and popping of corks, which greeted our ears when we walked up to the terrace of the chief's house, did not lessen our disgust. He was entertaining a superior member of the official confraternity, and thus keeping up the character for hospitality, called for by his possession of a Persian decoration, nominally acquired by the exercise of that virtue, but really by purchase, if Tabreez talk was true.

May 26th.—We did not pass a very comfortable night. I left before breakfast, to try and get our passports back from the officials who had them to examine. Returning

unsuccessful, I was just detailing my experiences to
Tucker, when he was suddenly seized with a fainting-fit,
which lasted for some time. When he showed signs of
recovering, I sent up Paul to the large house to say that
my friend was ill, and to ask for the loan of a saucepan to
make some soup in. He returned shortly, having seen the
head-officer, with the message that they did not keep
things to lend. Disgusted at our treatment the previous
evening, I had already written a complaint, which I purposed addressing to one of our friends at Tiflis. I now
added this detail to the catalogue of offences, and, on going
up a second time to receive our passports, handed over a
copy of it to the officer, whose guest I had learnt spoke
French, and would therefore be able to read it for him.
He did not then, or on our return, offer any excuse for his
conduct towards us; and we therefore carried out our
purpose of presenting the complaint, which, we were subsequently assured, met with attention at head-quarters.

When the Persian ferry-boat came over, about 8 o'clock,
Tucker was fortunately well enough to walk down to it.
We embarked with a miscellaneous company of six donkeys, two horses, some ragged peasants, and an Armenian
merchant, who rode with us up to Tabreez. There was no
rope to the boat, which was only propelled by two wretched
apologies for oars, and the current was so strong that we
seemed to run a chance of seeing the Caspian shortly.
The men in charge knew, however, how to take advantage
of the eddies and backwaters of the stream, and finally
landed us in safety on Persian soil, about three-quarters
of a mile below the custom-house. The banks had been
quite lately under water, and were now an almost impassable swamp; we picked our way along them some little
distance to a sandspit, on which a number of camels were
reposing, until the boat should be ready to return. In

the interval we hired some of them to convey us to the station, to which we rode in an imposing caravan, François leading the way with great solemnity. We found the head of the custom-house was the man to whom Monsieur Rulli had told us to apply for help. As he spoke a few words of French, we were able personally to explain our needs to him; and we were at once taken to his room, where he made a sofa for Tucker with his carpets, and shared with us his dinner, a most excellent mutton-broth. We found that our Russian paper-money would be useless beyond the frontier, and consequently had to change into tomans (a rude silver coin about the size and value of a franc) a sum sufficient for the next two days' journey.

From Djulfa to Tabreez is a distance of about ninety miles, which is divided into four stages. In Persia, as throughout the East where Russian influence has not yet introduced that doubtful luxury of civilisation, the 'telega,' wheeled carriages are the exception, and all the traffic and commerce of the country is carried on horse and camel back. The post-stations vary from sixteen to twenty-five miles apart, and the ordinary plan is to ride one stage in the morning, rest in the heat of the day, and start again with fresh horses towards evening. In this way from forty to sixty miles a day may be ridden without too much fatigue. We found no difficulty in obtaining animals to carry us the first stage, which Tucker now felt himself equal to undertaking. There was not, however, a spare horse for the postboy, who had in consequence to run along by our side, until we met some return-horses, one of which he mounted. Our saddles were in various stages of decay, and of most remarkable construction, with narrow seats and high peaks in front; the stirrups were ludicrously short, and incapable of being lengthened, so that our knees suffered no slight torture after we had been a short time on the road. The horse-track leads up into the elevated tableland of North-western Persia, by a gentle but continuous ascent of at least 3,000 feet. Leaving on our left the isolated cone we had remarked from the Russian bank, we rode along bare slopes in the direction of the only gap in the chain before us. The backward view was very wonderful; a shelf at the base of the hills on our left was dotted with the bright-green or chards of several villages, while below us lay the trough of the Araxes, bare, brown, and hideous, from which the eye sought relief in the pure snows and noble forms of the summits of Kanudschuch, which looked from here their full height of 12,854 feet. The narrow glen through which our road now lay was sufficiently picturesque. The mountains on either side were covered with grass; a little stream, with a line of trees along its banks, turned several watermills. Farther on the crags became bolder, and we noticed an extraordinary distortion of the strata. The narrowest point of the pass was defended by two old towers, now in ruins. We emerged at length on an upland plain surrounded by bare hills, a description applicable to the greater part of Persian scenery. The village of Datarzian soon came into sight under the hillside on the left, and the first house in it proved to be the post-station. The three we saw on the way to Tabreez were almost alike, and a description of one will serve for all. A square building of unbaked mud surrounds the courtyard, into which you ride; on three sides are the stables, on the fourth the rooms for the reception of travellers. There are generally two—one on the ground-floor beside the gateway, and the other (which is cool and airy in hot weather) built over it. These rooms we found quite bare of furniture, but had no difficulty in obtaining mattresses and pillows to lie on, which both looked and proved clean and free from insects. Datarzian is a small and poor village, but we succeeded in getting milk and eggs, and Paul sacrificed a fowl.

May 27th.—The morning was lovely, and the snowy chain north of the Lake of Urmia, towards which we were riding, was a fine object in the distance. Near at hand there was little to diversify the road, until we came to a large ruined khan, the doorway of which was handsomely decorated with tesselated tiles. Having traversed a watershed, we descended slightly, and crossed two streams, along which a few trees and watermills were scattered. Before us was a broad cultivated plain, like that of Cœle-Syria. A grove of trees on its opposite side was pointed out to us as Marand, the town at which the road from the Russian frontier joins the caravan route from Tabreez to Trebizonde.

A Persian town is a very curious sight when seen for the first time. A green grove appears in the distance; 'that is Marand,' says the postboy. As you approach the trees become distinct; you pass a few detached orchards surrounded by high mud walls, but it is not till you have fairly entered the place that any houses are visible. The main street of Marand is shaded by trees, and watered by a stream in which the juvenile population, mostly in a state of nature, were engaged in making mud-pies. The houses stand on either side, all but the poorest surrounded by gardens, vineyards, and orchards. They are of one story, and flat-roofed. The walls are built of grey mud, well smoothed and finished off (reminding Tucker of Devonshire cob), and often slope inwards towards the top. The windows are filled with very neat wooden lattice-work frames, the small interstices between which are plastered over with oiled paper, instead of glass. The women whom we saw struck us as exceptionally hideous; the men are an active-looking race, more akin to one's idea of Hindoos than to the more apathetic Turk. The common people

and children wear a kind of elongated nightcap—the upper
classes carry the tall Persian hat. We lunched off
'kabobs' (scraps of meat stuck on a stick and toasted),
and 'kaimak' (a kind of Devonshire cream); our greatest
success was the discovery of a bottle of wine, pure juice of
the grape, which owed little to any skill in its manufacture. When we were on the point of starting an unexpected question arose. At first we were told the next
stage over the hills was dangerous, on account of brigands,
and that we must take an escort; on our refusing to accede to this proposal, it was suggested we might go by a
longer and perfectly safe route, only we should have to pay
more for the horses. This we also declined, and finally,
after much talk and waste of time, were allowed to set
out. Our postboy led us, by a track running due south
into the mountains, up a valley terminating in fine rocky
cliffs. We now found out why the brigands had been
created; it was for the benefit of two foot-soldiers on the
march to Tabreez, who were naturally anxious to do a job
on the way by protecting us. They were fine raw material—active fellows and splendid walkers, unencumbered
by any uniform, save an old blue coat and a white belt, to
which was hung a flint-and-steel gun. Presently our path
turned east, and crossed a broad watershed, dividing the
stream which flows into the Araxes and the Caspian, from
one of the feeders of the Lake of Urmia. A bold summit
rose on our right in rocky slopes intersected by snow-filled
gullies; on the other side a village in a bleak situation,
and a serrated ridge of rocks, attracted our attention. A
long ride down a dull winding valley, between barren and
ruddy-coloured hills, brought us to Sofian, a village at the
foot of the mountains, overlooking the salt-plain which
extends to the Lake of Urmia. The people at the posthouse were very civil, and made us fairly comfortable for
the night.

May 28th.—We set out early for our twenty-four miles' ride across the plain to Tabreez, which was already visible, a dark green spot, in the distance. There was nothing of interest on the road, except in the parties we either overtook or met. Now a gentleman on his travels, dressed in a cool dove-colour or grey coat, bright silk shirt, and tall hat, his horse covered with a gaudy saddle-cloth, caracoled past us, followed by his servant with the saddle-bags; now we met a train of donkeys gaily decorated with many-coloured tassels, and bustling along as if they were all hurrying to a fête. To them succeeded a solemn train of camels, swinging, with every sway of the neck, enormous bells. As we drew closer to the city, we met more and more people on the roads, and crowds of donkeys carrying brushwood for the use of the brick-kilns we afterwards passed. A brick bridge of many arches crosses the considerable stream of the Aji Chaï, which, rising at the foot of Sawalan Dagh, waters the environs of Tabreez, and finally loses itself in the Lake of Urmia. We now entered the gardens of the suburbs; on either hand rose grey earth-walls, fifteen feet in height, with fruit-trees raising their heads over them, and vines pushing green tendrils through their upper and less solidly-built portions. The doors of the vineyards were of stone, exactly of the pattern of those in Bashan, and some of them nine feet high. We were amused to see the primitive mode of knocking for admittance, by picking up a stone from the ground and hammering it against the door. We rode at once to the English Consulate, which is the handsomest house in Tabreez, with a pleasant balcony and garden, and large cool-looking rooms. We found, as we feared would be the case, that Mr. Abbott had already left for Urmia. The dragoman of the Consulate (who, however, only spoke Turkish) soon appeared, and requested us to make ourselves at home until a lodging was found for us. I need hardly say that there is no hotel at Tabreez, and it is difficult for a European traveller to acquire sufficient familiarity with the manners and customs of the people to be able to put up with any comfort at a native khan. Dr. Cormick, the English physician, who is in charge of the Shah's son, was unluckily out for the day; but a German merchant came to our aid, and we consulted with him where we should lodge. After a good deal of doubt and delay on the subject, it was finally settled, towards evening, that we should take up our abode with a Nestorian Christian, who had been employed as dragoman by several embassies going to Teheran, and was said to speak English. We were greatly amused by our future host's manner of introducing himself: 'You come with me, all right; you know me? I Lazarus; find me 11th John in middle chapter; all missionary gentleman know me, all right.' 'Old All Right,' as we irreverently renamed this Scriptural character, led us off to his house, which was approached by a narrow lane between two high walls, and a downward flight of steps. The interior, however, was a pleasant surprise; we found a snug little room, furnished with European chairs and a table, ready for us. The walls were decorated, in the Persian style, with paintings of flowers. The windows, filled in with paper—the universal substitute for glass at Tabreez—opened out into a little garden, on the other side of which were some more apartments, in which glimpses of our host's wife and daughters might occasionally be obtained. At night beds were made up for us on the floor, and Lazarus turned out to be a good cook, with a special gift for rice-puddings; so that Paul for a time rested from his labours, while we 'fared sumptuously every day.'

CHAPTER VI.

TABREEZ, ARARAT, AND THE GEORGIAN HILL-COUNTRY.

The City — Brick Architecture — The Shah's Birthday — The European Colony — A Market Committee — Return to Djulfa — A Dust Storm — Ford of the Araxes — Aralykh — Start for Ararat — Refractory Kurds — A Moonlight Climb — Failure — A Lonely Perch — Vast Panorama — Tucker's Story — A Gloomy Descent — Return to Erivan — Etchmiadzin — The Armenian Patriarch — A Dull Ride — Hammamly — The Georgian Hills — Djelaloghlu — A Moist Climate — Schulaweri — Tiflis again — Moore joins us.

TABREEZ, May 28th to June 2nd.—We had come thus far to see a Persian city, which we might hope to find more beyond the reach of European influences than the towns of the Levant, or even Damascus, the romance of which is fast yielding to the frequent invasions of Cook's tourists. Tabreez far exceeded our expectations. In roaming about its bazaars, we felt the same sensation of unreality as on our first arrival at Cairo, which, considering what we had seen in the interval, is saying a great deal. I will begin by describing the view gained from one of the house-roofs, which form the favourite lounging-places of the inhabitants. Thus viewed, the city seems to divide itself into three portions. In the centre are the domed roofs of the bazaars; round these is a broad zone of dwelling-houses, the grey of their flat roofs and walls enlivened by the bright-green of the courtyards, and in the nearer ones by the woodwork of the window-blinds; outside stretches a ring of walled gardens, beyond which is the bare country, characteristic of this part of Persia. There are no minarets, and the only conspicuous building, which rises above the level of the low one-storied houses, is the massive tower of an old castle, in shape not unlike an Egyptian propylon. The horizon is mountainous on all sides. A range of red sandstone rises above the town on the north and east; in the south are seen the snowy summits of the Sultan-Dagh, which reach 13,000 feet; and to the west the eye sweeps over the plain, often beautified by mirage, extending to the shores of the Lake of Urmia, the mountains beyond which, fronted by those of the peninsula of Shahi, close the view in this direction. Owing to the height of Tabreez (4,000 feet above the sea-level), and the neighbourhood of snowy ranges, the heat of the sunshine is frequently tempered by cool breezes, and during our stay the temperature was delicious.

The bazaars were our favourite haunt, and where we spent the largest portion of our time. The principal ones, where the most expensive goods are sold—the Bond Street and Regent Street of Tabreez—are brick arcades, the roof composed of a series of small domes, through an aperture in the crown of which a column of sunlight falls on the goods exposed for sale below. In the same stall you see the fabrics of Lyons and Manchester, lying side by side with those of Shiraz and Ispahan. Here you may buy a gaudy French silk, or a cotton, in which the Eastern colours and designs are more or less faithfully reproduced by a Lancashire firm. We cast but a passing glance on such fabrics, but a gorgeously-embroidered tablecloth from Rescht, or a beautiful piece of Persian shawl from Shiraz, often made us linger to chaffer with its owner—generally, happily for our purses, without result, as the Persians are shrewder men of business, and harder to drive a bargain with, than even the merchants of Cairo or Damascus. Out of these arcades open halls or khans, covered by large and very flat brick domes, in the building of which the Persian architects greatly excel. The patterns introduced into the brickwork, and the variety and shape of the arches which support these halls, add much to their effect. They are the principal places where the wholesale trade of the city goes on; round the sides are the shops; over them again are offices, or rather dens, where

> 'Above their merchandise
> The merchants of the market sit,
> Lying to foolish men and wise.'

Each of these shops, with the den over it, is encased by a brickwork arch, the space inside the arch intervening between the lower and upper rooms being filled in with neat woodwork. The walls are often hung with very effective black-and-white cotton drapery, on which the favourite device of the Persian lion, with the sun rising out of his back, is displayed. The centre of the hall is generally occupied by a pile of carpets, on the top of which their owner is often to be seen performing his prostrations. There are other halls, entirely built of wood, and supported by roughly-finished poles; which are as gaily decorated and as quaint, if not so handsome, as the brick buildings. There are besides, in the business quarter of the town, numerous open squares, some of them very large; a row of shops extends all round them, and at the centre of each of the four sides is usually a brick apse, in the form of a gigantic alcove. In the middle is a fountain, and the court is planted with flowering shrubs, such as guelder-roses and lilacs.

The outer circle of bazaars, for rough retail goods, is mostly covered by picturesque wooden roofs. Here the grocers' stalls, tastefully decorated with devices in tinfoil, and the provision-shops, set out with nosegays of buttercups, poppies, and clover, are a perfect blaze of colour. In one of the brightest and liveliest alleys in the whole bazaar, you watch the shoemakers, all stitching as if the pair of shoes in hand must be sent home in five minutes; close by are the forges, where eight men, standing in a circle, hammer out, with alternate strokes, a mass of iron, how each hammer keeps clear of the next being a mystery to the uninitiated. Then there is the carpenters' row, the bookbinders' row, the old clothes row, the knicknack row, where you may buy anything, from revolvers to Persian ink-trays, and last, but not least worthy of notice, the saddlers' row. The Persians are decidedly a horsey people, and have studied all the requisites for a long ride. The roof is bright with saddle-cloths, some covered with the most beautiful Rescht embroidery; while in the stores you find gay girths, and tasselled bridles, carpet saddle-bags, and leather salt-pouches, heaped together in picturesque confusion. It was not the season to see the fruit bazaar in its glory, but the quantity even of nuts and dried fruits was extraordinary; they were piled up in ten rows of baskets, one behind another, on an inclined plane.

The crowd which fills the streets of Tabreez is purely Eastern; you do not meet two men in European dress during the day, nor do you see the red fez which lends such life to the cities of the Levant. The ordinary head-dress is the tall Persian hat; this is now generally made of cloth, and of moderate dimensions, as the Shah has published an edict against the steeple-like edifices of Bokhara lambswool which were formerly the fashion. Turbans, dark-blue or white, are however frequent; they are of tremendous dimensions, and resemble those in the Museum of the Janissaries at Constantinople, far more than any to be seen in the streets of the Sultan's capital. The colour of the dresses is quiet compared to the bright hues worn by the Turks. The women are clothed in dark-blue sacks, and have a sort of open crochet-work window in their veils for the eyes to peep through. There are said to be many Georgian and Circassian beauties in the harems of the wealthier merchants, but the chance traveller has, of course, no opportunity of admiring their charms. We were impressed with the busy air of the street crowds. Everybody walked fast; there were comparatively few fat men; friends met and told one another a good story, and passed on, and occasionally some of the younger sort indulged in the innocent amusement known at the universities as 'bear-fighting.'

We took several strolls through the outer quarters of the town, where long winding lanes, with a watercourse running down the middle, and shaded by trees, lead between high earthen walls; at the crossings is often found a coffee or rather tea-house, for the Chinese drink supplants the Arabian in Persia. We noticed many carved blocks of black basalt strewn about, which must formerly have belonged to handsome buildings. There is only one interesting edifice of any antiquity now left in Tabreez—the ruin known by the name of the Blue Mosque; the walls were coated inside and out with encaustic tiles, the prevailing hues being Oxford and Cambridge blue. The effect is still very beautiful, and before the destruction of the central dome—caused, as we were told, by an earthquake—must have been superb. The largest modern building is the custom-house, which covers a great extent of ground, and consists of large storehouses supported on columns, and roofed with small domes.

On Sunday we took tea in one of the gardens of the suburbs. They are, in fact, vineyards with walks round them shaded by trees, and planted with rose-bushes. The owner charges a small sum for admission, and your own servant brings the materials for your picnic. Though there is no attempt at the refinements of European gardening, they are pleasant places enough to while away an afternoon in, and are much resorted to for this purpose by the townspeople. The Persians are great lovers of flowers: while we were at Tabreez the single roses, red and yellow, came into bloom, and were hawked about in large nosegays; the pipe-stems at the eating-houses were wreathed with them, and they were made use of in the most tasteful way by the common people. We met a boy carrying round a dish of small trout for sale; the fish were laid out in a pattern on green leaves, with lumps of ice and roses placed between.

The day after our arrival was the Shah's birthday, and we were much interested in witnessing the departure of the high officials from a levée held by the Crown Prince in honour of the occasion. They were mounted on gorgeously-equipped steeds, but were themselves dressed in quiet although richly-coloured robes. We also saw a turn-out of troops in the courtyard of the palace—scarecrows of soldiers, scarcely worthy of a minor theatre. We were told that the men are personally brave when well led, and are capable of enduring great fatigue on very little food. They have only a pretence of uniform, are little drilled, and in consequence do not present a very imposing appearance. There are now, it is said, 500 percussion muskets in the whole force, and the conversion of flint and steel pieces is regarded with the same interest as that of muzzle-loaders among ourselves.

The little European colony at Tabreez soon found us out, and showed us the greatest hospitality; indeed, we dined out every night but one of our stay. The mixture of nationalities was most curious. There were only two English subjects—one, Dr. Cormick, the physician in charge of the Shah's son, the other a Maltese. There were, besides, some pleasant Swiss gentlemen, a Frenchman, and a young Italian, an ex-Garibaldian; the latter two kept up a constant fire of good-natured chaff on European politics in general, and the relations between the Pope and the French Emperor in particular. We gathered a great deal of information, which was new to us, about the internal condition of Persia and its Government. Crime seems to be repressed with a strong hand, and with the indifference to human life common in the East; 1,200 executions have taken place at Tabreez in the last nine years. We were told that death is inflicted in the quietest way, and that both the headsman and his victim behave like perfect gentlemen. The one lights his long 'kalian,' smokes a little, and passes it to the other, who has his whiff; and after hobnobbing thus for a while, the agent of the law remarks, by some Oriental periphrasis, 'Time's up,' and chops off his companion's head with neatness and despatch. Detected coiners still suffer the penalty of having their ears nailed to a post, and an instance of this punishment occurred during our stay, but we missed seeing it. Of public amusements Tabreez has few; we were just too late to see a kind of Oriental miracle-play, which had been performed in one of the squares, in which Jacob, Joseph, and Solomon, who are equally revered by Mahommedans and Christians, had been brought on the stage. One day we stopped to look at a tame lioness in the street; the poor animal had been partially blinded, and her performance was not of a very lively character—to us at least, who could not understand the jokes of her showman. We should like to have visited the Sultan-Dagh Mountains, where the European residents are in the habit of camping-out during the hot weather; to have explored the peninsula of Shahi, said to afford the best sport in this part of Persia; and to have ridden round the southern shores of the lake to Urmia, which was described to us as a very pretty place. A colony of American missionaries—who make it their special aim to encourage and assist the Nestorian Christians found in this part of Persia—has settled there, and established schools and a church. Time, however, was against us, and we made up our minds to return by the same road to Erivan, and not to risk delay or stoppage at one of the less-frequented fords of the Araxes.

Before we left, we were anxious to obtain some mementoes of our visit to Persia, and, afraid of coping unaided with the native merchants, gave a list of things we had remarked in the bazaars, and wished to purchase, to one of the European residents, who promised to put us in the way of getting them at fair prices. We went by appointment to his office, where we met a sort of market-committee, each of whom had brought something to show us. One man dealt in Shiraz shawls, another in Rescht tablecloths, a third in Bokhara lambskins. There were some half-dozen in all. They sat in a semicircle before us, and when one brought out any 'choice article,' all the rest nodded their heads, and uttered a deep 'wa-ah' of admiration. One toothless old fellow, a dealer in swords and antiques, who might have sat for a picture of Avarice, was most demonstrative in praise of his neighbours' wares, and altogether upset our gravity by his pantomimic expressions of delight. Our bargains were at last concluded, and a box was packed and directed to London, not without some fears (unfounded as it turned out), as to whether we should ever see it again.

June 2nd.—We ordered post-horses to be ready at 7 A.M., and succeeded in getting off an hour later, amidst a hubbub of claims for 'backsheesh' from various employés, who had, or pretended to have, done something for us. We lunched and rested at Sofian, and then pushed on for Marand; but having discovered that there was a second road between the two places, shorter but steeper than that by which we had come, we insisted on taking it. We soon turned off, with the telegraph-wires, up a glen to the right of our old track. Near its head we reached a village, surrounded by verdant pastures, and guarded by huge and very handsome dogs, which barked furiously at us as we passed. Openings in the hills on our right were closed by ridges of splintered crags. Our pass was already in view—a notch in the range before us, to which the road mounted by short steep zigzags; on our left was a wide pasturage, covered with sheep, and backed by the snowy peak which we had seen from the other road. The view from the top burst on us with unexpected beauty. Looking back, the foreground consisted of fine rocks and ruddy hills; beyond them lay Tabreez, now thirty miles off, and appearing as a dark-green spot on the plain, from which a long and uniform slope led up to the base of the mountains of Sultan-Dagh; on the other side we looked down on Marand, and towards the broken ranges which separated us from the Araxes valley, the gap through which, leading down to Djulfa, was plainly visible. At a wayside spring, just below the watershed, we met a native of artistic tastes, who, with an appreciation of landscape beauty very rare in the East, asked us which view we preferred, the Tabreez or the Marand side? The posthouse at Marand was full, but we were taken in by a hospitable native, whose house stood in a pretty garden, from which he brought us some red and yellow roses. There were some well-trained standard guelder-rose trees in front of the house. We had a comfortable room, and plenty of coverlets for the night. Our host enquired of

Paul if we had come to arrange for the construction of a railroad to Tabreez, and hoped we should make it pass through Marand, and that it would soon be begun. One was scarcely prepared for so keen an appreciation of the advantages of quick communication in this remote Persian town.

June 3rd.—We rode down to Djulfa, and as we passed the defile in the range, that shuts in on the south the Araxes valley, the rock-tower on the Russian side looked very imposing. We were already descending towards the Araxes, when we noticed a black cloud gathering on the mountains behind us, similar to, but more dense and inky than, that which had pursued us on the Erivan and Nakhitchevan road ten days previously. Feeling that Djulfa, the scene of our inhospitable treatment and of Tucker's illness, was not a place to arrive at drenched to the skin, we increased our pace, and reached a village about three miles from our destination, still uncaught. The storm was gaining fast, and looked so bad that we set off at a gallop. The race that followed could only be described by a mixture of the styles of Mayne Reid and De Quincey, to which I am wholly unequal. The ground was tolerably flat, the storm coming up rather on our flank. On making a sudden dip over the bank that bounds the actual trench of the Araxes, we saw that we were too late. The dust-cloud, which rides on the front of these steppe-storms, had crept round and cut us off from Djulfa. In a moment it was upon us, borne along by a wind which swept twigs and tufts of grass along the ground, and nearly blew me out of the saddle. The air was so thick that I could only see the two horses nearest me—one a riderless animal the postboy was taking back with him. Our beasts seemed to dread the storm as much as we did, and galloped at the top of their speed. In a minute our eyes were choked, and every particle of clothing covered with dust. Then came a second blast, and the darkness grew deeper, so that we could barely see the ground under our horses' feet. I remembered that a watercourse cut the plain close to the station, and managed to hit the plank-bridge over it, passing my companions, whose voices I could hear, on the right. In another minute the first heavy drops fell, but at the same moment the Persian custom-house loomed through the darkness, and we were in shelter. Paul was the only one of the party who got thrown; he almost rode against the building without seeing it, and, his horse suddenly swerving, he fell off, but fortunately without hurting himself. François, as usual, turned up among the first, serene and unruffled. At the posthouse the accommodation was wretched, and the people most inhospitable; but we had a resource in our former friend, who entertained us with kabobs, and a bottle of excellent Persian wine.

June 4th.—The only acknowledgment of his hospitality our host would accept was a scrap of paper, on which we wrote our names, and recommended all English passers-by to his care. We had to wait some time while the ferry-boat was towed up, by buffaloes, to the proper starting-point. The river had fallen considerably, and there was no difficulty in the passage. Our reappearance was evidently more of a surprise than a pleasure to the Russian officials, who had probably solaced themselves with the idea that they would never hear or see anything more of us. Believing that the jolts of the 'telegas' were mainly responsible for my companion's previous illness, we made a successful effort to obtain saddle-horses. We were, however, able to retain them only for the first stage, and no entreaties or threats would induce the stubborn and dirty postmaster of the halfway station to furnish us with any other means of conveyance than the hateful carts. We arrived at Nakhitchevan in the middle of the day, and I wasted the afternoon in an ineffectual struggle to hire some kind of spring-carriage. Having called on the governor, who expressed his readiness to do anything for us, I told him that my friend had been knocked up by telega-travelling, and that we were most anxious to hire a carriage; but in this particular he could afford us no aid. There were only two carriages in the town—one his own, the other an old 'tarantasse,' which we at first thought might serve our purpose, but which proved to have suffered so much from exposure and neglect that it was practically useless. Its owner was a curious character. He had been at one time in business in London, spoke French, and a little English; but having failed, as he gave us to understand, through his own extravagances, he had returned to Russia, and was now fixed as a sub-official of the custom-house in this remote corner of the Empire. The poor man complained bitterly of the dulness of his situation and the barbarism of his companions, and sighed after the theatres and diversions to which he had been once accustomed. There was a heavy thunderstorm in the afternoon, and the continued uncertainty of the weather made us rather despondent about Ararat.

June 5th.—We started late, in the usual 'telegas.' The big river, which had been so formidable ten days before, was now much lower; but we were nearly upset in it, owing to the stupidity of our driver, who let one of his horses flounder into a hole and break the harness in the middle of the stream. At the next station there were no post-horses, so we hired some peasants' animals, and sent on our men, with instructions to procure and cook some supper for us at the next station. We followed two hours later, and before we had finished the long stage of twenty-two versts the moon had risen, and the steppe looked wonderfully weird in its light, with Ararat and Alagoz looming like ghosts in the background. We slept at Sudaruk, where Paul and François had got some supper ready for us.

June 6th.—We drove two stations to Kamirlu, where we were to turn off for Aralykh, the village and Cossack station nearest Ararat, separated from us by ten versts and the Araxes. I had a note for the postmaster, and he made no difficulty about giving us carts, although Aralykh is not on a regular post-road. We soon saw we were approaching the river, by the swampiness of the ground, from which the water had but just retreated. The Araxes is here divided into three branches: the main stream is crossed by a ferry-boat; the two smaller branches are generally easily fordable, but now offered considerable difficulty. At the first our baggage was carried on men's heads, who waded across whilst we plunged in, cart and all. Though our small ark was nearly floated away in the struggle, we came out in safety on the other side. A hundred yards further we reached the second branch, the only means of crossing which was a leaky old boat, and the delay in the traffic caused a most picturesque scene of uproar and confusion. This ferry, though on no great caravan route, is much frequented by the nomad Kurds, who are constantly changing their pasture-grounds. These people are the Arabs of the mountains, and are nearly as striking as their better-known relations. Hundreds of them now lined the bank, in their bright and picturesque costumes, with their tents rolled up ready for crossing, their wives and daughters seated on the heaps of baggage, and their camels and flocks lying down or straying around. The men, who were fine-looking fellows, wore gay-coloured dresses, and carried queer old weapons in their broad belts. The girls were almost all pretty nut-brown maids, with bright eyes and plaits of beautiful brown hair, which streamed out from under a handkerchief, and reached down to the waist. They wore bright-colonred jackets and short petticoats. The men are said to be dangerous customers, but we always found them very civil and friendly, like most Mahommedan country-people, and they seemed pleased to discover that we were English. In the present instance they helped our baggage into the boat, and after some delay we got across. Meantime the passage of the flocks was amusing. Hundreds of sheep and goats were forced to face the stream by shouts and pistol-shots, while boys and girls dashed into the water to meet and land them safely. Camels lined the bank, waiting their turn with an air of patient resignation, and two huge sheep-dogs, off duty for a time, beguiled their leisure hour with a fight; three hundred yards off the ferry-boat moved backwards and forwards over the main river. All this made a lively foreground; and in the distance stood Ararat, as usual wrapped in his afternoon cloud, and the two peaks of Alagoz relieved against a bright-blue sky. It was a picture one longed to see transferred to canvas, by some painter equal to the occasion, and we lamented once more our own incapacity to use brush or pencil to any purpose. The ferry-boat, which is large enough to take on board a carriage, and works on a stout rope, floated us easily over the main stream.

The question now arose, how we should get on to Aralykh, still three miles off, as of course our carts had been left behind. There were some Cossacks in charge of the ferry, and on showing our letter, addressed to the Colonel at Aralykh, their chief found us a horse to carry the baggage. As no more animals were to be had, we were compelled to walk, in the full heat of the afternoon sun, over the bare sandy waste between us and the village, which is situated at the very foot of the mountain, at the commencement of the long uniform slope, which serves as a pedestal to the upper and more precipitous cone. On our arrival, despite our hot and dusty appearance, we were most cordially received by the officer in command of this out-of-the-way post, who has a comfortable house and the company of a wife to console him for his banishment from civilisation. He insisted on giving up his study to our use, and at once set about making the arrangements for our intended expedition, which were somewhat complicated, as the natives are as yet little accustomed to travellers visiting the mountain. The Colonel entertained us most hospitably, and insisted on our allowing him to procure the provisions we should require for the ascent. Unfortunately, neither he nor his wife could speak any language but Russian, which made our communications rather laborious, especially for Paul, who was in constant request as interpreter.

On the morning of June 7th, we set off from Aralykh on our expedition against Ararat. We were accompanied by a Kurd chief, in the Russian service—in whose charge the Colonel had placed us—and his servant; four Persians, the owners of the horses, and three Kurds, who were supposed to be mountaineers, and capable of acting as porters.

Starting on horseback for a 'grand course' is not quite in accordance with Alpine ideas; but when it is remembered that Aralykh is only 2,600 feet above the sea, and that the lower slopes of Ararat are perfectly uniform, bare and stony, we shall be excused for avoiding the dreary grind up them, under an Araxes-valley sun. At first we kept a course parallel to the river, but soon turned towards the great mountain, and began to ascend sensibly. We next skirted the base of a green bastion commanding the lower slopes, in the hollows and shelves of which several groups of Kurd tents were pitched. A somewhat steep ascent led up to the green plain which fills the space between the bases of the two Ararats. The Little Ararat rose immediately before us in an unbroken slope of about 4,000 feet; it is a typical volcano, uniform on all sides, but least steep on the Turkish, from which a Russian General is said to have ridden up on horseback.

On our right the base and upper portion of the cone of the Great Ararat were visible; the lower part being masked by buttresses, and the whole mass most deceitfully foreshortened. On a knoll about 300 feet above the plain we found the group of huts which have been used as a resting-place by most of the explorers of Ararat. These queer dwellings are underground burrows, constructed like the villages on the Georgian steppes. A door of twisted twigs, on being opened, reveals a hole in the hillside, which forms the mouth of a long, winding, dark passage leading into two or more chambers lighted by holes in the roof. The floor of these horrid caverns is the natural soil, and their atmosphere is earthy and tomb-like, while the darkness that pervades them adds to their depressing effect. The roofs are formed of branches covered with turf, and as there is nothing outside to distinguish them from the solid ground, it is easy to walk over them unawares. One of our horses, while grazing, suddenly sank into one of these dangerous traps, and was left, with only its forequarters emerging from the ground, in a position from which it was extricated with great difficulty.

On the way up we halted, and discussed our arrangements with the Kurd chief. We had been told below that we should find all we wanted at the huts; but they now proved to be uninhabited, and it was therefore necessary to get a further supply of bread at one of the encampments lower down. The porters wished also to borrow a

tent. We remonstrated at the delay this would occasion.
It was quite early in the day, for we had reached the huts at
11.15 A.M., and, being anxious to sleep as high as possible,
we proposed to the men that they should go on with us at
once, carrying our rugs, in which we were prepared to pass
the night. We had provisions enough for ourselves, and
we pointed out to them, that if they were afraid of sleeping
out at such a height, or had not sufficient food to last till
morning, they would have time before nightfall to return
to the huts, where they might sleep, and remount next day
to fetch down our rugs and other baggage. The men, not
unnaturally, were averse to the double toil and trouble involved in this plan, and utterly declined to carry it out, or
to join us in sleeping out without shelter and a further
supply of food. We were therefore obliged to remain at
the huts, and await the arrival of the tent and provisions.

Our position was curiously like and unlike many old
Alpine bivouacs. The surrounding pastures might have
been on the Riffelberg, and it was delightful to see again
many well-known Alpine flowers. The rhododendron indeed was sought for in vain, and we were too low for
gentians [*]; but their lack was partially compensated by a
new friend, a dwarf wild hyacinth, white delicately streaked
with blue, which grew in great profusion. Little Ararat,
however, was sufficiently unlike a Swiss mountain to dispel
any illusion, and if that had not sufficed, one glance down
his side into the brown, bare, burnt-up trough of the Araxes
would have been enough to recall to our minds the fact
that we were in Asia, far indeed from the old haunts.

After midday, clouds gathered, and Ararat indulged in
his usual thunderstorm. Some hours passed, but the men
did not reappear, and we were getting more and more
impatient, when about 3.30 P.M. they came into sight,
followed by a cow, carrying one of the regular Kurd tents,

---

[*] We afterwards found a solitary plant higher on the mountain.

too large to be useful for mountaineering, and too heavy to be carried by our men over rough ground. The three porters now professed themselves ready to go without a tent, but a second thunderstorm delayed our start till 4.30. With fuel we had load enough for four men; but as the owners of the horses declined to be of any service, François and Paul had to carry one of the bags between them.

Striking up the spur behind the huts, we made our way as directly as possible towards the mountain, traversing a good deal of rough ground, and crossing several hollows, by which we lost time, and partially deceived ourselves as to the progress made. The porters halted constantly, and our pace was slow; in about an hour and a half after leaving the huts, we found ourselves in a hollow between two spurs, and nearly at the snow-level. Here the porters stopped, and declined venturing upon the snow. It was a good place for a bivouac, and, although probably 500 feet lower, we thought we were at a height of at least 9,000 feet. We knew that the moon would allow us to start at midnight, and anxious moreover to save our men the fatigue of acting as porters, we agreed to halt. The weather looked promising, so we supped on 'Liebig' cheerily over a bright fire, and then rolled ourselves up in our rugs, with little misgiving for the morrow, despite the hindrances of the day.

After a sound sleep (at least I speak for myself) we were awake and stirring at 11.30 P.M. We had a glass of hot wine all round, and started at 12.10 A.M. The first *contretemps* was the discovery that François had, despite my warning, allowed Paul to leave Tiflis without proper boots, and that it was impossible he could come on with us. He had set his heart upon ascending Ararat, and therefore very reluctantly turned back.

After climbing two snow-slopes, we gained a ridge commanding a view of the ground between us and our mountain. In front lay a deep hollow, such as in the Alps would be filled by a glacier; the ridge along which we were proceeding appeared to be connected with, or rather to form a continuation of others, by which it was possible to make the circuit of the hollow, and reach the foot of the great rocks, which we had, for convenience and old acquaintance' sake, named 'les Grands Mulets.'

In the first hour and a half we had cleared a good deal of ground, and I remarked to Tucker how well we were getting on, and how 'fit' I felt. Nemesis was at hand. In another half-hour, though the ground was easy and the inclination trifling, I began to feel unwell, and experienced all the sensations of mountain-sickness, generally ascribed to the rarity of the air. In the present case, too much telega-travelling and want of training supplied a sufficient cause. Meantime the moon was shining gloriously in a cloudless sky, lighting up the huge white cone above us, and the distant ranges beyond the Araxes. Unluckily, I got worse instead of better, and was obliged to delay our progress by frequent halts. We were now beginning to climb the actual cone, and the rock-ridge, though still easy, became steeper. When fairly on the face of the 'Grands Mulets,' after three hours of feeble and intermittent progress, 'the force of nature could no further go,' and I sadly succumbed, leaving François and Tucker to go on and, as I hoped, to prosper. This was about 6 A.M. The sun was already high, and the air was pleasantly warm.

I was left on a shelf of the rock with a cup of wine and some food. For the latter I felt no inclination; as for the wine, it was soon disposed of by my dozing off and upsetting it with my arm, leaving barely a wineglassful of liquid as my provision for the morning.

After the first doze I made an attempt to follow my companions, but soon found it useless; so I resigned myself to fate, and lay down, now in one nook of the rocks, now in another, sometimes dreaming oddly, as one does in odd places, sometimes gazing drowsily over the top of Little Ararat (12,800 feet) into Persia, or over the Kara Dagh ranges to the white line of the Eastern Caucasus. The sun got very hot, and my head ached horribly; so I scrambled round the rocks to a shaded shelf, whence I could see far into Kurdistan, a region of snowy mountains and bare valleys. A streak below me was the infant Euphrates, but I did not feel much the better for seeing it. Of the Garden of Eden no tradition seems to linger even in this land of old stories, and if these barren hills were ever clothed by the groves of the earthly Paradise, the change has been complete indeed. My state of mind at the time scarcely made me a fair judge of the view, but I will try to give the impression it produced upon me, as compared with European mountain panoramas. Most people have seen in a sculptor's studio a block of marble hewn down to the rough outline of the group which he has it in his mind to produce. From a distance the eye catches a certain grandeur of effect which closer inspection destroys, by revealing that the parts are in themselves but rough and shapeless masses. So it is with these mountains of Kurdistan. On them the great sculptor, Nature, seems to have 'tried her prentice hand' before she had learnt how to chisel out with her graving-tools, frost and heat, the torrent and the glacier, those exquisite outlines of peak and valley which are a distinguishing feature of the Alps and the Caucasus. The first impression I received was,—what a wilderness of mountains!—in every direction nothing met the eye but snowy masses, lying in heaps instead of ranges. The general effect was exceedingly grand and impressive; but when the details were examined in search of some beautiful peak, the search was in vain. The slopes were characterised by dreary monotony, and the summits were without form or beauty. One distant mass (Bingol Dagh?) alone deserved to escape the general condemnation.

Time wore on, and at length, about 1.30 P.M., a shout above me announced Tucker's return. I augured ill from it, for it was not a cheerful 'jödel,' but I retained a hope that he might, for my sake, be subduing his feelings. To my surprise, the next shout came from below, and I knew that my companion must have descended by another route. Through the light cloud that was hanging on the mountain, I soon saw the two figures, and before long had joined them and heard their story.

The rocks above my halting-place turned into an arête, cut into towers, separated by deep gaps. The climbing here was exceedingly difficult, and the passage of some of the gaps required both care and steadiness. Fortunately, the ridge was not very long, and in an hour and a half from the place where I had stopped, a snowy saddle connecting the rocks with the upper mass of the cone was gained. Here they rested for half an hour, at a height probably of 13,800 feet. Above them stretched interminable snow-slopes, seamed here and there by rocks, but, unluckily, rocks of an utterly useless description to the climber. They were not ridges, but disconnected crags of lava, suggesting by their fantastic shapes the idea that half the animals, after leaving the Ark, had been petrified as they came down the mountain. Here was an elephant, glissading elegantly, using his trunk for an alpenstock; there a tapir, or some antediluvian-looking beast, by whose untimely fate, now for the first time discovered, naturalists have lost a species.

Before long the snow took the form of hard névé, and it was necessary to cut steps. François was by this time so exhausted that he could do no more; Tucker, however, pushed on alone, and by cutting about 1,000 steps, succeeded in reaching a point a little under 16,000 feet.* Such work, at a height equal to that of Mont Blanc, cannot be continued for ever, without long training; his breath began to fail, and his head to throb painfully, so that he was obliged to rest every twenty or thirty steps. The tremendous staircase required to reach the summit was not to be accomplished single-handed, and at 12.10 p.m.—after nearly four hours' solitary work, the top looking as far off as ever, and clouds collecting rapidly round the mountain—Tucker turned to descend. Having rejoined François, they returned quickly together down the tracks made in the ascent, avoiding the rocky arête, by slithering down the snow-slope on its left, which had been hard-frozen in the morning.

We plunged gloomily through soft snow, and over the tiresome rough lava-crags, and, despite the mists, found it easy to follow our old track to the spot where we had left the Kurds. They now shouldered with ease the burdens under which they had groaned and staggered the evening before, and led off at a quick pace for the huts, where we arrived about 6 p.m., having halted often on the way. The last part of the walk was in rain, Ararat having succeeded in his daily task of collecting a shower in otherwise fine weather. We regained the huts at 6.30 p.m., having been 18½ hours out.

We slept in the Kurd tent—I badly, but Tucker, as he deserved, soundly enough. These tents have black roofs, like those of the Arabs, from which they are otherwise very different. Stakes three feet high are driven into the

* We estimated it afterwards, carefully, as between 1,000 and 800 feet below the top.

ground in a circle, and to the ends of these the top of the tent is loosely fastened. It is afterwards forced up and made taut by poles inserted underneath. The sides are then filled in with a roll of matting or reeds, through which the winds penetrate far too easily. The next morning was fine, and the Kurd porters, to gain several more days' pay (their object throughout), were willing to carry the baggage anywhere. Our provisions, however, were exhausted, and we must have waited a day for a fresh supply. I did not feel able to try the mountain again with any chance of success, without rest and good food, and Tucker had done nearly enough. After an hour's debate, we decided that we had no choice but to descend to Aralykh.

On the way down, we had a good deal of talk (through Paul) with the Kurd chief, who was a good fellow in his way. Though otherwise an intelligent and well-informed man, he shared the superstition prevalent among the natives at the foot of the mountain, that its top never has been, and never can be, trodden by mortal foot. This belief is maintained, despite the two recorded and undoubted ascents of Herr Parrot in 1829, and General Chodzko in 1850. Neither of them are open to the slightest doubt. Parrot positively asserts that, on his third attempt, he gained the actual summit, of which, moreover, he gives an intelligible description. General Chodzko led a regular military expedition against the mountain, advancing slowly, but surely, until he pitched his camp a few hundred feet below the top. There he remained for a week, engaged in scientific observation. Both these ascents took place in the early autumn, when, owing to the diminution of the snow, the summit is most accessible. Our Kurd also knew Alagoz well, having been in the habit of feeding his flocks on the great upland pasturage, which lies to the south of the peaks; and he confirmed

General Chodzko's account of the existence of a small glacier near the top, by telling us that there was a river there which stood still on account of the cold. General Chodzko ascended the highest peak of Alagoz, when employed on the military map of the Caucasian provinces, in 1847. He describes the top as exceedingly small, and the final scramble as more fatiguing than difficult. The second summit he pronounces altogether inaccessible.

The Colonel welcomed us back most cordially, and invited us to stay and rest; but we were anxious to get to Erivan, and so, after paying off our Kurds, wished him good-bye, and rode on to the ferry. The Persian owners of the horses had demurred to being taken on to Kamarlu as part of the day's work, and we had compromised the question, by promising them a 'backsheesh' on our arrival there. At the ferry they refused to go any further; we surprised them by paying no heed to their noise, and taking the horses over with us. At the second branch we found the leaky boat replaced by a rude log-raft, buoyed on inflated skins, on which we crossed, some Kurds swimming our horses over. The scene was otherwise unchanged, except that the Kurd girls were even prettier specimens of their race than those we had seen before. The sun was scorchingly hot as we rode into Kamarlu, and we were glad to throw ourselves down for half an hour on the wooden benches, in a cool room, while our 'telegas' were prepared. The Persians having persisted in their resolve not to cross the river, we left their horses in the charge of the postmaster, and set out to drive the two terribly rough stages into Erivan, where we arrived, sore and sorry, about sunset.

June 10th to 12th, Erivan.—We sent out Paul to make enquiries, and endeavour to conclude a bargain for horses, to enable us to ride back to Tiflis, by country-roads, for we were quite determined to have nothing more to do, for the present, with the post-carriages of the Caucasus. Our plan was to visit Etchmiadzin, and take from thence a track leading along the flanks of Alagoz (which we entertained thoughts of attacking), and then passing through the hill-country of Georgia, and across three ridges, varying between 6,000 and 8,000 feet in height, before it finally rejoins the post-road, three stations out of Tiflis. Paul was successful in finding a man in the bazaar, who agreed to furnish us, at a reasonable price, with horses and men who knew the roads we wished to follow. In the evening of our first day's rest, I was attacked with a violent pain at the back of the head, got no sleep, and sent next morning for the Russian doctor. He said there was nothing the matter with me, and prescribed a mild solution of peppermint, which was neither nice nor useful; in the evening he came again, and ordered leeches and a mustard-plaister. This vigorous treatment was effectual; I slept fairly, and got better next day, so that by the morning of the 13th, I was ready to make a short stage of three hours, on horseback, to Etchmiadzin. Tucker, meantime, had been busy in the bazaar, in getting a light mattress stuffed and made up, and in making other small provisions for our week's ride across the hills to Tiflis.

June 13th.—The ride to Etchmiadzin, despite the distant view of Ararat, is on the whole a dull one. The road passes under the fortress of Erivan, which might perhaps be formidable to Asiatic troops, and crosses the stream from the Gotchka lake, which here flows in a picturesque ravine. There is an untidy botanical garden on its further bank. For some little distance out of the town the country is cultivated, but the greater part of the twelve miles is over a bare and stony plain, broken halfway by a village, and further on by a Druidic-looking ruin on the left of the road. The village of Etchmiadzin is conspicuous, from a distance, by the number of its churches; they are of the usual Armenian style of architecture, lofty for their size, with circular towers capped by stumpy steeples. The village and bazaar are poor, and the place is in a very uninteresting situation, on a broad plain, watered by the stream, which has its sources on the eastern flanks of Alagoz. The convent and cathedral are within a large fortified enclosure, which has in its time resisted many attacks from the infidels. We were assigned a room in the convent, and the monks did what they could to make us comfortable. They all wear the Circassian hood, or 'baschlik,' which is far more graceful than the square cap of the Russian priest, or the cowls of European Orders. The cathedral is a quaint old building, covered with elaborate but somewhat barbaric sculpture, and decorated internally with fine wood-carving and numerous pictures of saints. The greatest sign of progress about the place is a large reservoir, which has been lately constructed.

We were invited in the evening to take tea with the Patriarch. He is a fine but not intellectual-looking man, with a splendid beard; he was dressed in robes of purple silk, and wore magnificent Orders, some of which had been presented to him on his recent visit to St. Petersburg. We were introduced by an Armenian merchant, whose acquaintance we had made on the Black Sea, and whom we now again most opportunely met. 'His Holiness,' who quite plays the Pope amongst his countrymen, was very affable, and, could he have spoken any Christian language except Russian, would doubtless have given us a good deal of interesting information. As it was, he spent the best part of an hour in proving to his own satisfaction how much more charitable and tolerant the

Armenians were than the Greeks, and how much they sympathised with the English in ecclesiastical matters. Toleration is a virtue often found in the weaker party, and the poor Armenians need at present all the sympathy they can get, as their Church is divided against itself—one party, headed by the Patriarch, acquiescing in Russian supremacy and interference; while the other resents it, and urges the removal of the seat of the patriarchate to some spot outside the Czar's dominions. The room in which we were received was hung with a long series of portraits of (to us) unknown Kings of Armenia, headed by the present Czar of All the Russias. On our departure the Patriarch presented each of us with his 'carte-de-visite.' After we had returned to our room, a secretary appeared with an English document, which he wanted me to copy; it was the receipt of a Calcutta firm for some money paid in to the Patriarch's account by an Armenian missionary in 1814. The man was very anxious to know if he could get the money now by sending to London, but we thought it best to decline giving an opinion on that delicate point.

June 14th.—In the morning we were shown the convent library, which is small, but contains some magnificently-illuminated manuscripts. A Bible, with numerous and quaint pictures of Old Testament history, was I think the handsomest I ever saw. Our day's ride lay up the valley on the eastern side of Alagoz; the country in general is distressingly bare, and the track led us over stony downs, until it came suddenly to the brow of a cliff, under which flowed a stream in a verdant trough, with the village of Oschagan on the opposite side. An old bridge formed the foreground to a picture which perhaps struck us more than it would have done in a country less generally monotonous. The long gentle slope from here up to Aschtarak was a perfect Eden contrasted with the bare wastes beyond. Careful irrigation had clothed the soil with a rich mantle of vegetation, vineyards and orchards lined both sides of the path, and even the dividing hedgerows seemed to share in the general luxuriance. The village stands on a brow above the sunny slope which supplies it with corn and wine, and its inhabitants have a more prosperous air than most of the Armenian peasantry. We made our midday halt here, and I was glad to rest for two hours in a clean room, for my Erivan attack had left me somewhat weak and lazy.

The only interesting features in the latter part of the day's ride were the river-beds—picturesque troughs, almost gorges, sunk from 100 to 200 feet below the general surface, with rugged volcanic rocks cropping out from their sides. We ascended all day, and towards evening reached a high plain partially cultivated, and dotted with dismal villages. The first we halted at offered such bad accommodation that we rode on three miles further to Alekujak, which stands on the left bank of the torrent descending from the Alagos glacier. Our quarters here were about the nastiest we met with during our whole journey. To avoid the winter-cold at this height (about 5,000 feet), the houses are all constructed on the principle of a molehole—one passage leading into the family apartments, which no stranger can enter on account of the presence of the womankind, another into the stable. In the latter we had to lodge on a sort of dais, provided with a fireplace and two sleeping-mats, slightly railed off from the horses and cows that occupied the rest of the apartment. The only fuel obtainable was cowdung, so that the fire did not add much to the cheerfulness of the situation. What with the stiflingly pungent smell of the stalls, the noise of the animals, and the determined inroads of fleas and other insects, we never passed a more miserable night.

June 15th.—The hills were covered with a wet blanket

of mist, and our last hope of Alagoz—the summit of which (13,436 feet), a rock-peak of the Piz Languard type, had shown for a moment the previous evening—was extinguished. We rode on over intensely green upland pastures, surrounded by, if possible, greener hills. Mists swept over all their tops, and rain fell pretty steadily. We forded the stream three times before reaching Kondaksaz, a small village inhabited by Mahommedan Kurds, where we halted for lunch in a stable, a shade better than our sleeping-quarters. Alagoz now lay well in the rear, and the track leading over to Alexandrapol turned off on the left. A plain, on which large herds of horses were pasturing, was soon crossed, and we entered a long and narrow glen; the scenery and the weather were both bad Scotch, and we could not look forward with any pleasure to the passage of the watershed between the Kur and Araxes, which we were now approaching. After passing two villages, one on either hand, the glen narrowed, and the track finally made a sudden dash up the hillside on the right, bringing us very quickly to a grassy ridge 7,828 feet above the sea. We were surprised to see a rapid and long descent before us on the northern side.

The valleys of the tributaries of the Kur are everywhere much deeper cut than those of the Araxes, and the Georgian highlands are consequently more picturesque than those of Armenia. Snow lay heavily on the pass, and we had some little trouble with our animals. The wild flowers were lovely, many of them old English friends—such as cowslips, primroses, and violets. We also found a gentian, and saw again the dwarf hyacinths of Ararat. Two hours below the pass we came to Hammamly, situated on the banks of a torrent, at the junction of three valleys. It is on the road from Delidschan to Alexandrapol, and there is a post-station in the village, to which we of course went.

It was pleasant to find a clean room and a good fire, over which some mulled wine was quickly brewed, and proved most acceptable after our wet and cold ride.

June 16*th.*—The mists clinging to the bare hillsides around reminded me irresistibly of Scotland, and the first part of our ride was through scenery very like that of the Grampians. We followed the post-road for some distance down the valley; occasionally, near a village, a clump of trees broke the hillsides, but the general character of the country was unchanged. Several versts before reaching Kischlak, the next station, we turned over a brow on the left, and entered a side-valley, which runs up into the hills in a direction at right-angles to our previous course. We soon came to a village, tidier and more habitable-looking than the wretched places we had seen since leaving Etchmiadzin. Above this the hillsides were thickly wooded, a fact we appreciated the more from having seen no natural timber either in Armenia or Persia.

A rough cart-road led us over a grassy ridge, the summit of which was covered in mists, and it was not until we had descended some distance that we gained our first view of Gergeri, a large village situated in a secluded basin, and surrounded by finely-timbered hillsides. Our horses waded with difficulty, through horrible mud, into the military cantonment, which is a short distance from the village. We found shelter from the incessant rain in a small cottage, built after the Russian style, and bearing evidence, in its fittings, of inhabitants more civilised than the Georgian peasantry. Pictures, mostly of a religious type, were pasted on the walls, and there was an old family Bible on the table. Having learnt with satisfaction that Djelaloghlu, our sleeping-place, was only ten versts distant, we, after a short rest, proceeded on our journey. The road ascended a small valley, with bold hills on the left, for some distance, and then crossed a low steep ridge, from the top of which we overlooked a green tableland filling up the space between the Bezobdal and Lelwar ranges, both of them offshoots of the Anti-Caucasian chain. The rain was falling in torrents, but, happily, Djelaloghlu was at hand. It is a place of some size, laid out in the straggling style common in the Caucasian provinces. Detached cottages are set down in rows on either side of a broad street of mud. The houses, individually, are quite as good as an ordinary English cottage. Here there are, besides, large government stores, and officers' quarters, with some pretence of a garden in front of them. We were directed to the village shop, and found shelter in a sort of back-kitchen, opening out of it, which would have been comfortable enough, but for the chilly look given by a damp earth-floor. We shared our quarters with some enormous wine-skins, which were stowed in a row along one side of the room. The presence of the military ensured us fair food, and we spent the evening in writing letters, and working ourselves up into an unusual state of patriotism, by drawing comparisons between a Georgian and an English June—all in favour of the latter. Calculations showed that rain had fallen on twelve out of the fifteen days since we left Tabreez, and it fell on each of the three following days up to our arrival at Tiflis.

June 17th.—We meant to have started early, knowing we had a long day's journey before us; but in the morning Paul complained of being ill, and would do nothing but groan. It turned out that he had neglected our injunction to change his wet clothes the previous afternoon, and had consequently caught a chill. A strong 'pick-me-up' cured him for the time, and at 8 o'clock we set out once more, to face the rain and mists. Djelaloghlu stands on the brink of a curious cleft, the bottom of which is at least 100 feet below the level of the plain, and the sides almost perpendicular. The road dips down, by a steep zigzag, to a bridge over the Debeda. At the top of the ascent on the other side stands a fine old stone cross. For several miles we rode over a plain covered with the most luxuriant herbage, and then entered a long valley between bare green hills, one of them crowned by a tall wooden cross. At a point were two streams meet was a large village, where our men wanted to halt; but we, wishing to reach Schulaweri before nightfall, refused to let them. We forded one branch of the stream without much difficulty, and followed the other up to its sources among the hills. The height assigned, in Kiepert's Map of the Caucasus, to the pass we now crossed, is 5,805 feet. It is probably picturesque, but we saw nothing but fog and mist, till we had descended several hundred feet on the northern side, when we found ourselves on a wooded slope, high above the recesses of a deep valley. The neighbouring mountains were clothed in the most beautiful park-like timber. The glades and grassy knolls were enlivened by Kurd encampments, sheep and horses were grazing on the fresh herbage, and the bright costumes of their owners gave colour to the scene. At this height the trees were still in their spring tints, and the white-thorn was coming into full blossom. We noticed a great many wild fruit trees, especially pears and apples. After many windings, the cart-track succeeded in descending to the side of the stream, which we followed for some distance down an exquisitely wooded valley. It was our first introduction to Caucasian forest scenery, and we were constantly halting and calling each other's attention to some wall of verdure, built up of gigantic beech-trees, or a glade where gnarled old trunks and luxuriant underwood afforded a subject for the artist or photographer. The brown torrent, encouraged by the recent rains, ventured on some remarkable falls; at one spot a tributary leapt suddenly out of the foliage, and tumbled in a sheet of foam into the larger stream, forming one of the most effective 'water-meets' imaginable. At last the valley opened out a little, and we came upon cornfields, showing that habitations were not very far off. We halted at the village-store, a roadside hut soaked with rain, where we had difficulty in finding a dry corner to eat our lunch in. Here, nevertheless, our horsemen wanted us to stop for the night, and told the usual lies to induce us to accede to their wishes. It was said that we had only ridden halfway, and therefore could not arrive at Schulaweri till long after dark, and that there were wicked people on the road, which was moreover barred by an impassable torrent. We were by this time pretty well used to these bogies, and persisted in starting again as soon as possible. The track at once crossed, by a bridge, the stream we had been following, and then a short but steep climb led to the summit of a low watershed, the valley on the other side of which was broader and more open than that we had just left.

An utterly-deserted village contrasted strangely with the smiling landscape and frequent cornfields, and the hedgerows, gay with flowering shrubs, often reminded us of England, to the hillier parts of which the features of the country bear some resemblance. Where the valley bent round to the north, and contracted into a defile, we encountered the terrible torrent. The old man in charge of the horses was much alarmed, and declared the water was rolling down big stones, and that the passage was too perilous to be attempted; but we rode through with perfect ease, scarcely finding it necessary to lift our feet in the stirrups. We had to cross the stream three times, but familiarity, as usual, bred contempt, and even the leading old man did not hesitate twice. The hills gradually opened as we drew near Schulaweri, the situation of which is very beautiful; the town stands in the centre of a richly-wooded basin, and is surrounded by walled vineyards and groves of fruit-trees. The ground on the north falls in a long slope to the Khram, and the eye sweeps over the plain-country to the chain of hills that surrounds Tiflis. A curious natural arch, in some castle-like crags on the top of one of the hills that overlook the town, is a conspicuous feature of the view. We were first shown into a gloomy den, but in a short time got possession of a clean though bare room in a two-storied house, which was but just finished, and still unoccupied. We were not sorry, after our forty miles' ride on tired horses, to spread our mattrass on the floor, and lie down to sleep.

The next morning (June 18th) we rode out through a fairly-furnished bazaar, and crossing, for the fourth and last time, the stream of the day before, left the vineyards behind, and found ourselves on cornland, where teams of sixteen oxen were ploughing furrows, six inches deep, to the monotonous chaunts of their drivers. Under ordinary circumstances it is an easy day's ride from Schulaweri to Tiflis, but a flooded stream now barred the direct road, and we were obliged to make a long circuit to the west to find a bridge. The way was enlivened by the frolics of two half-tipsy Georgians, both riding on the same horse—a form of cruelty to animals to which the people of this country are much addicted. They narrowly escaped drowning, in an attempt to ford the stream, half a mile below the bridge. On the bank stood a comfortable farm-house, surrounded by some fine trees, which might have been made into a very pretty place. After crossing a second stream, by a new bridge, we at last passed, some way off on the right, a large building, apparently an old caravanserai. We halted at a village, meaning to lunch; but, though there were many vineyards in the neighbourhood, no wine was to be had. The puzzle was explained when we found that the people here were Mahommedans, and those of the next hamlet Christians. We rode on, and in half an hour found ourselves again on our old track, at the third station from Tiflis on the Erivan post-road. It was ten versts on to Kody, where we were obliged to sleep, for our horses,

> 'Hollow pamper'd jades of Asia,
> Which could not go but thirty miles a day,'

were completely played out by their previous performances, and plodded on at a pace which was painful to everyone concerned.

There was scarcely anything eatable to be found at the station—indeed, the postmaster's only object seemed to be to get rid of us. We consoled ourselves with the thoughts of Tiflis, and hotel luxuries on the morrow. Our sleep was soon broken by the howling of some miserable dogs. The concert was so prolonged that my friend finally lost patience, and broke it up by firing a revolver into the middle of the performers, unluckily without fatal effect.

June 19th.—From this point we intended to strike straight for Tiflis across the hills—a route which had been described to us as both shorter and more picturesque than the tedious approach by the valley of the Kur. During the first part of our ride we were still following the post-road; the mud was something indescribable, and the ground ordinarily driven over so heavy that the carts had been taking lines of their own through the fields. The postmaster at Kody told us, with apparent satisfaction, that if we had wanted 'telegas' he should have given us five horses to each, as a less number could not pull even that light weight through the slough of the highroad to Persia and the Caspian. Presently turning off at right-angles, we struck up the hillside on our left, by a steep horse-path mounting beside a gully, in which a quaint little village sheltered itself. On reaching the brow of the hill, the whole of the great city of Tiflis and the course of the Kur for many miles burst upon us with startling suddenness, at least 2,000 feet below. The view is very striking, and when the snowy chain of the Caucasus is clear, it must be still more so. The descent was long and steep, down a hillside covered with brushwood and broken by crags. We met strings of donkeys carrying out goods from the city, and passed others, laden with firewood, going in the opposite direction, as we rode down through a suburb of gardens into the Persian quarter. After Erivan and Tabreez, the streets seemed wonderfully European, with their tall houses, shops with plate-glass windows, and smartly-dressed ladies in Parisian costumes. The Russians have spent a great deal of money to establish a handsome European city south of the Caucasus, and they have effected their object. Tiflis is undoubtedly a success. It is polyglot, but not Asiatic; and the Persians, like the foreigners in Leicester Square, keep their own quarter, and even there look shady and dull compared with their countrymen at home.

Tiflis, June 20th to 26th.—We were delighted to rejoin all our luggage at the comfortable Hôtel d'Europe, and to find the missing tent and portmanteau arrived from Kutais. My time during the next few days was spent principally in visits to the governor and postal officials, which did not produce any very great results. I took pains to explain our plan, which was in itself sufficiently simple—namely, to go to the foot of Kazbek by the post-road, ascend if possible that mountain, and then cross, by two passes laid down in the Russian maps, into the valley of the Rion. To the official mind, however, the unknown and the impossible are coextensive terms; and while I was met with the greatest personal civility and desire to aid us, I could get no definite information, or promise of assistance, beyond the posthouse of Kazbek, where the Governor of Tiflis told me he hoped to meet us in a week's time. There seemed even to be a question whether we should get thus far, for at the post our 'crown-podorojno' was laughed at, and we were told no carriage could be promised us for an indefinite period.

The Grand Duke Alexis (the son of the Emperor) was expected from St. Petersburg on a visit to the Caucasus, and consequently all the official world were in motion to meet him, and no one without epaulettes and a band round his cap had a chance of meeting with the slightest attention. After several efforts I gave up the post in despair, and sought out a German carriage-master, who agreed to let us a 'tarantasse' with four horses, to travel voiturier-fashion. We had also to make several visits to the police to enquire about François' passport, which the officials at Poti had promised to forward. The authorities would hold out no certain prospect of its restoration, and seemed to wish him to buy of them a Russian document, costing two roubles, in its place; so we commissioned the master of the hotel, who was going back to Europe, to stir up the Poti police, and he succeeded in recovering the missing passport. Travellers anxious to avoid that fever-stricken swamp, Poti, often go straight through, trusting to the promise of the police to send their passports after them—a promise which, in the cases which came under our personal knowledge, was invariably broken.

On the 20th, the day fixed on for a rendezvous with my friend Moore, who was to come out straight from England by the Danube and Constantinople, a telegram from Kutais announced to us the welcome news of his arrival in the country, and on the 21st he appeared in person, having been most fortunate in getting brought on from

N

Poti by a Russian lady, who was coming to live with her daughter at Tiflis. He described the state of the road as something awful: for half the distance they had found no post-horses, and had been obliged to pay high prices for peasants' animals, brought in from the fields to meet the demand. The mud was very bad, but not so deep as it had been a few days before, when a Russian family, whom he met at the third station from Kutais, had been obliged to have their carriage dragged for one stage by bullocks, and had taken twelve hours to accomplish sixteen versts. These unlucky people, who did not care to pay for extra horses, had taken five days to get over the hundred miles between Tiflis and the station where Moore found them. With our previous experience, we were rather dismayed to hear that our friend had left one of his portmanteaus at Kutais in charge of an official, to be forwarded, and our fears were justified by its non-arrival for three days after it was due. On its appearance our preparations were quickly made, and our 'tarantasse' was ordered to come to the hotel at 1 o'clock on the afternoon of the 26th June.

Two days after we reached Tiflis Paul had declared himself ill, and, to our great embarrassment, had taken to his bed; he had never entirely recovered his wetting at Djelaloghlu, and was now suffering from a kind of intermittent fever. We felt sure that if we could get him well enough to go up with us to the Kazbek posthouse, a week's rest in mountain air would restore his strength; but his illness was a great discouragement, just at the moment when we were starting for the portion of the journey in which his services were most indispensable. The doctor, whom we sent for, had recourse to the Russian panacea, leeches, which in this case did not do much good, and it was by frequent doses of our own quinine that the patient was finally brought into a condition to travel.

CHAPTER VII.

THE KRESTOWAJA GORA AND ASCENT OF KAZBEK.

Start for the Mountains—The Pass of the Caucasus—Kazbek Post-station—The Governors—A Reconnaissance in force—Legends—Avalanches—The Old Men's Chorus—Men in Armour—Our Bivouac—A Critical Moment—Scaling an Icewall—The Summit—The Descent—A Savage Glen—A Night with the Shepherds—Return to the Village—Caucasian Congratulations.

WHEN our vehicle drove into the courtyard, we, ignorant still of the utter uncouthness of all Russian conveyances, were surprised to find a mere shell of a carriage without any fitments inside. However, by making use of our own luggage and rugs, we soon succeeded in heaping together seats, which, if they had a tendency to collapse, were luxurious in comparison to those of our late 'telegus.' Amidst the good wishes of the hangers-on of the hotel, we started on the journey which was to carry out the object long and anxiously planned, and throughout all our wanderings steadily kept in view, as the centre and chief aim of our travels—the exploration of the *terra incognita* of the Caucasian range.

We slowly jolted over the badly-paved streets of Tiflis, now about to be strewn with earth to spare the bones of the expected Archduke—a proceeding which, if the weather held fine for a few days, would be certain to throw dust even in imperial eyes. Our coachman, a regular Russian peasant, stupid, obstinate, and good-humoured, crossed the Woronzoff Bridge and took a road along the left bank of the Kur, which passed through several villages, and, though hilly, was more direct than the line taken by the post-road. Before reaching Mscheti we had to cross the Aragui, a large tributary of the Kur, one of the branches of which the Kreuzberg road follows almost to its source. On our return, two months later, the long wooden bridge had met with the usual fate of bridges in this country, and was so much damaged by floods as to be rendered impassable.

Mscheti, if we may believe Georgian chronicles, is one of the oldest towns in the world. It is asserted to have been founded by Mtskethos, son of Karthlos, who lived in the fifth generation after Noah, and who chose this site on account of its beauty and natural strength. A little below it, on the top of a green hill, are the remains of an extensive church and convent, from which it is said that a mystic chain used once to extend in mid-air to the cathedral tower of Mscheti, and serve as a means of mutual communication for the saints of either church. We drove close under the walls of the old fortified cathedral, where we joined the post-road which crosses the Kur a mile higher up, and has to return some distance to the town. The Aragui here flows at the base of high bluffs, along the sides of which the road is carried. For this stage and half the next workmen were employed on the construction of the new and still-unfinished macadamised roadway, which, as soon as the river allows it, descends to the level ground. Sukan, the second station, is situated in the centre of a fertile basin, encircled by well-wooded hills, purple as we saw them in the fading sunset. Although not travelling with post-horses, our 'podorojno' gave us the right to lodge in the stations. Our reception, however, was at first anything but hospitable; we were even told to turn out, until the master found we were willing to pay for rooms and to order supper, when he became less bearish in his manners.

The stations on the Dariel road are very different to the

ordinary type of Caucasian posthouses. They are substantial stone buildings, with verandahs, bow-windows, and sometimes a billiard-room. Their internal fittings by no means correspond with their pretensions. Downstairs the rooms are furnished only with square stools, and the usual wooden bedstead. The salle-à-manger is usually large, with, in one corner, a cupboard containing a motley collection of delicacies, mostly liquid—a sort of museum of various shaped bottles labelled with the names of the choicest brands. I have seen in a row 'Veuve Clicquot,' 'Château Lafitte,' 'Allsopp's Pale Ale,' 'Guinness' Stout,' and 'Old Madère' (sic); there is very seldom more than one bottle of each. The champagne is generally five roubles, and the English beer one rouble fifteen copecks, a bottle. A few boxes of sardines and a plate of stale cakes form a set-off to this tempting array. The samovar and tea are always forthcoming; 'borsch,' or cabbage-soup, a national dish in Russia, is usually to be had very quickly, and sometimes a beefsteak will be cooked if ordered; but, as often as not, there is nothing more solid than eggs in the house. Upstairs are a set of rooms provided with mattrasses, which are charged for extra, such arrangements being considered quite unnecessary luxuries. It was our readiness to pay for these reserved apartments which smoothed away the difficulties at first made to our reception.

June 27th.—We got off at 5 A.M., and enjoyed the beauty of a fresh clear morning. After a straight stretch of several versts, the road left the valley of the Aragui, and turned up a narrow glen; a long and gradual ascent brought us to a green tableland, where a little tarn appeared amongst the meadows. The posthouse of Duschet stands by the side of a hollow, but the town lies on a sloping hillside, at some distance to the right; a good many

Russians live here, and we had a letter for the commandant of the district, who, we were informed, would probably be able to aid us in our preparations for attacking Kazbek. We found, on enquiry at the station, that he, like everyone else, had gone off to meet the Grand Duke, so we pursued our journey without delay.

The next stage was across a ridge, wooded to the summit with fine park-like timber, and down a long and narrow glen on the other side to Ananour. On the tongue of rock projecting at the mouth of the glen stands a most picturesque group of buildings, consisting of two old churches and a belfry, enclosed by battlemented walls and towers. The larger and more modern church is decorated externally with large and elaborately-curved crosses, and sculptures of trees with animals feeding on their branches. The village clusters round the foot of the fortified mound, in a very pretty position at the junction of two torrents. The road now led up a narrow valley, the wooded slopes were frequently dotted with castles and towers, and the vegetation was richer than that of a Swiss, but the rocks not so bold as those of an Italian, Alpine valley. At Pasanaur, remarkable only for a church in the most gingerbread style of Russian architecture, the river forks, and the road, following the western branch, enters a defile, above which the upper valley, lying at the foot of, and running for some way parallel to, the main chain, opened before us. Scattered hamlets and noble trees studded the slopes; the lower wooded buttresses of the mountains were beautifully shaped; the higher ridges (9,000 to 10,000 feet), 'up to their summits clothed in green,' often ended in peaks of bold outline, and picturesque glimpses of the snowy chain opened from time to time up side-glens. The horses, which had done seventy-five uphill versts in the day, required a great deal of persuasion to trot the last half-hour into

Mleti, and we were amused at the difference between the long guttural grunts of the Russian driver, and the sharp tones used by an Italian voiturier in like circumstances. The station at Mleti is one of the most frequented and best provided on the road.

June 28th.—The ascent from Mleti, up a slope broken by cliffs, is steeper than any Alpine carriage-pass I remember, except the wonderful zigzags beside the Madesino Fall, on the south of the Splugen. At one picturesque corner the road is seen on the top of a cliff overhead, and anyone unused to mountain engineering might well wonder how it got there. A little fountain, spurting up a jet by the wayside, is an incongruous bit of civilisation in the wilderness. In mercy to the horses we walked, and having scaled the rocky mass, which, during the latter part of our drive the evening before, had seemed to block the valley, we found ourselves on grass slopes covered with azalea-bushes and smaller flowering plants. From this part of the road the view of the head of the valley beneath is very striking. A thin waterfall leaps down the opposite cliffs; a village, close beside a curious isolated rock, occupies the last habitable spot in the valley, and higher up a mere ravine runs under the base of a pointed peak, which rises above it in grand precipices. A group of houses—consisting of barracks, a station, and a wayside inn—stands on the mountain-side about 1,000 feet below the pass, filling the place of the 'hospice' on an Alpine road. I had slightly rubbed my foot during the ascent, and therefore waited for the carriage, but the rest of the party walked on as far as Kobi. We now traversed, at a level, a steep hillside cut into terraces, and staked up to prevent avalanches from gathering impetus enough to sweep over and carry away the road. The old horse-path crossed the ridge at a point slightly to the east of the course now followed. The grass and flowers were most luxuriant, owing to the quantity of springs which burst out of the ground on all sides. There is little distant view from the summit, on which is a stone refuge. The Krestowaja Gora (or Kreuzberg, as translated on German maps) is the real name of the pass over the chain of the Caucasus leading from Asia into Europe;[*] the ordinary name of 'Dariel' road is only so far appropriate that the defile of Dariel is the most striking natural feature between Vladikafkaz and Tiflis. If the pass of the Splugen from Chur to Chiavenna was ordinarily termed the 'Via Mala road,' it would be an exactly parallel case.

The descent on the north side into the valley of the Terek is one of only 1,500 feet, but it must be very dangerous in spring, as the way lies down a deep glen choked at the bottom with the remains of enormous avalanches, which in more than one place still buried the track, obliging a passage to be cut through them. The slopes are terraced, to protect the road; the idea of building covered galleries has either not occurred to the Russian engineers, or was considered too expensive by the Government. It must be adopted if the pass is ever to be kept open at all seasons. Kobi, the first village on the northern side, is strikingly situated, at the point where the glen joins the valley of the Terek. A high cliff shelters the posthouse, from whence the summit of Kazbek is not in view, being hidden by massive buttresses. The postmaster here was tipsy. As an English traveller mentions the same fact in 1837, and as he was in a similar happy condition upon

[*] I follow the most eminent modern geographers in considering the Caucasian watershed as part of the boundary between Europe and Asia. Though this conclusion has been for many years generally adopted, the public and their instructors are, as yet, scarcely awake to the necessary corollary that Mont Blanc and Monte Rosa must be regarded as usurpers, and that Elbruz and Kazbek, Koschtantau and Dychtau, are entitled to precedence on the roll of European mountains.

our two subsequent visits, a week and two months later, it is fair to suppose that the complaint is chronic. The scenery of the valley of the Terek is entirely different from that on the south side of the pass: treeless valleys, bold rocks, slopes of forbidding steepness (even to eyes accustomed to those of the Alps), and stone-built villages scarcely distinguishable from the neighbouring crags, but for the one or two towers of defence which rise above the clustering hovels, are the main features of the sixteen versts' drive from Kobi to Kazbek. A bold pinnacle of rock on our right reminded me of a Tyrolese dolomite, while the trough-like character of the valley, and the stern barrenness of the scenery, carried Moore's thoughts back to Dauphiné.

We passed, halfway, a hamlet bearing the familiar name of Sion, behind which a few trees had been planted, the only ones in the vicinity. Clouds as yet prevented our catching any glimpse of the snows of Kazbek, but did not hide the lower mountains. The village is in a fine position, backed on the east by very steep grass and rock-slopes, the supports of a massive rock-peak of at least 12,000 feet in height. Soon after our arrival, the clouds, which up to this time had filled the glen opening opposite the posthouse, rolled away, and revealed at its head Mount Kazbek, a magnificent mass of rock and snow, towering thousands of feet above all the neighbouring summits. The form of the mountain-top is that of a steep-sided dome; the uppermost crags, which break through the ice, are of a horseshoe form, and are curiously prominent in all views of the mountain from the east, and even from Vladikafkaz. We were glad to find the posthouse in the charge of a civil couple, a man and his wife, the latter of whom spoke a little German. The charges were high, but we had no other ground of complaint, and enjoyed during our stay plentiful food, fair wine (selected from the

the usual medley in the cupboard) and much civility. Although a very cursory inspection of the mountain suggested several routes, offering a fair chance of reaching the summit, yet it was felt that to make the assault without a previous reconnaissance would be unadvisable, bearing in mind especially our utter want of training. Paul was therefore told to find a native who would accompany us, in the morning, to some point of sufficient elevation to command an uninterrupted view of the mountain, and at the same time to accustom our muscles to the work before them. In due course he reappeared, with a good-looking man known as Alexis, who, he said, was a mighty hunter, and knew more about the mountains than anyone else. This worthy seemed to our eyes a feeble creature, but as no one else was forthcoming, and it was not probable that we should put his ability to a very severe trial, he was engaged to be our pioneer on the morrow.

June 29th.—We were up betimes, and starting before 5 A.M., on as fine a morning as ever rejoiced the heart of a mountaineer, climbed to an old church perched on a lofty brow 1,500 feet above the village. This building is regarded with great reverence by the inhabitants, and is made an object of pilgrimage; but their religious feelings do not prompt them to keep it in repair, and the interior is in a very desolate and ruinous state. In Klaproth's time it was the practice to open it only once a year, but the attendant, who had joined us on our way up, made no difficulty about admitting us, although to open the door he had first to gain admission by getting in himself through one of the windows—no easy task. From the enclosure round the church we could see the ground between us and the base of the great mountain. Just opposite, and easily accessible from where we stood, a snow-clad peak, evidently commanding a view of Kazbek, offered itself

as a suitable goal for our morning walk. The way
to it lay up a broad grassy ridge adorned by rhododendrons with large white flowers, several kinds of gentians, and many other plants which lack of botanical
knowledge prevents my naming. We had not underrated
Alexis' capacity: so long as the way lay over grass he went
well enough, but on reaching the snow he stopped
abruptly, and declined to go any further, so we left him
with Paul at the foot of the final ascent. A climb up
steep snow-slopes succeeded by easy rocks led to the
summit, which was more of a ridge than a peak, and over
10,000 feet in height. Kazbek was now directly opposite
us, a long glacier streaming round its south flank, and
ending at our feet. From this point of view we saw the
second or western summit, which (totally invisible from the
station) here appears equal in height to the eastern.
This was a source of perplexity. Opinions were divided
as to the relative claims to superiority of the two peaks;
and although the majority were inclined to award the
palm to the eastern summit, there was sufficient doubt
about the matter to leave us all well pleased at the discovery, that from the glacier on the southern flank of the
mountain, the gap between the two peaks appeared to be
accessible by a series of crevasse-broken but easily sur-
mountable slopes, merging in a steep wall of snow or ice,
only partially visible, and as to the exact character of which
it was difficult to judge accurately. As any mistake with
regard to the real culminating-point would be very annoying, and it was clear that, once on the ridge, we should
have only to turn right or left, as might seem advisable, it
was unanimously agreed that this route should be tried—
an additional argument in its favour being supplied by
the evident existence, high up on the left bank of the
glacier, of several excellent sites for a bivouac.

With the great mountain full in view, I may now briefly advert to the position it holds amongst Caucasian summits, and to the legends with which it has been connected. From the earliest times Kazbek has taken a place in history, and has somewhat unfairly robbed its true sovereign, Elbruz, of public attention. Situated beside, and almost overhanging, the glen through which for centuries the great highroad from Europe into Asia has passed, it forces itself on the notice of every passer-by. The traveller—who, even if blessed with a clear day, sees Elbruz only as a huge white cloud on the southern horizon, as he jolts over the weary steppe—is forced to pass almost within reach of the avalanches that fall from his more obtrusive rival. It is not difficult, therefore, to see why Kazbek has become thus famous, why the mass of crag on the face of the mountain, so conspicuous from the post-station, is made the scene of Prometheus' torment, or why a later superstition declares that amongst these rocks, a rope, visible only to the Elect, gives access to a holy grot, in which are preserved the Tent of Abraham, the Cradle of Christ, and other sacred relics.

We were told by Mons. Khatissian, an Armenian gentleman, who has spent many months in examining the vicinity of the mountain, and in making scientific observations on its glaciers, that the Ossetes occasionally call Kazbek, Beitlam and Tscristi Tsoub ('Christ's Mountain') —names which seem connected with these traditions. On the top of Kazbek is said to stand a splendid crystal castle, and near it a temple, in the middle of which hovers a golden dove. The mountain has undoubtedly been held in reverence for many centuries by the neighbouring population, and it is not only the native inhabitants who have associated it with superstitious legends. A traveller in 1811 breaks forth, on reaching the station of Kazbek, into the following rhapsody: 'Alternate sensations of awe and rapture quickly succeed each other in this ancient land of enchantment: it was assuredly in these abodes that Medea compounded her love-potions and her poisons; here it was that Prometheus received the reward of his bold impiety; this is the very birthplace of magic; and it is from these lofty peaks that the immense roc used to take its flight, intercepting the rays of the sun.'

Mons. Khatissian also informed us of the existence of human habitations, now deserted, at a height of 11,000 feet, on the eastern flanks of the mountain. These consist of cells, half hewn from the solid rock, half built up of the rough boulders which abound in the neighbourhood, amongst which a cross of white porphyry still remains. Here, according to tradition, once lived a band of monks. The superior was renowned for his austere life and stern piety, and a daily miracle proved his claim to the title of saint. At daybreak a ray of light penetrated through an aperture in the wall, and illumined the darkness of his cell. In the centre of this ray the holy man was accustomed to lay the volume he was studying, which remained suspended in the air without any apparent support. The high claims of their superior to their reverence could not, however, reconcile some of the younger monks to the severe discipline he imposed upon them. By the machinations of these wicked men, the saint was exposed to a temptation similar to that of St. Anthony, but unhappily with a different result. The suspension of the miracle followed; the heavy volume, when laid in its accustomed place on the sunbeam, fell with a crash to the ground. The Abbot, overcome by the malice of his enemies, retired to a cave still higher on the mountain, to pass the remainder of his life amidst perpetual snows. The monks, freed from all restraint, gave themselves up to the license for which they had schemed, until at last the anger of Heaven was aroused by their misdeeds. A fearful storm fell on the mountain, the cells were destroyed, and nothing more was ever seen or heard of their inmates. So firmly is this story still believed, and so great is the reverence felt by the peasants for the once holy place, that Mons. Khatissian had the greatest difficulty in persuading anyone to conduct him to the ruined cells; and his guide, when induced to venture, fell on his knees at every other step, imploring Heaven to overlook their presumption. A heavy rain-storm the following evening, which threatened destruction to the hay-harvest, was attributed by the villagers to the Divine wrath at Mons. Khatissian's explorations, and he was recommended by the late Prince Kazbek to leave at night, if he wished to escape personal violence.

The accuracy of the above legend is, I fear, rather impugned by the fact that a lady, who published her 'Letters from the Caucasus' in 1811, actually saw one of the last of these recluses, of whom she does not seem to have formed a very favourable opinion. I quote her own words: 'I had often heard of hermits, but had never seen one. Learning, while at Kazbek, that I could satisfy my curiosity, I went to visit, in a cell not far from that place, one of these sloths, who are such vast pretenders to piety. I was surprised to find a healthy young man: his hermitage is hollowed out of the rock, where, thanks to the superstition of the people, who look upon him as a saint, he lives in abundance. Should he ever be canonised, I shall not indulge much hope from his mediation; for I saw nothing in this recluse but a cunning rogue, and that sort of address by which the lazy feed on the simplicity of others.'

The name by which the mountain is now known, and

which has been apparently accepted by geographers, to the exclusion of several more or less unpronounceable native titles,* is, like Elbruz, of Russian origin. A certain Prince Kazbek, or Kasibeg, who lived in the village of St. Stephen (the present Kazbek), was one of the first of the mountaineers to perceive that his best policy was to recognise a *fait accompli*, to embrace Christianity, and to acquiesce in Russian supremacy. He received his reward; the conquerors have given him immortality, by conferring his name upon the village in which he lived, and upon the great mountain by which it is overhung.

Even with the Russians—who, as a race, have no feeling for mountains, and regard them more as barely tolerable eccentricities than as admirable beauties of nature—Kazbek has, during the last twenty years, excited a good deal of attention. The creation of an ice-barrier across the torrent issuing from the great glacier of Devdorak, on the north-eastern flank of the mountain, has from time to time caused calamities wrongly attributed by the Russians to avalanches. On our arrival in the Caucasian provinces, the first thing we were told was, 'Oh, you are just in time to see the great avalanche from Kazbek.' Some years ago the Dariel road was swept away, and a similar catastrophe was considered probable during the coming summer. Everyone in Tiflis was talking of it, but happily it never came off, and we learnt from Mons. Khatissian that some, at least, of the historical avalanches are apocryphal. The record of one (in 1842) is preserved in the official archives at Tiflis, where the reports of the officers stationed at the Dariel fortress, and commissioned by the then Viceroy to ascertain the immi-

* Mquinvari is the best known.

intervals, and the supposition that the injury was caused by avalanches, are equally ridiculous.*

Attempts to ascend Kazbek have not been numerous. Klaproth claims to have got halfway up, but, as he admits that he did not reach the snow-level, the halfway did not amount to much. In 1811, the well-known German traveller Parrot made a series of most determined attempts to reach the summit, by the same route we adopted; but he was compelled to retreat from the foot of the icewall by bad weather, and the fears of his companions. About 1844, Herr Moritz Wagner ascended 'to the lower limits of eternal snow,' to use his own words—a very moderate measure of success, upon which some German and English newspapers lately claimed for him the honours, such as they are, of the first ascent. Several half-hearted attempts to climb the mountain have been made of late years by Russian officers, but with very little success, owing to the

* Mons. E. Favre, of Geneva, a well-known geologist who visited the Devdorak glacier a few weeks after ourselves, came to the following conclusion as to the nature of the catastrophe. No avalanche, he says, could without the aid of water traverse the space between the end of the glacier and the Terek, and he accounts for the disasters which have taken place in the following way. He believes the Devdorak glacier, to which he finds a parallel in the Rofen Vernagt glacier in the Œtzthal Alps, to be subject to periods of sudden advance. During these the ice finds no sufficient space to spread itself out in the narrow gorge into which it is driven, and is consequently forced by the pressure from behind into so compact a mass that the ordinary water-channels are stopped, and the whole drainage of the glacier is pent-up beneath its surface. Sooner or later the accumulated waters burst open their prison, carrying away with them the lower portion of the glacier. A mingled flood of snow and ice, increased by earth and rocks torn from the hillsides in its passage, sweeps down the glen of Devdorak. Issuing into the main valley it spreads from side to side, and dams the Terek. A lake is formed, and increases in size until it breaks through its barrier, and inundates the Dariel gorge and the lower valley.

Mons. Favre has also printed a paper, entitled 'Les Causes des Avalanches du Glacier du Kasbek, par le Colonel Statkowski, extrait du Journal du Ministère des Voies et Communications, 1866,' which contains an explicit statement as to the most recent catastrophe. The Colonel says: 'The last avalanche of the Glacier of Devdorak fell in 1832. In 1842 and in 1855 similar disasters were expected, but did not take place.'

O

exaggerated fears of their native guides, and their own lack of proper mountaineering gear—such as rope, ice-axes, and spectacles. Hence we found in the Caucasus a widespread belief in the inaccessibility of the peak, and we were regarded at Tiflis with a mixture of amusement and pity, as 'the Englishmen who were going to try and get up Kazbek,' and had the audacity to expect to succeed, where Captains, Colonels, and even Generals of the Imperial Russian Service had failed.

We spent a pleasant hour on our lofty perch, and then, by a 'rapid act' of what may be called 'snowmanship,' rejoined Paul and Alexis. The snow being in excellent order, we sat down, one behind the other, at the foot of the rocks, and letting go, slid with great velocity to the base of the peak, where our companions were waiting for us. They, never having seen such a performance before, were horror-struck at our apparently headlong descent, and could scarcely believe their eyes, when the confused heap, in which we landed, resolved itself into its component parts, apparently none the worse. By 2 P.M. we were back at the posthouse, and were delighted to find that the Governor of Tiflis had arrived, accompanied by Colonel Soubaloff, the Commandant of Duschet. They had come thus far to welcome the two Grand Dukes, who were about to pass on their way to Tiflis. The acquaintance of the Governor of Tiflis we had already had the pleasure of making, and both he and the Commandant entered heartily into our plans, and rendered us all the aid in their power in making our arrangements.

The most experienced mountaineers of the village were at once summoned—to wit, three aged men, all more or less lame or blind, who in the way they nodded their heads together, and by their occasional outbursts of eloquence, reminded us forcibly of the old men's chorus in 'Faust.' We at last settled with them to take four men as porters, at two-and-a-half roubles (seven shillings each) a day. They were to follow where we led, and to pitch our little tent where we directed. I must do them the justice to say that they carried out their part of the bargain with an honesty

Mountaineers in Armoury.

and good-humour which led us to form an unluckily premature estimate of the general character of the people with whom we should afterwards have to deal.

In the evening, through the kindness of the Commandant of Duschet, we had an opportunity of witnessing a sword-dance, performed by some mountaineers, habited in complete suits of chain-armour, who had come down from a neighbouring village to greet the Grand Dukes. They carried small round shields, like those of the Kurds, which they used very cleverly to parry the blows of their assailants; the principal feat seemed to be for one man to defend himself against the assault of two enemies.

June 30th.—Having marshalled our porters, who had a horse to help in carrying the luggage as far as possible, we started on our ascent of Kazbek, receiving a parting benediction from the two officials, who came out into the balcony to see us off. Instead of climbing to the old church, we took a path to the right, which led us into the glen opposite the station, and we then passed, over rough ground beside the torrent, to the point where the streams, coming respectively from the Ortzviri glacier, and from the smaller ice-stream which descends from the east face of the mountain, unite. A narrow track mounted, by zigzags, the bluff which projects between the two branches of the glen. A long and steep ascent, which was beguiled by the variety and beauty of the flowers, led up to a gently-sloping meadow, such as in the Alps would have been occupied by a group of châlets, a little beyond which the horse was left, although he might have gone farther without difficulty. We were now close to the snout of the Ortzviri glacier, which, as before mentioned, sweeps round the southern flank of Kazbek, and, despite many remonstrances from the porters, already getting beyond their beat, we climbed on, up the steep slopes on its left bank, until at 2.30 P.M.—at a height of 11,000 feet—we found a most suitable spot for a bivouac. It was a mossy plot, in a hollow protected on one side by the moraine, on the other by the great southern spur of Kazbek. Here we pitched our tent, and under François' superintendence established our cuisine, which turned out some excellent soup, broiled ham, and a brew

KAZBEK, FROM THE SOUTH.

of mulled wine. We should have been happy enough, but for the very doubtful appearance of the weather. Soon after our arrival there was a sharp shower of rain, followed by hail, succeeded in its turn by a violent wind, which, when we retired for the night, about 7 o'clock, was roaring in a way suggestive of anything rather than an ascent of Kazbek next morning.

July 1st.—The cold in the night was not excessive, and we slept in a broken sort of way till 1 A.M., when we rose, and began to prepare for a start; but it was not until 2.45, after more than the usual petty delays, that we—that is Moore, Tucker, and I, with François—were fairly off on our adventure. Before leaving the tent we had by pre-arrangement fired off a pistol, to give notice to the porters, who had retired to lairs at some little distance, and out of sight; but no one answered, and we heard nothing of them until we were just starting, when there was a distant howl, to which we in our turn made no response, the fact being that we were not anxious for the company of our friends, who in any serious difficulty would probably have been more of an hindrance than help. We therefore started alone, carrying only our rope, and sufficient provisions for the day.

Our camp must have been very close to the deserted cells, afterwards described to us by Mons. Khatissian, and it is quite possible that the porters, who, we remarked at the time, went off with the air of knowing what they were about, and did not waste time in looking for holes among the rocks close at hand, may have sought shelter in them. Such conduct would not agree with the superstitious fears the natives are said to feel of the spot, but our men may have thought that, having gone so far already, it did not much matter what they did further.

The morning was calm and lovely, and we fully enjoyed

the moonlight view of the great glacier and ice-mailed peaks around, and the glorious sunrise-flush which soon succeeded it. We mounted along gentle snow-slopes between the glacier and the mass of Kazbek, and gradually rounded the base of the eastern peak of the mountain. Arrived at some rocks, beyond which the tributary glacier from between the two summits joined the main stream, we halted to put on the rope, and Moore left his new Cardigan waistcoat under a rock, intending to pick it up on our return. As will be seen we never did return.

We now began to climb the face of the mountain—at first by rocks, afterwards by broken slopes of névé—and gained height rapidly, bearing somewhat towards the base of the western summit. At 6.30 A.M. we were at an altitude of 14,800 feet, only 1,800 feet below the top. At this time the view was magnificent and perfectly clear; some fine snowy peaks, which we afterwards knew better as the Adai Khokh group, were conspicuous to the west; to the south the eye already ranged over the main chain of the Caucasus, and across the valley of the Kur, to the hills beyond; while behind the rugged ridges which rise on the east of the Terek valley, the peaks of Daghestan raised their snowy heads. From this point our difficulties began; the crevasses became large, and had to be dodged. François resigned the lead to Tucker for forty minutes, during which the favouring snow-slope was exchanged for blue ice, covered with a treacherous four inches of loose snow. The work of cutting steps became laborious, and François presently resumed the lead. An incident soon occurred which might have been serious. A bergschrund, a huge icicle-fringed crack in the ice, three to four feet wide, of which the upper lip was about five feet above the under, barred our progress. François was first, I followed, Tucker was behind me, and Moore last. We had all passed the obstacle without serious difficulty, when the rope, which in the passage had got somewhat slack, was discovered to have hitched itself round one of the big icicles in the crack. Tucker, having, from the position in which he was standing, in vain tried to unhitch it, began to cut steps downwards to the upper lip of the crevasse. At no time is it an easy thing to cut steps in ice beneath you; try to do it in a hurry, and what happened in this case is almost sure to occur. The step-cutter overbalanced himself, his feet slipped out of the shallow footholds, and he shot at once over the chasm; of course the rope immediately tightened with a severe jerk on Moore and myself, who, though very insecurely placed, fortunately were able to resist the strain. Tucker had fallen, spreadeagle-fashion, with his head down the slope, and we had to hold for many seconds before he could work himself round and regain his footing.

The escape was a very narrow one, and we had reason to be thankful that neither the rope nor our axes had failed us at so critical a moment. So startling an occurrence naturally shook our nerves somewhat, but little was said, and our order being re-established, we attacked the exceedingly steep ice-slope, which separated us from the gap between the two summits. For the next four hours there was scarcely one easy step. The ice, when not bare, was thinly coated with snow. A long steep ice-slope is bad enough in the first state, as mountain-climbers know, but it is infinitely worse in the second. In bare ice a secure step may be cut; through loose incoherent snow it cannot. François went through the form of cutting, but it was of little use to the two front men, and none at all to those in the rear. In many places we found the safest plan was to crawl up on our hands and knees, clinging with feet and ice-axes to the slippery staircase. It has always remained a mystery to us how we got from step to step without a slip. The difficulties of the feat were increased by a bitter wind, which swept across the slope in fitful blasts of intense fury, driving the snow in blinding showers into our faces as we crouched down for shelter, and numbing our hands to such a degree that we could scarcely retain hold of our axes.

Time passes rapidly in such circumstances, and it was not until 11 A.M., when François was again exhausted by the labour of leading, that we gained the saddle between the two summits. There was no doubt now that the eastern peak was the highest; at this we were well pleased, as, in such a wind as was raging, the passage of the exceedingly narrow ridge leading to the western summit would have been no pleasant task. Snatching a morsel of food, we left François to recover himself, and started by ourselves, Tucker leading. The final climb was not difficult; a broad bank of hard snow led to some rocks; above lay more snow, succeeded by a second and larger patch of rocks (where François rejoined us), which in their turn merged in the final snow-cupola of the mountain. A few steps brought us to the edge of the southern cliffs, along which we mounted. The snow-ridge ceased to ascend, and then fell away before us. It was just midday when we saw beneath us the valley of the Terek, and knew that the highest point of Kazbek was under our feet. The cold, owing to the high wind, would not allow us to stop on the actual crest; but we sat down half a dozen feet below it, and tried to take in as much as possible of the vast panorama before us.

Clouds had by this time risen in the valleys, and covered the great northern plain, but the mountain-peaks were for the most part clear. The apparent grandeur of the ranges to the east was a surprise. Group beyond group of snowy peaks stretched away to the far-off Basardjusi (14,722 feet), the monarch of the Eastern Caucasus. Nearer, and therefore more conspicuous, was the fine head of Schebulos (14,781 feet). On the western horizon we eagerly sought Elbruz, but it was not to be found; whether veiled by clouds, or hidden behind the Koschtantau group, we could not say. We fancied afterwards that we recognised Kazbek from Elbruz: of course in this case the converse is possible. Except in the immediate vicinity of Kazbek, there seemed to be but few and small glaciers nearer than the Adai Khokh group, on the further side of the Ardon valley.

After a stay of about ten minutes, we quitted the summit, where it was impossible to leave any trace of our visit. We could not spare an ice-axe, to fix upon the snow-dome, and the rocks were too big to use for building a stone man. In a quarter of an hour we regained the gap, and then held a council. From the commencement of our difficulties our minds had been troubled about how we should get down, though, fortunately for our success, they had been more pressingly occupied with the business of the ascent. Now, however, the question had to be fairly faced — how were we to descend the ice-slope we had climbed with so much difficulty? With a strong party— that is, a party with a due proportion of guides, and when good steps can be cut—there is no more delicate mountaineering operation than the descent of a really steep ice-slope. Our party was not a strong one, and on this particular slope it was practically impossible to cut steps at all. A bad slip would result in the roll of the whole party for at least 2,000 feet, unless cut short by one of the numerous crevasses on the lower part of the mountain. The exact manner of its termination would, however, probably be a matter of indifference when that termination came.

We were unanimously of opinion that an attempt to return by our morning's route would end in disaster, and that a way must be sought in another direction. This could only be on the northern flank of the mountain, and it was satisfactory to see that, for a long distance on that side, there was no serious difficulty. A steep slope of snow (not ice) fell away from our feet to a great névé-plateau, which we knew must pour down glaciers into the glens which open into the Terek valley below the Kazbek station. A very few minutes' consideration determined us to follow this line, abandoning for the time our camp and the porters on the south side of the mountain. The first hundred feet of descent down the hard snow-bank were steep enough; I was ahead, and neglected to cut good steps, an error which resulted in Moore's barometer getting a jolt which upset it for several hours. Happily, the little thing recovered during the night, and told us our approximate heights for many a day afterwards. Very soon the slope became gentle enough to allow us to dispense with axe-work, and we trudged straight and steadily downwards, until we were almost on the level of the extensive snow-fields upon which we had looked from above. Here we again halted, to consider our further course. We were on an unknown snow-plain, at a height of 14,000 feet above the sea, and it was most undesirable to hazard our chance of reaching *terra cognita* ere nightfall by any rush or hasty move. One plan suggested was to turn to the left, and cross a pass we had good reason to believe connected the plateau we were on with the névé of the glacier by which we had ascended. This course, if successfully carried out, would have brought us back to our tent and baggage, but its probable length was a fatal objection. Eventually we determined to keep nearly due north, across the snow-field, towards a ridge which divides two glaciers flowing into different branches of the glen of Devdorak. We des-

cended, for some distance under the rocks, along the left
bank of the most southerly of the two glaciers, until the ice
became so steep and broken that further progress promised
to be difficult; we therefore halted, while François climbed
up again to the ridge, and made a reconnaissance on its
northern side.

After some delay, a shout from above called on us to
follow, and we rejoined François, after a sharp scramble,
at the base of a very remarkable tower of rock which
crowns the ridge, and is visible even from the Dariel road.
It will be useful as a finger-post to future climbers.

The view of Kazbek from here is superb; its whole
north-eastern side is a sheet of snow and ice, broken by the
steepness of the slope into magnificent towers, and seamed
by deep-blue chasms. We were glad to find that there
was a reasonable prospect of descending from our eyrie to
the lower world without too much difficulty. The crest
of the ridge between the two glaciers fell rapidly before
us, and offered for some way an easy route. We followed
it—sometimes crossing a snowy plain, sometimes hurrying
down rocky banks—until we saw beneath us, on our left, a
series of long snow-slopes leading directly to the foot of
the northern glacier. Down these we glissaded merrily,
and at 5.30 halted on the rocks below the end of the glacier,
which was of considerable size, and backed by two lofty
summits. The view of the lower part of the glen was shut
out by a rocky barrier, and before we reached its brow,
mists, which we had previously observed collecting in the
hollow, swept round us, and for the next two hours we
were enveloped in a dense fog. A long snow-filled gully
brought us to the bottom of the gorge, of which we could
see but little, owing to the unfortunate state of the
atmosphere. It must be of the most savage description.
The torrent was buried under the avalanches of many

winters; huge walls of crag loomed through the mist, and pressed us so closely on either side, that, but for the path afforded by the avalanche snows, we should have been puzzled to find a means of exit. This aid at last failed us, the stream burst itself free, and tumbled into a gorge. After a laborious scramble for some distance over huge boulders, we found it impossible to follow it any farther, and therefore made a sharp but short ascent to the right, when François happily hit on a faint track, which led us by steep zigzags into the same glen again, at a lower point. After more than once missing and re-finding the path, we rounded an angle of the valley, and, the mists having lifted somewhat, saw that we were close to the junction of our torrent with that from the main Devdorak glacier. On the grassy brow between the two streams cows and goats were grazing, and as it was now 7.45 P.M., we debated on the propriety of stopping here for the night. The question was decided by the information we got from the herdsmen, an old man and two boys, who proved to be very decent fellows. All communication, except by pantomime, was of course impossible; but necessity sharpens the wits, and we gathered from them, without much difficulty, that the Devdorak torrent was bridgeless and big, and that they had fresh milk, and would allow us to share their shelter. It was only a hollow under a partially overhanging cliff surrounded by a low wall, which was but a poor protection against the attacks of inquisitive sheep and goats, who invaded us several times during the night, and succeeded in carrying off and eating some gloves and gaiters. Despite these inroads, and a Scotch mist, which fell pretty heavily from time to time, we managed, with stones for pillows and our mackintoshes spread over us, to snatch a good deal of sleep.

July 2nd.—As we had not even taken off our boots, the preparations for our start in the morning did not occupy long. Our aged host accompanied us to the Devdorak torrent, which at this time of day, before the heat of the sun had melted the upper snows, could be waded without serious difficulty; and one of the boys volunteered to accompany us to the post-station, and relieve François of some of our traps. A well-marked path led us over grassy knolls considerably above and to the right of the united torrents. On a brow near stands, we were afterwards told, a pile of stones resembling in shape an altar, and covered with the horns of chamois and bouquetin. This is a spot held sacred by the pagan inhabitants of the neighbouring village of Goslet, and once a year they all repair hither, sing strange chants, and make their offerings to the *genius loci*. His name, according to our informant, is Duba, and that of the tribe who worship him is Kists. Before very long the defile of the Dariel opened beneath us, and a short descent brought us to the Terek. We kept for half-a-mile on the left bank, along a meadow covered with old tombstones, and then crossed by the bridge close to the stone hovels of Goslet, situated in a most savage nook at the mouth of a ravine. We had still a long uphill pull of eight versts (5¼ miles) to the village of Kuzbek; but towards the end we were able to cut short the zigzags of the road, and about 9 A.M. aroused, with our best 'jödels,' the people of the post-station. Our arrival did not at first create much excitement; everyone seemed to take it as a matter of course that we had not been to the top of the mountain, but equally as a matter of course that we should say we had. The first thing to be done was to rout up Paul, who, still unable to shake off his fever, was in a very stupid and gloomy mood, expecting death hourly. Through him we sent up a messenger to look for our porters, whom we had left encamped, at a height of 11,000 feet, the previous morning. The commission was promptly executed, and in the course of the evening the porters returned, bringing in safety all our belongings. Even a pair of spectacles, mislaid in the hurry of a start in the dark, had been picked up, and were now restored to their owner. The men, who naturally had supposed us lost, and felt uneasy as to what the authorities would say to their having allowed us to go on alone, were overjoyed to see us again, and now simultaneously talked, kissed, and hugged us all, including François. The excitement among the villagers grew intense; the porters told them that we had disappeared up the mountain, and that our tracks were visible to a great height on the southern face; the shepherd-boy, who had arrived with us, was a witness to our mysterious appearance on the other side the same evening. The two facts showed that we must have crossed the mountain very near the top, and been, at any rate, thousands of feet higher than those before us, and we suddenly found ourselves installed as heroes, instead of humbugs, in the public opinion of Kazbek village. Two of the porters even thought it worth while to allege that, searching for us on the second day, they had followed in our footsteps to the top; but this bold fiction was only intended to raise their reputation at home, and they did not press it on our acceptance, or make it the ground of any money-claim.

The old men's chorus, by whose help our first arrangements were made, came in during our supper, when more kissing and hugging had to be endured. The chief of the party was very excited and enthusiastic in his congratulations, and dilated at length on the Generals and Colonels, who, with companies of Cossacks to aid them, had desired to do what we had done, and had failed. We tried to explain to him the use of the rope and the ice-axe, and to show that such aids were much more useful on a snow-mountain than any number of Cossacks. The Grand Dukes had passed during our absence, and had carried away the officials with them; we had promised to let them know how we fared, and accordingly wrote a short account of the 'happy despatch' of Kazbek, which we sent to the Commandant of Duschet, leaving it to his discretion to publish it in the *Kafkas*, the official journal of Tiflis.

In that publication it finally appeared, and contributed in no slight degree to the reputation of modern Munchausens, which before leaving the country we had succeeded in establishing.

CHAPTER VIII.

THE VALLEYS OF THE TEREK, ARDON, AND RION.

A Geographical Disquisition—The Upper Terek—Savage Scenery—Ferocious Dogs—Abano—A Dull Walk—Hard Bargaining—An Unruly Train—A Pass—Zacca, on the Ardon—A Warm Skirmish and a Barren Victory—An Unexpected Climb—The Lower Valley—A Russian Road—Teeb—The Ossetes—The Mamisson Pass—Adai Khokh—A Shift in the Scenery—Glarachavi—The Boy-Prince—An Idle Day—View from the Rhododendron Slope—Glola—The Pine-Forests of the Rion—Chiora.

July 3rd.—It was less than a week since we had left Tiflis, and already the first piece in our programme was accomplished, and the most formidable of the two great peaks we had pledged ourselves to attack successfully disposed of. We had now to turn our thoughts to the less imposing, but really far more difficult, task of making our way along the foot of the main chain of the Caucasus, from Kazbek to Elbruz, a distance, as the crow flies, of 120 miles. Before leaving England we had studied German maps, which, although shown, by better acquaintance with the country, to be often inaccurate, yet gave a sufficiently correct idea of the disposition of the upper valleys, on either side of the watershed, to enable us to form a plan for our proposed 'high-level route.' Since landing at Poti, we had learnt that the Mamisson, one of the passes we intended to cross, was well known to, and occasionally used by, the Russians, as a route between Vladikafkas and Kutais. Beyond this we could gain from the officials little information, and the plan of the journey we had worked out was scouted by them as impracticable. A volume given me by Herr Radde, containing the account of his explorations in the higher valleys of Mingrelia, showed us that he had traversed, at different times, all the country west of the Mamisson, to a point south of Elbruz, with the exception of one short link, between the valleys of the Rion and Zenes-Squali. It is one thing to make excursions from a base to which you can return for supplies, and where you can leave much of your baggage, and another to push on from point to point, carrying everything with you, and harassed by the constant difficulty of engaging fresh porters. We saw no reason, however, to give up our original plan, despite the small encouragement it had received from others, and accordingly were ready on the morning after our return from the ascent of Kazbek, to drive back to Kobi, where we purposed to bid farewell to post-roads and such civilization as they carry with them, and to adventure ourselves among the primitive paths, and native inhabitants of the mountains.

Before I enter upon the account of our journey, and its various adventures, I must ask my readers to open the map, and to look at the disposition of the ridges and valleys amongst which we are about to wander together. It will be seen that the watershed of the Western Caucasus, from a point south of Elbruz to the Adai Khokh group, on the west of the Ardon valley, is an uninterrupted and tolerably straight ridge, which nowhere sinks below 10,000 feet, and is traversed only by glacier-passes, some of them practicable indeed to Caucasian horses, but even those equal to the well-known Theodule in the extent of snow and ice to be crossed. This central mass, according to the testimony of recent geologists, confirmed in most parts by our own unskilled observation is mainly composed of granite. On either side, but more especially on the south, the upper valleys are troughs running parallel to the central chain, and thereby aiding the traveller who wishes to explore it. These upper basins are enclosed between the main chain and the lower but very considerable limestone ridges, which guard both its flanks. The rivers rising in the glaciers of the central mass are consequently compelled to make their way to the low country by deep gorges cut through the lateral ranges. In this part of the chain, that is from Suanetia on the west, to the eastern source of the Rion, the relations of the watershed and the two lateral ridges, though sometimes interrupted or rendered indistinct (as by the sources of the Zenes-Squali, on the south, or by the great promontory of Dych-Tau on the north), are on the whole easily traceable. The next section eastwards presents at first sight, on the map, a curiously changed aspect; the watershed having for so large a space run from north-west to south-east, bends suddenly due south, and sinks to the comparatively low gap of the Mamisson Pass. After a few miles it resumes its former direction, but entirely fails to recover its former grandeur, and although the peaks rise frequently to heights of 11,000 and 12,000 feet, they support but few and small glaciers, while the passes between them vary from 7,500 feet, the height of the Krestowaja Gora, to 9,000 feet. North of this insignificant watershed, we find a line of summits averaging at least 14,000 feet, and terminating in the noble outwork of Kazbek, 16,540 feet. A second glance at the map shows that these grand peaks are in an exact line with the glacier-crowned chain which forms the watershed further west, and that the ridge which now divides the basins of the Kur and the Terek is, in fact, the continuation of the southern lateral range. I have only further to point out that the head-waters of the Terek and the Ardon are divided by a low ridge, which connects the Kazbek group with the watershed. If thus much of the geography of the Western Caucasus has been made clear, my readers will be as well able to see, as we were when we left Kobi, the obvious line of march for a party who wished to follow as closely as possible the foot of the main chain, where the finest scenery might be expected to be found. Our plan was to ascend the Terek to its source, cross to the Ardon, descend the eastern, and mount the western branch of that river, traverse the main chain by the Mamisson Pass, and then work across the upper basins of the Rion and the Ingur, between which several ridges separating the sources of the Zenes-Squali barred the way, and enclosed glens seemingly without inhabitants.

I have here attempted to give some idea of the physical configuration of that part only of the Caucasus which we visited, and have not entered into details of the complicated system of mountains and river-basins of Daghestan, famous as the last refuge of Schamyl and the scene of his final capture.

We started from Kazbek station, on July 3rd, in grand style. Our turn-out consisted of the best pair of telegas we met with in Russia, with good horses, which had drawn the Grand Dukes two days previously, and had, in consequence, their harness still intertwined with gay ribbons. The day was gloomy, and before long the rain, of which during the month we were destined to have more than our share, began to fall in torrents, so that, despite mackintoshes, we arrived at Kobi wet through. The postmaster was in his usual state of intoxication, but we succeeded in getting a fire lighted, and then sent for the head of the Cossacks stationed there, who had been ordered by the Commandant of Duschet to have horses ready to carry our baggage. We found that two animals had been procured, but that they could not go with us beyond Res, the highest village in the Terek valley, the pass from which into Dwaleth, as the Upper Ardon valley is called, was said to be impracticable for laden animals. The rain-storm having passed over, we set out on foot, with our baggage packed on the two horses, which were accompanied by their owners.

The portion of the valley immediately above Kobi is bare and uninteresting; long and steep grass-slopes shut in the view, and no snowy peaks are visible. We walked along swampy meadows as far as a spot where the valley forks, and the main torrent comes out of a narrow opening on the left. Our path then followed the left bank of the Terek, through a long and savage but scarcely picturesque defile. Huge avalanches had fallen in spring down the gullies, and in many places still covered the path; from the traces we saw here and elsewhere of their ravages, far exceeding the devastations caused by similar agency in the Alps, we were led to suppose that the winter snow-fall is heavier in the Caucasus than in Switzerland. Mineral springs abounded, some of which were impregnated with iron, and coloured the ground for many yards round their source. An abominable stench which pervaded one part of the defile probably arose from a sulphur spring, although Paul tried to persuade us it was caused by the decay of the vegetation lying amongst the débris of the avalanches. We emerged, after a time, into the upper valley—an open basin perfectly bare, and surrounded by uniform slopes capped by rock-peaks of a very commonplace character. The nearer beauties of nature were more conspicuous, and the carpet of flowers, which almost hid the grass under our feet, consoled us for the rather disappointing tameness of the general scenery. As we suddenly turned a corner, we came upon a group of natives sitting on a bank of turf.

and amusing themselves with music and singing. They were a handsome set of men, tall and military-looking, dressed in the usual long frock-coat and high fur hat of the Ossetes, and carrying about their persons the indispensable variety of swords, daggers, guns, and pistols. They rose to meet us, and, after a few minutes' friendly

An Ossete Village.

conversation, we passed on our way. After a walk of three hours from Kobi, we came in sight of Kektris and Abano, two villages about half a mile apart, and both on the left side of the valley. There being no wood in this district, the houses are entirely built of stone: they are generally gloomy-looking masses of rough masonry, in which small holes are left for the windows; but the peculiar character of the villages is given by the number of towers, which are often found in the proportion of two towers to three houses. There is nothing picturesque in these primitive fortresses, which, from their walls sloping inwards towards the top, closely resemble, from a distance, a collection of exaggerated brick-kilns; many of them are in ruins. In passing through Kektris we were put in bodily fear by the dogs—a magnificent race, as big as the St. Bernard, and of the same colour, but with shaggier coats and even more sagacious faces. The narrow lane wound along between the houses, on the roofs of which our enemies took their stand, greeting us with savage barking and every demonstration of a desire to rush down and eat us. I believe, however, that this ferocity is more apparent than real. At Abano our horsemen selected a lodging for us at the house of the wealthiest man in the village, where we found a clean upper room with two bedsteads. Supper was promised, and we had nothing to complain of in our reception, as a samovar was quickly brought and a fowl slaughtered for our benefit.

July 4th.—In the morning a dispute arose with our host as to the payment we should make, and we were obliged to resist his excessive demands. The valley did not increase in interest as we mounted it. There are few duller walks in a mountain country than that from Abano to Gumara; the trough of the Terek is bare, and destitute of any natural attractions, and a glimpse of the fine snowy head of Gumaran Khokh, up a side glen, forms but a momentary relief to the general dulness. This part of the Kazbek group deserves exploration; its glaciers and ridges are laid down in the vaguest way on the Five Verst Map, and the only fact I can state concerning it is, that it sends out a large ice-stream, known as the Gumaran glacier, the head of which probably abuts on that of Orzviri. From hence to Res the distance was not great, and the change in the scenery showed that we were drawing close to the head of the valley. The slopes became less uniform, while bolder and loftier summits rose around us. The hamlet of Res, where our baggage-horses were to be left, is a cluster of stone hovels, perched one above the other on a steep hillside. We unladed our packs in the middle of it, and sitting down on some stones began our lunch, while the question of porterage was discussed with the inhabitants, who of course soon gathered round us. They were a handsome but ruffianly-looking lot, but we had become too much accustomed during the last six months to find ourselves among queer company to think much of their appearance. The first demand made was that we should hire ten men to carry our luggage to Zucca, the highest village in the Arlon valley, and that we should pay them two roubles apiece, which would have made the whole sum twenty roubles, or 2*l*. 15*s*. We offered them half, which they at first contemptuously refused, but finally accepted, when we, as a stratagem, ordered the horses to be reladen, and pretended to be about to return the way we had come. The packs, which were ludicrously light (not above one-third of the weight ordinarily carried by Swiss peasants), having been with much difficulty and loss of time adjusted, we started for the pass, which was now visible in front of us. A strong stream, flowing out of a snowy hollow in the northern chain, had to be crossed, and gave some trouble to those who attempted to perform the feat dryshod. The men made the passage an excuse for a long delay while they rearranged their shoes.

The sandals of the mountaineers of the Caucasus are too peculiar to be passed over without a description. A tangle of leather bands is stuffed with dry grass and bound round the foot, so that the sole is renewable at pleasure; these remarkable boots seem to be everlasting, and at the same time to afford the feet sufficient protection from rocks and cold. For a long time we thought they would fail when brought into contact with snow and ice, but the way in which the men of Pari crossed the steep snow-slopes between the valleys of the Nakra and the Baksan in them, quite disabused our minds of this prejudice. Such being the ordinary style of shoe of the country, it may be imagined what surprise our double-soled and heavily-nailed English boots created, and we used often to hold up our feet, as a show, in the villages, while some arithmetical genius endeavoured to count the nails in the soles. The last sandal having been satisfactorily strapped and re-arranged, our train moved on.

The path, a fairly-marked one, steadily rose above the Terek, the highest source of which was now in sight, issuing from a small glacier at the base of Zilga Khokh, a fine peak at the point where the ridge over which our pass lay joins the watershed. Numerous springs burst out of the hillside, and their channels were bright with masses of the yellow blossoms of the ranunculus. The final climb to the pass was up a steep slope of shale, on which a good deal of snow was still lying. Our native companions were silly enough to prefer a straight course to the well-made zigzags of the path, the pains expended on the construction of which caused us some surprise; the rest of the party, however, stuck to the zigzags, except Moore, who kept with the porters, in order to have an eye on their dealings with our goods. As they soon lost breath, and wanted every minute to sit down, he had enough to do to drive them before him, and his difficulties suggested to our minds a comparison between his present position and that of Enid when driving the unruly steeds before her through the waste. By our several routes we all arrived at nearly the same point on the ridge, which is over 10,000 feet in height. The actual crest was bare, but plenty of snow lay around; there was nothing, however, to prevent horses, so accustomed to snow-work as those of the Caucasus, from crossing the pass. The view looking back towards Kazbek, and forwards to what must, I suppose in deference to the Five Verst Map, be called the Adai Khokh group, ought to have been fine; but unluckily clouds hid all the more distant summits, and we saw little more than the bold mass of Zilga Khokh close at hand on the south. This summit (12,645 feet) was ascended, in 1852, by General Chodzko, who spent several days near the top for the purpose of the government survey. He describes the expedition as a difficult one, and seems to have encountered considerable glacier obstacles. The path, on the western side of the pass, first bore away to the right, and then descended rapidly into a green basin, such as is familiar to all Alpine travellers; a pass lower than that we had just crossed led out of it on the south, across the watershed, immediately to the west of Zilga Khokh.

We looked forward with mingled pleasure and dread to the necessity of making fresh arrangements for the transport of our baggage: on the one hand we were only too delighted to be rid of the Res men, who had been most provokingly insolent during the descent; on the other, we dreaded a prolonged wrangle before a fresh bargain could be concluded. On reaching Zacca we succeeded in finding a house, the owner of which was willing to get us something to eat, and on a raised terrace outside, we sat down and collected together our luggage. A crowd immediately surrounded us, and soon, not content with staring, pushed in and jostled us so roughly, that we asked the man who had promised to secure us some bread whether he could not also find us a room to rest in. He pointed out one close by, and by stationing François at the door, we managed to free ourselves from the inquisitiveness of the mob, and to confine our visitors to a select few of the elders, whom we entertained by displaying some of our European knick-knacks, such as knives, telescopes, and portable drinking-cups. As soon as we had got all our goods into our own hands, Paul was given the 10 roubles to pay to the porters. This was handed over, and at first quietly accepted, but they soon began to clamour for an extra rouble as backsheesh, or trinkgeld, or whatever is the Ossete synonym for those well-known terms. We having just sought refuge from the jabber and jostling of the outside crowd, were not drawn out again by the every-day sound of angry voices, and it was not till the row became serious that we sallied forth, Moore and Tucker going first. They found the Res men hustling Paul, who was sputtering with rage, while the villagers looked on and laughed. When my friends appeared, one of the scoundrels snatched at Paul's sheepskin cloak, and then they all hastily retired, carrying it with them. This was the state of the matter when I came upon the scene, and saw Paul frantically excited, and our late porters standing in a knot on the path, fifty yards off, with our cloak in their possession. Knowing nothing of what had gone before, and remembering the effect any decided course of action generally has with Easterns, I fancied a prompt move would settle the question, and accordingly ran up to the men of Res, and, taking hold of the cloak, motioned to them to drop it. They had no such intention, and began instead to pommel me in their own way, which fortunately was a very harmless one, consisting of roundabout pats on the top of the head. This, no doubt, is an effectual mode of bonneting an adversary who wears a tall sheepskin, but it is singularly harmless to a man with a hard wideawake. In self-defence I was obliged to let go the cloak, and in a few seconds my friends came to the rescue, Tucker hitting straight

into the eyes of the thieves, while Moore charged down
the hill with the point of his ice-axe directed full at their
stomachs, and François lent the weight of his elephan-
tine bulk to the united onset. After Tucker had been
rolled down the embankment on which the skirmish took
place, and some dozen blows had been planted fairly in the
thieves' faces, the foe suddenly fled, and did not stop till
they had put the river between themselves and us. We
thus remained masters of the field, but the enemy had all
the fruits of victory, as they got clear off with their booty;
we consoled ourselves, however, in the smallness of our loss,
and in the fact of our retaining a very fine staff which Paul
had borrowed, and which afterwards served him as an
alpenstock during our whole journey.

Our next move was to turn to the chief of the village
and ask how it was that he stood by and allowed strangers
to be robbed, whilst his own people aided and abetted
the thieves? The only reply of this specimen of nature's
nobility was, that if we would give him something for his
trouble he would get us back the cloak, an offer which I
need hardly say we declined to accept. The looks of the
population were not friendly, and we came to the conclusion
that it was better to submit to extortion in engaging horses,
than by delay to run any risk of further robbery. We
consequently agreed with two handsome smartly-dressed
fellows to start down the valley at once, with two horses.
We were heartily glad to shake the dust of Zacca off our
feet, and to feel ourselves once more on the road with
only two, instead of ten, of these impracticable mountaineers
to deal with. The valley is treeless, but the scenery is far
superior to that of the Upper Terek. The slopes are
varied and broken, jagged peaks show at the head of
lateral glens on the south, and clusters of houses, each
dominated by one or more towers, are perched on every

defensible rock-knoll. We climbed on to a level-topped green brow at some height on the left bank of the stream, then made a dip into a lateral ravine, on the opposite bank of which we passed another hamlet. The map showed that this was the last of the upper cluster of villages, and after some discussion we halted a few minutes further on at a solitary house by the wayside. A large empty barn was our quarters for the night, and as we were able to add eggs and milk to the provisions we carried with us, we did not fare badly for supper. The position of affairs during the evening was not pleasant, as the manner of our horsemen was insolent and suspicious, and led us to apprehend an attack in the night. In order, therefore, to let them see that we were prepared to meet it, we had a grand review of our forces before retiring—that is to say, we ostentatiously fired and reloaded our three revolvers, a performance which excited considerable astonishment. The baggage was all collected at one end of the barn, and we slept lightly; but the night passed without disturbance, and I hope our suspicions of the men may have been unfounded.

July 5th.—We had only engaged our horsemen for the previous evening, but being unable to find others, we were obliged to retain their services at their own valuation, which was of course an extravagant one. We expected to have an easy stroll down one branch of the Ardon and up the other to the foot of the Mamisson Pass, and meant to sleep at one of the villages on its eastern side. Our first intention had been to leave Paul and the heavy baggage at Dalla-Kau, at the fork of the torrents, and ourselves to descend the main valley for some distance, and then turn up a lateral glen, which appears from the Five Verst Map to be well wooded, and to contain at its head the largest glacier of the Adai-Khokh group, over which we might have found a way back across the mountains to our luggage. After the specimen the Ossetes had just given us of their gentlemanlike behaviour, it seemed imprudent to separate our party, and to leave our goods for an uncertain length of time at their mercy; so this idea was given up, and we determined to push on, in the hope that the inhabitants of the Rion valley would prove more friendly than their neighbours, and that from it Adai Khokh might be accessible. We crossed the stream by a narrow footbridge immediately below our night-quarters. The sheep and goats were at the same time starting for the pasturage, and it was amusing to watch the way in which they hustled one another in their eagerness to pass. The sheep would follow peaceably enough for a minute, until an old goat made a dash into the crowd, upset a lamb or two into the water, and not unfrequently overbalanced himself and got a ducking. The stream was strong enough to give the poor lambs a good tossing before they got on their legs again, and came out dripping and bleating from their morning bath.

Nothing is more annoying than a mountain in your way when you have no reason to expect it, and it was not without careful enquiry into the necessity of the exertion that we consented to leave the valley, which our horsemen assured us contracted below into an impassable gorge, and set our faces against a mountain-side of 2,500 feet. A good horsepath, mounted at first by very steep zigzags, and then gradually crept along the top of a grassy ridge, and round the head of a hollow, to the summit of a spur about 9,800 feet in height, whence we looked down into another side-glen of the Ardon. This point commanded an admirable panorama of the extraordinary chaos of mountains and network of ravines which form the upper eastern basin of the Ardon. This river, like the Ition, is formed by two torrents running parallel to the main chain as far as their junction, whence their united streams turn suddenly at right-angles to their former course, and force a way through the deep cleft which divides the Adai Khokh and Kazbek groups. The range between the Mamisson and Zilga Khokh was clear, and presented a line of bold rocky summits separated by deep gaps, offering passes of from 8,000 to 10,000 feet in height into the southern valleys. The mountain-range on the north is on a far grander scale, but clouds unluckily hid all the tops of the Adai Khokh group, and we could see only the tail of one glacier.

At the base of the projecting mass on which we stood, was a deep valley terminated by a rocky cirque, above which a remarkably-pointed peak, called Tau Teply, showed itself through the mists. When the clouds blew off, we saw that the sharp rock-cone was supported by a long icy ridge, depriving the mountain of some of its apparent boldness of outline. We were, not unnaturally, in the constant habit of comparing Caucasian with Swiss scenery, as the best means whereby to confirm or correct our first impressions. Thus far we were agreed that in form the Caucasian peaks were at least as bold as the summits in the most serrated portion of the Alpine range. The features missing in the valleys of the Terek and Ardon are large glaciers and forests. The earth's surface must be wonderfully broken to render a district absolutely bare of trees anything but monotonously savage; despite therefore some striking views, at points where lofty peaks close either end of the valley, the scenery of the Upper Ardon must be characterised as on the whole dull.

There was a good deal of snow on the path, but it was tolerably hard, and did not cause any difficulty to the horses. After a last glance at the mountain-encircled den from which we had just made our escape—probably one of the most out-of-the-way corners of the Caucasus—we commenced the long but pleasant descent which led down into the lower valley. The hillsides were gay with flowers; near the snow we found gentians of two sorts, the common Alpine variety, and one of a duller blue; further on masses of the white Caucasian rhododendron, interspersed with pink ox-eyed daisies and orange-coloured poppies, made us remark the curious difference in hue of the same flowers at home and in the Caucasus.

The village at which we determined to make our midday halt is built on a narrow hog's-back, projecting between two streams. The people seemed a shade more civilised than those we had left, and we were soon received in the house of one of the villagers. A large and dark entrance, in which all sorts of implements were stored, led to a more cheerful room, one side of which opened on a balcony overlooking the torrent. The articles of furniture in the Ossete houses are few but quaint; the greatest amount of pains is bestowed on the cradles and armchairs. The former are elaborately ornamented; the latter are broad and shallow, with a low carved back suited for Darby and Joan to sit in together, but quite incapable of being used as places of rest. The tables are in shape something between three-legged stools and the low velvet-covered pieces of furniture now in fashion in London. In an inner room there were two raised couches, over which the arms of the master of the house were hung up against the wall. A large herd of horses was feeding in the meadows on the opposite side of the river, where we also noticed a cluster of men whose number gradually increased during our stay. We met with nothing but civility from our hosts, and our horsemen were treated most liberally; when one of them had tossed off his fourth tumbler of 'vodka' as though it had been water, without being apparently in the least the worse for it, we thought it about time to be off. We descended to the stream, and crossed by a bridge to the meadow on its opposite bank. The group which we had before noticed now advanced towards us, and a grizzled old gentleman asked to see our permit to travel. Thinking that a British Foreign-office passport might be beyond his comprehension, and at the same time not wishing to raise a needless difficulty, I offered for inspection an old 'crown-podorojno.' We were surrounded for some minutes by a curious crowd, but in due time the paper was restored, the chief professed himself perfectly satisfied, and we went on our way unmolested.

After crossing a tributary stream, and passing another gloomy-looking village, we had a dull but easy walk along level meadows to the fork of the valley. The numerous villages, alike in their rude stone houses and frequent towers, are invariably perched on the hillsides, and often on the isolated promontories of rock which form one of the peculiar features of this district. The defile through which the Ardon flows out to the north seemed to be wooded in its lower portion, but the western arm of the upper valley was as bare as that we had just traversed; we crossed its torrent by a bridge, and mounted the further bank to reach the track of the projected carriage-road from Vladikafkaz to Kutais over the Mamisson Pass. The road has been traced, and partly cut along the hillsides, but as wherever a mass of rock required blasting, nothing has been done, it is of course impassable for vehicles: moreover, in many places torrents and earthslips had already half destroyed the track, which appeared to have been abandoned to its fate. Roadmaking is not a Russian virtue, and the authorities are so little accustomed even at home to see anything which would be called a road in

Western Europe, that they are naturally slow in the
appreciation of the necessity of good highways in the
Caucasus. If military purposes demand a means of communication, soldiers are set to work, and one sufficient for
the momentary need is constructed; had all the roads
which have been traced and cut, at immense cost both of
money and labour, been finished and kept in repair, the
Western Caucasus would now be very fairly provided with
routes practicable for light carriages, and much more
would have been done towards the civilisation of the
country. The road now in question has some chance of
completion, owing to its obvious importance as the shortest
line from Vladikafkaz to the Black Sea coast. The Viceroy
of the Caucasus passed this way in September last, and
his visit may perhaps have the effect of giving the needed
impulse to the local authorities. There was absolutely
nothing to look at during the walk up the western arm of
the Ardon to Teeb. The track mounted gradually along
the northern side of the valley, passing above several
villages surrounded by fields of barley enclosed by untidy
fences. We met a drove of colts being taken southwards
for sale; the Kabarda, a district of which I shall presently
have more to say, is celebrated for its breed of horses, and
exports large numbers annually to the markets of Tiflis and
Kutais.

Teeb consists of several hamlets scattered on the
hillside above and below the road; we sent Paul to reconnoitre, and waited to learn the result of his enquiries,
which proved satisfactory, and we were installed in a clean
little room on the housetop. The people, living on a
frequented path, and having had troops quartered near them
for many months, were more accustomed to see passers-
by, and less churlish than those of the other branch of the
valley. There was even a priest in the village, who talked

Russian, and assisted Paul in his search for fowls and
eggs, and his enquiries after fresh horses. The men who
had come with us from Zacca had evidently got beyond
their home-circle, and did not find anyone to treat them
to 'vodka'; they consequently wanted to take up their
quarters with us, but we told them plainly that we thought
them no better than thieves, and wished to see no more of
them. Having received their pay, they loitered about the
place for some time, casting longing glances at our
numerous belongings; but finding we were on the watch,
and that there was no chance of carrying off a field-glass
or revolver, the objects which they looked at most cove-
tously, they took their departure before nightfall.

July 6th.—Paul had found two honest-looking men,
who were willing to come with us for three roubles (eight
shillings) a day, for man and horse. This was much
above the price of the country, but was only half of what
we had given the Zacca men, and we gladly concluded the
bargain; our new attendants turned out pleasant and
obliging, and we kept them with us for several days.
Teeb was one of the few places we had halted in since
leaving the Dariel road, where we had no reason to complain of churlishness or extortion of some sort; and the
friendliness of the villagers caused us to modify the otherwise universal condemnation we felt disposed to pronounce
against the Ossetes, of whom we now took leave for the
present. This tribe, one of the most famous of the
Caucasus, was converted at a very early period to Christianity, which they continue to profess, although they
trouble themselves little about either its letter or spirit.
Their worship is mixed up with sacrificial feasts, apparently of pagan origin, and the doctrines they hold are
compatible with a severe law of vengeance, resulting in
long and bloody feuds between families and villages.

There seems to be no poor class among them; all the men we saw were well and even handsomely dressed. The tall sheepskin hat is universal, and great attention is bestowed on the numerous ornamental details of their costume. The cartridge-boxes on the breast are often inlaid with silver, and when they go abroad they invariably wear a belt (generally silver), to which is attached a double-edged

An Ossete.

dagger like the Roman short-sword, enclosed in an ornamental sheath; on the other side hangs a heavy flint and steel pistol, in addition to a variety of smaller necessaries, such as a leather case for tinder and flints, a knife, and a little box of oxidised silver prettily worked, in which they keep the grease to anoint their bullets. Their dresses are usually in good condition, and a shabby or poor-looking man is hardly to be met with. Altogether it is impossible not to admit that their external appearance is some excuse for the title of 'Gentlemen of the Mountains,' which Count Leverschoff gave them.

We were still at some distance from the head of the valley, the scenery of which continued to be of the same monotonous description. About half an hour above Teeb there was a fine view, looking back towards a great snow-crowned mass, a western outpost of the Kazbek group. The track, gradually ascending by a uniform gradient above the torrent, made long and frequent circuits round lateral ravines, until, after passing several villages, the head-waters of the Ardon opened before us, and the long straight valley broke up into several glens, running up into a semicircle of peaks, several of which were remarkable for their bold pyramidal forms. Our road turned up the northern of these glens, and wound along its side for some distance, almost at a level, until a huge snow-drift, which rose in a wall across the track, capped with an overhanging cornice, forced the horses to descend into the bottom of the glen, while we kept along the line of the intended carriage-road. The snow was just melting off the turf, and the flowers were exceedingly beautiful. We were pleased to find the homely cowslips and primroses, mixing with gentians and other alpine plants; but the newest sight to us was the mass of snowdrops which whitened the ground, in many places proving their claim to their French name of *perce-neige*, by pushing their green leaves and clustered blossoms through the still unmelted snowdrifts.

We saw beneath us a large troop of natives, who had crossed the pass in an opposite direction, and were making their midday halt. The ridge was now in view, and over it a bold peak, evidently belonging to the mass designated Adai Khokh in the Five Verst Map, shot up in the most alarming way through the clouds. Before beginning the final zigzags the road makes a wide sweep to the right, to cross the stream flowing from a small glacier which fills up the angle between the ridges at the head of the glen. The snow had entirely covered all the excavated track near the top, and had not a path been by this time trodden out of the steep drift which had accumulated under the actual ridge, our horses might have had difficulty in getting up.

Owing to the position of the pass, there is little distant view to the west, and the Rion valley is still hidden; but the head of a glen, containing one of the sources of the Glola-Squali (one of the feeders of the Rion) was at our feet, and above it rose the stupendous eastern peak of Adai Khokh, towering above several neighbouring summits. A very steep and much-crevassed glacier, the largest we had seen since leaving Kazbek, poured down into the valley, and we agreed that there was little prospect of any successful climbing in this direction. A heavy shower soon blotted out the view. The road descended in a series of very long and gentle zigzags, now obliterated by snow; the winter-fall had been heavier this year than usual, but it is probable that, should the carriage-road ever be established, this part of it will have to be roofed over with galleries, which there would be no difficulty in making with so much wood close at hand. We jumped across the small stream, and on its opposite bank passed a well-built house, erected for the accommodation of the officers in charge of the soldiers who traced the road. I met one of these officers afterwards, and he descanted eloquently on the hardships he had endured while living for three months (as he phrased it) on a glacier. The stream tumbled quickly down into a deep ravine; the road followed it more leisurely, sweeping over fine pasturages. Suddenly we came to the corner, where the hillsides trended away to the west, and looked down for the first time on a large portion of the upper Rion basin, in which term I include the valley of its first considerable tributary, the Glola-Squali. Few people who have not seen an absolutely treeless district can appreciate the magical effect of coming out of one, suddenly, into a densely-forested region. Below us was the head of a deep valley, the slopes covered with birch and ash, mingled lower down with noble pines, the dark green of which came out in strong contrast to the lighter foliage. Spur behind spur, ridge behind ridge, carried the eyes up to a cluster of finely-shaped peaks on the southern side of the river, which, like the Ardon, is enclosed by mountain-ranges, and finds an outlet through a narrow gorge. We stood for some time in delighted surprise, and agreed that we had never seen a landscape more beautiful, lit up as it was by the afternoon sun, which had burst through the clouds, and was shining with that special brilliancy so common in the interval between heavy storms. We soon found ourselves among the trees. Scattered birches first hung their graceful branches over the path; the mountain-ash next appeared, accompanied by many varieties of flowering shrubs, and by flowers (such as campanulas and wild roses) the presence of which betokened a more genial soil and climate.

From its position on the map we had counted on Gurschavi as a desirable resting-place, and when the hamlet came in sight, its lovely position determined us at once to make it, if possible, our headquarters for a day or two. A dozen wooden cottages, more resembling an untidy Swiss village than the stone fortresses of the Ossetes, were perched on the edge of a triangular meadow projecting

from the base of the mountain. No less than three glens
opened up behind it, all more or less tempting to an
explorer, and in front the position commanded a wide view
of the basin of the Rion and the peaks on its southern
side. The main chain was hidden by the intervening
buttresses. The road makes an immense zigzag down the
valley to reach the bottom of the ravine under the village;
but after running down a short cut, and climbing some 200
feet on the other side, we found ourselves close to the
houses, which are surrounded by a remarkably fine plantation of stinging-nettles. There was no one loitering outside, so we put our heads into the nearest cottage, and
found a large low room with a few benches and stools,
which opened into another with a fireplace in the centre,
occupied by two old women, to whom Paul addressed
himself. At first there seemed likely to be some difficulty,
as Caucasian etiquette prevented our lodging in the same
house as the beauties before us; but we had spied out a
very unexpected luxury, in some joints of beef hanging up
to one of the rafters, and were quite determined not to be
put off. Opposite the cottage was a well-built barn; on this
we set our eyes as a likely resting-place, and made our
way into it. A heap of hay filled one corner, and the
place looked quite habitable, although somewhat gloomy
from the want of a window. More natives soon turned up,
and, finding we should be contented with the accommodation of the barn, they set to work with a will to make the
place as comfortable as possible. One swept it out,
another fetched a bench, and Paul found everybody
willing to aid him in his culinary operations. While he
prepared a steak, we sent François to cut some young
nettles, which, when chopped up and boiled, make an
excellent vegetable, scarcely distinguishable from spinach.
The hamlet was a small one, and during all the time of

our stay there we saw scarcely more than twenty people. They were not dressed in the showy style of the Ossetes; their clothes were old and sometimes ragged, and their cartridge-pouches made of horn and wood, while their belts were of plain leather, and the daggers hung from them in sheaths equally unornamented. The 'swell' of the place seemed to be a lad of 14, a round-faced fellow, just like an English schoolboy, who wore a wonderful wideawake hat, with a broad brim swelling out into a circular crown, divided by braid, and shaped like an orange. He took a great interest in us and our doings, and 'fagged' several 'lower boys,' whom he kept in great subjection, to fetch us anything he thought we should want. In return we amused him by displaying our knives, field-glasses, and other knick-knacks, so that I believe our visit was a great source of enjoyment and enlightenment to him. To us it was a great relief to get among a colony of simple peasants, and to be freed from the numerous restraints of travelling among the 'gentlemen' of Ossetia. We passed a very comfortable night, though my mind was a good deal disturbed by discovering the loss of my revolver, which I now remembered I had unfastened, and must have left behind, near the house at the foot of the Mamisson Pass. We determined that, if the weather was fine, Moore and François should go the next day on an exploring expedition up to the foot of the chain, while Tucker and I (both having rubbed heels) should stop at home, and see to the preparation of an extraordinary banquet.

July 7th.—The morning was not very promising, but Moore and François set out, in the hope that the weather would clear up. However, it came on to rain heavily, and we stay-at-homes, rather congratulating ourselves on our superior position, settled down very contentedly to write up letters and read Shakspeare, a Globe edition of whose works formed the bulk of our travelling library. Our companions did not return till late in the afternoon, bringing with them my revolver, but without having gained much additional information about the mountains. They had taken shelter, during the worst of the storm, in the house before mentioned, and then climbed the ridge, between two of the sources of the Glola-Squali, to a height of about 11,000 feet. The ground was covered with snow of the most extraordinary pink or rather brick-red hue, a phenomenon we noticed frequently. It is of very rare occurrence in the Alps, and when seen there, the pink tinge is not generally so vivid. Clouds had hidden everything except the tail of a glacier on their left, so that we could form no definite plans, and had nothing to do but to wait till the weather cleared.

July 8th.—The rain was over when we awoke, and the bright morning sunshine poured down upon the rich basin below us, and brought out fresh beauties of colour and distance in its wooded slopes. The peaks overhead stood out boldly against the blue sky, and everything looked fresh and inviting. It was manifestly a day for a view. Our chief object was to inspect the southern face of the Adai Khokh group, and to ascertain if there was a reasonable prospect of effecting any high passes or ascents in it. The best way to go seemed to be up the hillside behind Gurschavi, as we knew that we must soon gain a sufficient height to see the great chain over the grassy buttresses which now hid it from us. At the back of the cottages is a burial-ground, marked by some tall tombstones, where the 'rude forefathers of the hamlet sleep' under the shade of fine trees. We kept along the edge of the little plateau on which Gurschavi stands, until we came to a bridge over a stream flowing out of a recess in the south-eastern angle of the ranges which enclose this end of the Rion basin. The scale of the scenery, the richness of the vegetation, but, above all, the ruddy colouring of a set of rock-teeth which sprang suddenly out of the slopes on the eastern bank, reminded me strongly of several similar scenes amongst the Dolomites of the Italian Tyrol. A woodman's path ran along beside the torrent, but, as it did not gain height rapidly enough for us, we turned straight up the slopes. For about 1,000 feet we scrambled up amongst the beautiful forest-trees, growing with a luxuriance and variety unequalled in the Alps. The underwood was so dense that we had often difficulty in pushing our way up through it, and were glad to help ourselves up by the tough branches of the white rhododendrons, which grew in great quantities, and were now in full blossom. Through the tree-tops snowy peaks were seen from time to time, and when we found a bank where no branches intercepted the panorama of the main chain, now full in view opposite to us, we thought it better to halt and make our observations, rather than to push further up the hillside. The chain was not cloudless, but thanks to a strong wind, which was blowing in the upper region, we got a view, at one moment or another, of every section of it, although the whole was never quite clear at the same time. The first and most striking of all the summits before us occupied the position assigned on the Five Verst Map to the peak of Adai Khokh.* Three long ribs of rock

---

* Tuilass Mta of Herr Radde. Caucasian nomenclature is at present in a state of hopeless confusion. It has seemed to me best to follow in most cases the authority of the Five Verst Map, which the traveller will probably have in his hand. Herr Radde, who frequently differs from it, has not as yet published the result of his researches in the form of a corrected map of the Central Caucasus. I have avoided, as far as possible, encumbering my pages with such unpronounceable names as Sagvbigora, Chrewlioto, Sarziwisdairia Mta, Sophitigoram Mta. All of these peaks look down on the sources of the Rion.

and ice ran up into a sharp point, and created one of the most striking mountain-forms I ever saw. The rocks on the left-hand or north-west rib, seen through a telescope, were of the most formidable character, some of them appearing actually to overhang; and the other sides of the mountain were so sheeted with ice as to be, if not absolutely inaccessible (a word which had, perhaps, best be banished nowadays from a mountaineer's dictionary), practically so for our party. Separated by a deep gap from its slenderer neighbour rose a double-headed mass, supported by huge and fantastically-broken buttresses of rock. Huge séracs hung in a curtain under its crest, and raked the lower snow-slopes. These two summits are probably nearer 16,000 than 15,000 feet; to the west of them the chain sinks considerably, and a succession of snowy eminences, none of them sufficiently marked to arrest the attention, are connected by icy ridges, steep and high enough to present a serious obstacle to anyone wishing to make a pass, and discover what lies beyond and behind them. Masses of rock abutting on the main ridge divided the basins of sundry small glaciers which filled the hollows at its foot. On the left, the isolated snowy tower of Tau Burdisula formed a striking termination to the group.

We were completely puzzled to know what use to make of the knowledge we had now gained of the neighbouring mountains. The first point settled was, that the two big peaks must be let alone, and we inclined to a suggestion made by Moore, that we should cross over to the northern side, and back, by two glacier-passes. The Five Verst Map showed, on the west side of Tau Burdisula, a pass called Por Gurdzieveesk, leading from Chiora to the valley of the Urach. We thought we detected a weak point in the mountain-wall before mentioned east of Tau Burdisula, from which, if it should prove accessible from the northern side, we could be sure of effecting a descent into the Rion valley. It was finally decided that we should return quickly to Gurschavi, collect our baggage, and go down to Glola to sleep. From there to Chiora must, we knew, be an easy day, and from that village, if the weather continued fine, we could cross the known pass to the Uruch. Its track on the north side was shown by the map as running along the side of a large glacier, the head of which must be behind the gap in the ridge we had already observed, as likely to give a passage to the south side. This scheme had the advantage of leaving us the alternative, in case of bad weather, or any other hindrance to its execution, of returning over the same pass we had crossed by, and regaining our base. The decision once made was promptly acted on, and we raced down through the wood to the village. Our Teeb horsemen, whom we still had with us, soon got the animals ready, and our goods packed on their backs. Before we left, a sickly-looking man, who was suffering, as far as we could judge, from consumption, was brought to us to be cured. Of course we could do nothing really for the poor fellow, but, willing to give him satisfaction, as well as to keep up our own credit, we unlocked our little medicine-case, and poured him out a dose of chlorodyne. The dram was carefully drained, and as soon as the patient felt its warmth, he gratefully rubbed his stomach, and, pouring a few drops of water into the cup, he drank them off, in the hopes of catching any lingering flavour. Our farewell to the boy-prince and the rest of the village was most cordial, and the payment we offered was this time accepted, with real demonstrations of surprise and pleasure at its amount. Amidst universal hand-shaking, and expressions of hopes that we should come again another year, we made our way out of the village, our regret at leaving which was only tempered by Paul's announcement that we had eaten up all the beef. The prince and a friend accompanied us down to the road, where they took a final leave, and we saw no more of the jolliest boy in the Caucasus. The road down the valley keeps on the left bank of the river, and has to make long circuits round

Adai Khokh from the Rion Valley.

the ravines which furrow the lower slopes, above which sharp snow-streaked summits* peered from time to time between the trees. The torrent falls very rapidly; the road descends more gently through the most magnificent pine-forest, varied with birch, poplar, and elm, and carpeted with moss and a variety of subalpine flowers. Before reaching

* The Wallatshibis Mta of Herr Radde; the names Dolomis-Zweri and Oeshi appear on the Five Verst Map.

the point where the largest tributary of the Glola-Squali flows out of a glen running deep into the heart of the main chain, road and river are again on a level. The afternoon was beautiful, and we enjoyed a superb view of the two peaks of Adai Khokh, which exactly fill the opening. The right-hand ridge of the eastern peak, seen from here, is a most exquisitely sharp and thin snow arête, and its sides are of a steepness appalling to anyone who has ever allowed the idea of climbing them to enter his head.

An artist might sit down at this spot within ten yards of the road, and paint a perfect picture, without putting in a foreground, or in any way improving on nature. The foaming torrent, and the rich foliage near at hand, the wooded slopes in the middle distance, and the gigantic mountain-forms which close and crown the view, are worthy of a master-hand, and the rough outline sketch, which was all we could carry away with us, can give but a faint idea of the scene. Every bend in the road opened some fresh vista of wood, water, and snow. The floor of the valley had now widened, and the forest soon gave way to hayfields, in which parties of women and girls were at work. Having brought on the scene the far-famed beauties of the Caucasus, this would, I feel, be the place for romance. Unluckily, like one of our American friends, who, being called on to admire an Egyptian sunset, declared 'skyscapes were not in his line,' descriptions of female beauty are not in mine, and I have the further plea that in this instance I should have, not to describe, but to invent. The forms and faces of the women who left their work to stare at the unprecedented sight of three English mountaineers, had lost, by exposure to weather and field labour, any traces of comeliness, and the group, but for certain details of dress, was just such as might be met with in any Swiss valley.

Before long Glola came into sight on the opposite side

of the river, built at the mouth of a tributary stream,* which had its source in a glacier of the main chain, a portion of which was for a few minutes visible, and with which we were destined in a few days to become better acquainted. The bridge above the village, which existed at the time of Herr Radde's visit, had gone the way of all Caucasian bridges; and we had to make a circuit, which cost an extra ten minutes, to reach its successor, built half a mile lower down the stream, there flowing in a wide stony bed, and then to return up the opposite bank. Glola is more like a Swiss village than any we saw before or afterwards; the houses are all built of wood, and have overhanging eaves, balconies, and roofs laid with stones, in the fashion of those of the Canton Berne, although without any of their elaborate carving and general air of finish. The place is sheltered behind a projecting cliff, and on a brow above it are the ruins of an old castle, which add considerably to its general effect. We were first led to a cottage, which had such an indescribable air of griminess and dirt about it, that we altogether declined to take up our quarters there. We soon settled on an apparently uninhabited outbuilding, attached to one of the larger houses, where we found a small room with some hay for our beds, and a broad balcony with a table and benches, which served us as a sitting-room. Paul had to do his cooking in the adjoining house, and I believe the presence and interruptions of certain well-meaning old women put him out; whatever was the cause, we had to wait till long after dark before we got our dinner. Such incidents seem almost too trifling to record, but they serve to remind one of what is often lost sight of afterwards—the difference between travel in a country organised for pleasure-visitors, and one entirely, so to speak, in a state of nature. We got

* The Scharula Squali of Radde.

our food at last, accompanied by a rare luxury, a bottle of wine. We had not seen such a thing since leaving Kazbek, and what was now brought us under the name of wine was a muddy liquor which owed little to the skill of its maker, but was, at any rate, unadulterated juice of the grape, a recommendation quite sufficient to Tucker and myself. The people of Glola were of the same type as those of Gurschavi—homely peasants, who wore the usual style of dress, crowned by felt hats of various and sometimes intensely comical shapes. We were regaled with new bread, one of the few delicacies of the country, which we found almost everywhere. The bread of the Caucasus is peculiar, and would be considered detestable by many people, but I must own to giving it a decided preference over the sour black loaves of the German Alps. The peasants never think of baking until they actually want food; sufficient for the day, or rather for the meal, is provided, and when more is wanted it has to be made afresh. The general shape of the loaves is round and flat, and a hungry man can eat two or three of the ordinary size at a meal. Some, however, are of a more substantial nature, and have a layer of melted cheese inside them, and these, when hot, are by no means despicable. Most of the varieties of cakes are made of barley, and are brown in colour, very close, and more or less heavy; they vary, of course, according to the quality of the flour used, and the skill of the maker. Here we found another kind, made of indian-corn, pleasant to the eye and palate, but very difficult of digestion.

July 9th.—The night destroyed an illusion which for a full week we had cherished most fondly. Hitherto we had been entirely exempt from the pest of insects, and we laid ourselves down to rest absolutely without suspicion of the misery in store for us. Tucker, famed for his sufferings in Swiss châlets, was the first to be attacked; the noise

consequent on the pursuit in which he was engaged
aroused me; once awake, to sleep again was impossible,
and we all lay tossing and growling until morning put an
end to our tortures.

The day was again fine. The view from Glola, down the
valley towards the Schoda chain, was very striking, and in
the opposite direction rose the two peaks of Adai Khokh,
grandly defiant as ever. We had no unpleasantness at
parting, and flattered ourselves—alas! how vainly—that
our difficulties with uncivil and extortionate villagers were
over, and that henceforth we should be free from those
petty vexations which destroy half the pleasure of travel.
Having recrossed the same bridge, and rejoined the new
road, which does not pass through Glola, we soon again
entered the heart of the primeval forest, where the
overhanging arch of foliage entirely shaded us from the
sunshine; the woodcutter's axe seldom thins these glades,
for the needs of the scanty population of the upper valley
are small, and the lower district of the Radscha, between
here and Kutais, is so richly wooded that no one has
occasion to come here for timber. The passer-by may
see illustrated the whole life of a tree, from its first stage
to the last: the cone just dropped on the ground, the
tender sapling, the forest giant sprending its branches in
every direction, and the trunk, broken and rotten, prostrate on the ground and gradually mouldering away into
the soil, from which a fresh generation will soon spring.

A sharp hour's walk below Glola brought us to the
mouth of the narrow defile through which the collected
waters of the Rion basin make their escape. Three
streams join to form the river just above the gorge: the
largest, the true source of the Rion, comes down the
western arm of the valley, from the mountains behind
Gebi; the Glola-Squali, known also by the less euphon-

ious name of the Dschandschachi-Squali, flows to meet it from the east, and between them a smaller stream cuts its way directly down from a glacier in the main chain, through a narrow opening in the lower hills. A bridge crosses the Rion, a short way below the double confluence, and on a plot of level ground close to the river-bank stands a house, evidently of Russian construction, but now falling rapidly into decay. Having made a considerable bend into the mouth of the defile, and crossed the united streams, we turned up a path which led along the right bank of the true Rion, through woods as dense as before, although the single trees were not so fine as those which grow on the opposite slopes. The path was level for some distance, until it mounted a spur of the Schoda chain, which nearly barred the valley.

The sunny slopes were converted into meadows and cornfields, and dotted with dark-brown hay châlets, scarcely a quarter of the size of those in the Alps. A solitary pine, of great size and perfect shape, marked the top of the ascent, beyond which we came in sight of Chiora and a stretch of the upper valley, in which the river, to judge by the width of its stony bed, is accustomed to commit great ravages, and to change its course very frequently. Arrived opposite the village, which is prettily situated on a gentle southward-facing slope of cornfields, we found that there was no bridge where one-half of the stream now flowed, and we had consequently to wait while our horsemen sought the best ford, and then to ride, one by one, through the water. It was a long time before we got the baggage fairly to the other side, and the heat of the sun, reflected from the stones of the river-bed, made the delay anything but pleasant.

Chiora is built in a totally different style to Glola and Gurschari. The houses are all of stone, two-storied, with sloping roofs, scarcely any eaves, and very small holes for windows. On our arrival, we were conducted to a shed, open on one side to the air, where we were requested to wait till quarters were prepared for us. In the meanwhile the whole village gathered round to stare at, and no doubt criticise, the extraordinary beings who had come to visit them. Paul went off to survey our intended lodging, and came back very disconsolate, for first appearances certainly were not cheerful. The interior of all the houses seemed the same—a couple of ill-lighted rooms with rough stone walls, with wooden pegs stuck into their crevices, from which hung clothes, sheepskin cloaks, and various household and field implements. It was too dark inside either to write or read, so we removed our mattrass to a rough wooden balcony projecting from the front of the house, whence there was a beautiful view down the valley, closed by the dark forests opposite Glola, and the serrated summits of the Wallatschibis Mts. A very sharp rock-peak, rising over the southern slopes immediately opposite the village, is in shape an almost exact model of the Rothhorn from Zermatt; we saw afterwards that it is an impostor, being in fact only the end of a long and narrow ridge running towards the valley from one of the chief summits of the range between Gebi and Oni.

In order to put into execution, as soon as possible, the plan formed on the rhododendron slope above Gurschavi, of effecting two passes across the main chain, we made enquiries among the villagers as to the Gurdzieveesk Pass, which proved well known to them: at first they asserted it took two days to reach the other side of the mountain, but on being pressed, they admitted that a good walker might do it in one. Our only difficulty was in settling the terms on which a peasant would accompany us to the snow-level on the southern side, and in making up our minds what instructions to give Paul, whom we meant to leave in charge of our baggage. Both questions were settled before nightfall; a native agreed to be ready to start with us at half-past 1 A.M., and we determined that Paul should hire a horse and go on to Gebi, which was only an hour's walk farther up the valley, there to await our return.

The people of Chiora were less simple and kindly than those of Gurschavi and Glola; but we had no reason to complain of any inhospitable conduct, beyond the usual desire to make a good bargain, and get as much as they could out of us. Their impression of our position was shown by a request, made to us in the course of the evening, that on our return to Kutais, we would represent to the Governor of Mingrelia the unfair distribution of the mountain pasturages, by which the neighbouring villages got more than their share, and Chiora had not enough for its flocks and herds. According to Paul, we were generally believed to be officials employed on some survey, or such-like mystery of civilisation, the existence of which was known to, though its benefit was beyond the comprehension of, the common Caucasian intellect.

CHAPTER IX.

THE GLACIERS AND FORESTS OF THE CENTRAL CAUCASUS.

Caucasian Shepherds—A Lovely Alp—Sheep on the Glacier—A New Pass—A Snow Wall—A Rough Glen—The Karagam Glacier—Bivouac in the Forest—An Icefall—A Struggle and a Victory—The Upper Snowfields—The Watershed at last—Chcek—A Useful Gully—An Uneasy Night—Gleba again—Pantomime—Gobi—Curious Villagers—A Bargain for Porters—Azalea Thickets—The Source of the Rion—Rank Herbage—Camp on the Zeura-Squali—A Low Pass—Swamps and Jungles—Path-finding—The Glen of the Scena—Wide Pasturages—The Nakasgar Pass.

July 10th.—We made a late supper, or early breakfast, soon after midnight, and having insisted on the peasant who was to accompany us sleeping in the same house, found no difficulty in starting at the hour appointed. Before separating from Paul, we told him to explain fully to our guide the part we expected him to perform, and the pay we should give him if he fulfilled it to our satisfaction. This was a necessary precaution, as we had no means, except signs, of communicating with our companion, who only knew the Georgian dialect commonly spoken on the south side of the chain. We climbed the hillside immediately behind Chiorn, soon leaving below us the cultivated fields, and finding ourselves on grass-covered slopes, adorned with clusters of trees in the manner of an English park. Shepherds' fires shone here and there through the darkness, and our guide took us round in order to pass near some of his friends, who were camping-out with their flocks. The peasants of the Caucasus do not take nearly so much pains as those of the Alps to provide themselves with a substantial shelter while spending the summer on the mountains. It is only rarely, and in certain districts, that huts at all resembling the Swiss châlets are met with. I only recall three instances—two in the Urach valley, and one close to the source of the Rion. In general the herdsmen are contented with a slight shelter, constructed of a few boughs and a sheepskin, which can afford very little protection in bad weather. Close at hand a forked stake is driven into the ground, on which, if the owner is at home, he hangs his gun. This and a milking-pail constitute nearly all the furniture of a Caucasian shepherd, who, as the flock under his charge consists mostly of sheep, oxen, and horses, is spared the delicate and complicated cares of a large dairy establishment.

Having passed the last of the shepherds' bivouacs, we steadily followed the somewhat steep zigzags of the sledge-path, until it surmounted a brow which had previously cut short our view. Dawn had not yet broken, and the graceful forms of scattered copses of birch and fir formed a fairylike foreground to a long moonlight vista up the Tchosura to the glaciers and snowcapped summits of the main chain. Deep below, in the dark shadow of the valley, the white towers of Gebi were distinguishable, and behind us the bold peaks of the Schoda chain stood out against a sky paling with the first approach of daybreak. A herd of horses, disturbed by our early movements, trotted off across the hillside, which now became more open.

The path still mounted, and soon even the birch, the tree always found nearest the snow in these regions, was left behind. A host of alpine flowers, amongst which the white rhododendron was again conspicuous, covered the ground, only just free from snow, which still lay in deep drifts in the hollows. The path for a long time followed a ridge, narrow at first, but gradually broadening into grassy undulations; on one side the ground broke away suddenly towards the Rion, on the other it sank more gradually into a barren recess, a branch of the Tchosura valley, above which rose a steep-sided range covered with small glaciers. The height of 9,500 feet we had already gained was sufficient to give us a good panorama of the Upper Rion basin, which served to confirm our previous estimate of its beauty. The ridge we were walking along now bent round to the northward, and separated the water flowing down into the Rion at Gebi from the upper basin of the stream, which joins the river close to its meeting with the Glola-Squali. Far below us, on our right, we looked down into a deep wooded defile, the outlet through which this stream escapes. Here the track began to descend, but first made a long sweep round the hillside, before finally plunging into the beautifully-timbered little plain, at the mouth of the narrow glen which leads up, due north, to the Gurdzievcsk Pass.

Knowing that this, the chief part of the day's walk, was still before us, we grudged bitterly the 2,000 feet of height thus lost, and, having now been five hours on the march, determined to stop, and open our provision-wallet. The beauty of the spot, and a spring bubbling up under a clump of alders, formed additional inducements to a halt. The level meadow in which we were sitting was partially covered with trees; the glades were filled with lush herbage, and bright with many flowering plants. Grassy ridges, rising above the level of the forest, but not reaching that of perpetual snow, shut off this sequestered nook from the lower valley, and immediately overhead, on the east of the narrow trench, which offered a way up to the crest of the mountains, the steep snowy sides and tower-like summit of Tau Burdisula caught the eye. The glen up which our path lay was soon terminated, by a steep glacier falling over in a long icefall from the unseen snowfields above. The rich pasturages of this beautiful plain, and the surrounding slopes, are not allowed altogether to run to waste; we passed herds both of horses and oxen, and saw smoke rising from the bivouacs of the peasants in charge of them. Steep walls of rock hem in the upper portion of the glen, and the glacier-torrent has covered the space between them with granitic boulders, amongst which we picked our way.

A long and gradual ascent to the foot of the glacier was followed by a very steep but easy climb up the slopes of snow and rock on its right. Halfway up we stopped, and while resting saw to our surprise a large flock of sheep, accompanied by their dogs and shepherds, descending towards us. The animals hurried and slid down the snow at a great pace, apparently anxious to finish their march, and reach the tempting herbage, already in sight below them. The dogs were fine animals, but somewhat savage, and not at all disposed to acquiesce quietly in our presence; they were called off by their masters, with whom we were, of course, unable to hold any communication. Our Chiora peasant now expressed by signs his wish to return, so, having given him a good day's walk, we paid him all he asked for, and let him go.

For some distance further the ascent was very rapid, still over alternate beds of shale and snow. At last we were on a level with the top of the icefall, and looked into the deep névé-filled basin which feeds it. We had not carefully followed the sheep-track, and on looking back saw that it had turned sharp up the slopes to the left, some distance behind. Although it would have been perfectly easy to regain it, the course up the glacier to its head was so far

the most obvious that we adopted it without much
thought. The upper snowfield was more extensive than
it looked from below, and rose in a succession of gentle
steps, each more or less broken by large crevasses. These,
with the safeguard of the rope, we found no difficulty in
turning, and came at last in sight of the point at which
we should hit the ridge—a well-marked and striking gap
between two rocks at the extreme head of the snowy
basin. The glacier scenery was wild, but not particularly
grand; the summits around us, which cut off all distant
view, were, with the exception of Tau Durdisula, of no
great height, and even that did not look very imposing
from this side. The trudge over the last snowfield was
heavy, and we began to count the number of steps, and to
wonder how many more would be necessary to get over
what seemed to the eye a very small distance. When (at
12.30) we reached the gap, and found shelter under some
rocks from the cold blast which was blowing through it,
we congratulated ourselves on having perpetrated that
delight of Alpine climbers, a new col, though whether
our notch would prove one was still a question. A snow
couloir fell away rapidly for some hundred feet between
splintered towers of rock, and then, the angle becoming
still steeper, was lost to sight, and the eye descended to a
tolerably level and smooth glacier backed by an icy ridge,
equal in height to that on which we were sitting. Up
the glacier a long procession of sheep was slowly wending
its way towards the regular pass, in the track of those
we had encountered in the morning. The northern side,
although easy to a mountaineer, seemed to be defended by
a crevasse large enough to form a serious obstacle to
sheep and dogs. The bold rock-shapes in the foreground,
and the wildness of the whole view, entirely confined to the
snow-region, and devoid of any touch of softness, reminded

us of those Alpine subjects with which all who are acquainted with Mr. Walton's drawings must be familiar.

Before attempting the doubtful descent we 'jödeled' and fired off our revolvers, to attract the attention of the shepherds, who halted for some minutes to watch us. Even a common snow-slope—and this was by no means a common one—looks remarkably like a wall when seen from any point nearly opposite it; I fancy, therefore, that our performance during the next half-hour must have been fully as exciting and gratifying to the spectators as those of Blondin and Leotard are to a London crowd. Fortunately, the snow was in perfect order, firm enough to hold without being too hard to dig steps into with the foot, and with ordinary care there was little risk in the descent on to the glacier, notwithstanding the really formidable angle of the slope, which was equal to that of the last piece of the Wetterhorn, but about 2,000, instead of 700, feet in height. François took exactly the right course, and by swerving to the left, about halfway down, avoided an overhanging mass of sérac, which would have brought us to a sudden check. Once on the level of the glacier, we had nothing to do but to follow its course, which led us in a north-easterly direction. The ice being covered with a tolerably thick layer of dirty snow, and almost free from crevasses, our progress was rapid, and we were enabled to make use of the smooth surface to the point where the glacier terminated, and the stream issuing from it struggled, with only partial success, to free itself from the snow-beds which still strove to bury it from the light of day. When we had once got below the snow-limit, which is comparatively low in this rock-encircled and sunless glen, the walking became very rough. Huge boulders, fallen from the cliffs above, strewed the ground, and offered under their sides lairs, evidently often made use of by shepherds or travellers desirous to cross the glacier while the snow was hard, a precaution almost essential for people to whom the use of a rope is unknown. We were too eager to gain a view of the ice-stream, which we knew from the map must fill the hollow in which the glen at last merges, to take advantage of any of these bivouacs. Scattered firs made their appearance, relieving the otherwise desolate character of the scenery, and the Caucasian rhododendron covered the ground, filling up the crannies between the rocks with its thick branches.

There were now some slight traces of a path, and we came suddenly upon two peasants (probably natives of Zenaga, the highest village in this branch of the Uruch valley) and a donkey, the object of whose mountain excursion we were unable to ascertain owing to our ignorance of the Caucasian dialects. The two parties having satisfied each other, by a close mutual inspection, that no harm was intended on either side, separated; the peasants taking a track leading towards the valley, while we went forward in search of a point which might overlook the great glacier, and afford some insight into the chances of our proposed venture on the morrow. An isolated grassy knoll, just in the mouth of the glen, seemed the spot most likely to offer the view we wanted, and the scene which burst upon us on reaching it so far exceeded and differed from our expectations that, at first, we could hardly realise its magnificence. The whole bed of the valley into which the glen falls is filled by an immense glacier only surpassed in the Alps by the Aletsch.* Its head was hidden behind nearer buttresses,

* Herr Abich alludes to this glacier in the following terms:—'A superb glacier of the first class descends on the north from the Adai-Khokh group between the ridges of Bordjoula and of Snourlaour. It is the Khaltschillon glacier. It is at least 1,500 feet broad, and traverses the forest region for a great distance. Approaching the village of Zenaga, it descends to a level of 6,700 feet, the lowest point known to be reached by any Caucasian glacier.'

but we had a good view of the ranges on its right bank.*
Opposite rose a high and steep mountain-wall; higher up,
looking in a south-easterly direction, an odd tower-shaped
rock appeared in front of a long curtain of ice, surmounted,
on the right, by a tall gracefully-shaped peak, and on the
left by a serrated ridge, in which tooth succeeded tooth in
the most formidable array. Following with the eye the
course of the ice-stream, to the point where it made a
sudden plunge into a branch of the valley of the Cruch,
we looked over the great waves which marked the commencement of the fall, and saw, beyond and far below
them, a tolerably wide valley. Its slopes were wooded
with firs, and we could distinguish some cultivated
land; the northern horizon was formed by rugged peaks,
too low to carry perpetual snow. The summit of one of
them bears a striking resemblance to a castle with a tall
turret at one of the angles.

Our position was much the same with regard to our
proposed pass as that of a traveller at the Montanvert
intending to cross the Col du Géant. We had ascertained that the great glacier marked in the map existed,
and was in reality far larger than it was represented;
but we could see nothing of its upper portion, and could
only be certain that it poured out of a gap just visible
on the right of the great snow-peaks. These offered
us a puzzle, which our united endeavours failed to
solve, as it was impossible to identify them with any of
the summits we had studied from the south, while we
found it hard to believe that such lofty peaks did not
form part of the watershed. This problem was left to
time to solve, and we came to the conclusion that,
though we had gained no positive information as to the
possibility of crossing the chain by the great glacier,

* The torrent issuing from it is called the Karagam, a name which seems the most appropriate for the glacier.

yet appearances were sufficiently encouraging to justify our
resolving to make the attempt. We accordingly set to
work to search for a suitable spot for a bivouac. There were
plenty of boulders strewn about, capable of affording more
or less shelter, but there was no water; therefore we
reluctantly decided that we must descend the steep fir-clad
bank below us to the side of the stream flowing from the
Gurdzievesk Pass, which here runs in a narrow channel
between the hillside and the huge lateral moraine of the
Karagam glacier. We had not gone far when François,
fortunately, hit upon a spring, and as there was a tolerably
level and sheltered spot of ground not far off, we at once
settled to remain there for the night. The first thing to
do was to cut a quantity of young fir-branches, to serve for
beds, and to eject sundry small boulders, which inconveni-
ently contracted our not over-large sleeping-place. We
next proceeded to unpack our provisions, to count over
and apportion our store of bread, and to hunt out certain
rare delicacies, which had been specially reserved for some
such occasion as the present. Our dinner was of the
most recherché description: a first course of sardines was
followed by chicken, and a box of *pâté de foie gras*, one of
two purchased at Tiflis, which, spread not too thickly over
slices of Caucasian loaves, proved a 'lingering sweetness
long drawn out' to all of us.

It was very difficult to realise, as we sat and chatted
round the log-fire which François had prepared, how far we
really were from home, and that our resting-place was not
some old Alpine haunt from which we should cross on the
morrow to Zermatt or Grindelwald. In reality our position was sufficiently strange, unable as we were to hold any
conversation with the people of the country, and separated
from our interpreter and luggage by a long day's journey
and a great range of mountains. Its discomforts and un-

certainties were not, however, sufficient to counterbalance the pleasure derived from the sense of novelty and adventure, and the only real subject of anxiety which disquieted us was the state of Tucker's heels, both of which he had rubbed raw during the day's walk. Moore was fortunately provided with plaister, and the failure of the natural was supplied by an artificial coating. When daylight faded away we arranged our side-bags, the only luggage we carried, as pillows, and soon fell off to sleep.

July 11th.—Our sheltered position, combined with the fineness of the weather and a good fire, which was kept up nearly all night, prevented our suffering from cold, and I have seldom enjoyed sounder sleep than I did in this bivouac in the fir-forest. We were, in fact, almost too comfortable, for no one stirred before daybreak, and it was not till half-past three that our preparations were concluded, and we were ready to start for the unknown region in which the great ice-stream flowing under our feet had its origin. We had previously discussed the question, whether it would be better to descend on to the glacier or to keep along its left bank, and had decided in favour of the latter course, notwithstanding the necessity it involved of partially retracing our steps, in order to cross the stream flowing from the Gurdzieveesk glen. The circuit necessary to effect this passage cost us an hour's most toilsome walking. The place at which we jumped the stream was one not to be generally recommended, as it was necessary to advance to the point of a smooth and slippery rock, and jump from thence on to the farther side. The jump itself was easy, but the 'take-off' was so bad, and the consequences of a slip into the rapid torrent would have been so serious, that no small care was necessary. After a tiresome scramble, amidst rhododendron bushes and over large and frequently

loose boulders, we were heartily glad to meet with a
distinct though narrow path leading in the direction we
wished to follow. After mounting for some distance, by
steep zigzags, up a bank broken by crags and covered
with underwood, we found ourselves on a sloping pasturage by the side of the Karagum glacier, the fall of
which towards the valley was here very considerable. In
about two hours from our bivouac we had gained a projecting brow, which had hitherto cut off the view of the
upper glacier. To our surprise, we found here a small
stone-built hut and enclosure, used by the shepherds
during the summer months, but not yet inhabited.

We now saw how the gap on the right of the great
snowy peaks was filled, and in what manner the glacier
descended from the upper snowfields. A second and
previously invisible peak appeared further to the west, and
through the deep hollow between it and the summits we
had admired on the previous afternoon poured the main
body of the ice-stream, in a frozen cataract of the greatest
and (I speak for myself) most repulsive beauty. So great
was the impression made on my mind by the tangled
web of crevasse and sérac, that I expressed some hesitation
as to the prudence of our attempting to force a passage;
my doubts, however, were promptly suppressed by François,
who gave a very decided opinion in favour of the practicability of the icefall. The shepherd's path still remained
faithful, and conducted us easily along the slopes above
the glacier, which are deeply scamed by numerous watercourses. Before reaching the foot of the icefall, a level
space is found between the moraine and the hillside,
where the ground is covered with soft turf, and a little
stream has space to dance along between grassy banks,
and to expand itself in places into crystal pools. The
saucy water-nymph played us a sorry practical joke, by

tearing Tucker's drinking-cup out of his hand, and hurrying off with it. The cup was of no value in itself, but we could ill spare it at the time. Two lateral glaciers coming from either side join the Karagam immediately below the icefall, but neither of them is of considerable size. The path, which had hitherto served so well, now came to an end, where low stone walls built under an overhanging boulder showed its object, and indicated the occasional visits of shepherds and their flocks to the grassy slopes which rose above the glacier on our right. We here took leave of vegetation and *terra firma*, which we were not again to tread until we reached the valley of the Rion. In crossing the moraine we did not take the best possible course, and had, in consequence, to walk along the narrow ridge of a pile of rubbish, which formed quite a typical arête, and would, if provided with proper precipices, have been most sensational.

The foot of the icefall was soon reached, and the way in which it could best be attacked became obvious. The lower portion promised to be tolerably plain sailing; but it was evident that, after a time, it would be necessary to bear to the left, on which side alone the upper maze of crevasses appeared at all assailable. The ice, which was almost level where we first entered on it, soon began to rise before us, and the surface, although not as yet seamed by any deep chasms, became uneven and slippery. Having reached the beginning of the real work, where it was necessary to put on the rope, we took the opportunity of halting to eat a second breakfast, to which a rivulet, just unloosed by the returning warmth from its night's imprisonment, contributed most usefully. A teetotaller would have a decided advantage over his winebibbing companions in Caucasian mountaineering; the general impossibility of obtaining drinkable wine or spirit in any of the upper valleys is a rather serious matter to men taking hard exercise, and accustomed to some such support.

The real work of the day now began. At first it did not seem likely to prove very serious, for we found little snow-valleys which led past and round the towers of broken ice, and enabled us to turn the huge chasms which ran across the slope to left and right of us. Our prospects of success began, however, to look very questionable when these chasms became more continuous, and cutting in half the snowy dells forced us to plunge into the intricate labyrinth of ice-towers and crevasses, in our endeavour to force a way through the tortuous mazes of the fall. The difficulties of a broken glacier have been often and well described by Alpine travellers, and those which we now encountered presented no particular feature of novelty. They were, however, the most numerous and complicated of their kind any of us had ever battled with. Once, after struggling through trenches, up walls, and under towers of blue crystal, fair to the eye, but liable at any minute to topple over, and therefore to be avoided or hastily passed by, we came to a great chasm, which at first sight seemed impassable. Behind us was 'clean starvation,' for our stock of provisions would not hold out over another day. The only alternative course was to descend to the village of Zenaga, and try by signs to procure food there, at the risk of being arrested as suspicious characters, and sent down to give what account of ourselves we could at the nearest Cossack outpost. Our situation, therefore, gave us every inducement to persevere, if not at all hazards, at least as far as prudence would permit.

A snow-bridge, which elsewhere might not have been approved of as fitted for public use, was, under the circumstances, voted worth trying, and François went ahead, to make such improvement in the footway as the axe could effect. This was not much, and the deficiencies of the frail structure were too serious to be supplied by any ingenious contrivances. We had to descend a bank of ice, six feet high, to reach the level of the snowy crest, fully twenty feet in length, which, like an arch of Al Sirat, was flung over the icicle-fringed chasms yawning to unknown depths on either hand. The top of this crest was uneven, and about the middle of the bridge an accurately-measured and delicately-managed jump was requisite to reach two pigeon-holes, cut by François for the feet, on the further side of an awkward gap. Then each walked carefully for several yards, like a cat along the top of an old and rotten wall, to the point where, instead of abutting against the steep snow-bank on the opposite side of the crevasse, the bridge broke down altogether, making a second and still more awkward jump necessary. The man in front, having made the leap and anchored himself in the snow-bank, turned round and grabbed tight hold of the arm of the next—a desirable precaution, as a little too much impetus might have thrown the jumper backwards into the chasm. Steps were then cut on to a promontory of ice separating the big crevasse from a smaller relation, which was not beyond a straight-forward jump, more formidable in appearance than reality, as the landing-place was good. We were now in a position to take advantage of a series of connected ridges, by which we made our way back into one of the snow-filled depressions we had found useful earlier in the day. For half an hour good progress was made; then again for half an hour we did little but wander up and down, seeking some exit from a fresh labyrinth, and almost despairing of final success. The séruc scenery throughout was of the grandest description, and we were constantly forced to admire the beauty of our stubborn foes, the icicle-fringed and blue-caved crevasses, and to wonder at the curious forms and grouping of the frozen towers and pinnacles.

As time went on we could see, by the diminished proportions of the ridges behind us, and the change from the clear crystal substance of the lower glacier to the half-formed ice, or rather 'névé,' of the upper regions, that we were slowly, but surely, drawing near the top of the fall. We were rashly congratulating ourselves on having achieved the victory, when a fresh obstacle appeared—a great split in the surface, with an upper lip ten feet higher than the lower. François made some foothold on the further side with his axe, jumped across, and attempted to work himself up the face of the perpendicular upper lip, while we watched his proceedings with some anxiety from an insecure situation on the lower bank. After several vain endeavours to wriggle or work himself up, he gave in. It was no easy matter to get back again, but he managed it by a skilful tumble; then with gloomy forebodings were traced our steps, until, several hundred yards to the left, where the crevasse was lost in a big hollow, we found a part of the wall of loose floury snow, up which there was no serious difficulty in forcing a passage. The steepest part of the icefall was now fairly below us, but we were still unable to see far ahead, as a line of broken waves of névé separated us from the unknown land above. Though our course was still necessarily zigzag, and occasionally subject to an annoying check, the first chapter of difficulties was overcome. The soft state of the snow, into which we sank at every step, now seemed likely to prove a less exciting but scarcely less serious hindrance to our progress. François having of late had more than his share of work, Moore relieved him by taking the lead, and soon enforced his request that we would keep the rope taut, by sinking up to his shoulders in a concealed crevasse. Not long after this incident, the slope before us lessened, and the view of the upper region, to which we had been anxiously looking forward for so many hours, burst upon us. It was now 1.30 P.M., and we had therefore spent six hours in fighting our way up the icefall, the height of which, from the rough measurements possible with our aneroid, we estimated at but little under 4,000 feet. The famous 'séracs' of the Col du Géant are child's-play when put in comparison with these Caucasian rivals, and I think it very possible that a party endeavouring to force this passage at a later period of the summer might meet with a signal repulse.

Our feelings, on viewing the new scene revealed to us, were those of mingled admiration, astonishment, and perplexity. Like Jack when he had climbed his beanstalk, we were a good deal taken aback by the strange region in which we found ourselves, and not a little puzzled what to do next. Before us stretched a vast reservoir of snow, which soon split into two bays, running respectively east and south. The eastern branch broke down upon the other in a grand fall of séracs; its head was surrounded by a number of magnificent rock-peaks, including amongst them the mass which had presented so imposing an appearance from our bivouac, not at all dwarfed by nearer approach. The surface of the southern bay was almost a dead flat, hemmed in by icy ridges, the summits of which scarcely attained the dignity of separate mountains. The extent of these snowfields was, to us who had to make our way to the other side of them, almost appalling; how much of the effect they produced on our minds was owing to the exciting struggle we had gone through to attain them, it is of course impossible to say. I suppose, if the first men who saw the upper regions of the Aletsch glacier had come upon them by the Jungfrau Joch, they would have been likely to exaggerate their effect. A similar allowance must be made for us. The principal cause of perplexity was our inability to recognise, in any of the peaks now in sight, the two summits of Adai Khokh, so conspicuous from the eastern branch of the Rion basin, and which we had hitherto confidently reckoned on as landmarks. Each had a different theory, but there was no time to stop and argue it out; so we agreed, by a majority of voices, to put our trust in the compass, and push up the southern bay, as that direction must, we believed, ultimately bring us to a point overlooking the Rion valley.

There are no incidents to relate in a three-hours' tramp across a soft snowfield, but such an operation is not to be lightly estimated, because it occupies but a brief space in the narration. Each in turn took the arduous task of leading—no slight exertion, when the leader sank at every step nearly up to his knees. The mountain we had seen from the bivouac now towered grandly in our rear, its western summit offering a striking resemblance to the Matterhorn from Breuil. In front the monotonous snow-plain seemed more endless the further we advanced; on the left hummock after hummock was passed, while on the opposite side some projecting rocks were for long our goal. They were reached and left behind, and still there was no change in the sameness of the view, except the more prominent appearance of a considerable mountain on the right, which now revealed itself as our old acquaintance, Tau Burdisula, under a new aspect. At last, about 4 P.M., we reached an almost imperceptible watershed, first indicated by the appearance of the blue ridges of the far-off Achaltzich mountains over the neighbouring snows.

For some distance the fall was as slight as the rise had

been on the other side; then the slope suddenly steepened, and we recognised for the first time our exact position. From our feet a glacier, a small portion of the outflow of the 'shining tablelands' we had been traversing, poured down into the glen, the torrent of which joins the eastern Rion at Glola. We were, in fact, standing on the very snows of which we had caught a glimpse when entering that village a few days before. The natural course was to find a way down this glacier; but after a few hundred feet of easy descent, it toppled over in a tremendous icefall, apparently as long as, and a good deal steeper than, the one we had ascended with so much difficulty. Once entangled in this complicated labyrinth, there seemed little prospect of getting free of it before dark, even if the passage proved practicable at all. We took therefore what, although its adoption disgusted me at the time, was, I believe, the only sensible course, and deliberately returning to the Col mounted the slopes to the east, and crossed a snowy head which projected above the top of the icefall.

There was still a moot point to be decided—whether we should endeavour to get down the mingled rocks and snow-slopes on the left of the icefall into the glen leading directly to Glola, or whether we should bear still more to the east, and descend into one of the branches of the wide valley which leads up to the base of Adai Khokh. Whichever course we decided on, it was necessary first to traverse diagonally a series of steep slopes, overhanging ground above the glacier, which might for all practical purposes be considered as a precipice. The surface was soft, and we had no trouble in step-cutting; but the snow more than once showed a disposition to crack and slide downwards in masses, leaving bare a substratum of ice, which, to anyone versed in mountaineering craft, was unpleasantly suggestive. We instinctively held our axes with a firmer grasp, and congratulated ourselves that there was no weak brother, or lumbering porter, likely to test the power of a slip in carrying down with a run the whole upper layer of snow. We were well pleased to reach a rib of rocks, and to clamber down them for some little distance; and having by this time decided at any rate to examine the descent into the eastern valley, we made our way over some safer slopes to the crest of the ridge in that direction.

The fall of the ground on the eastern side was undeniably steep, and an extremely ugly-looking 'couloir,' opening immediately at our feet, did not offer a tempting exit. François promptly pronounced against this side of the ridge; I was not satisfied with the grounds of his verdict, and untying the rope scrambled down a few feet to gain a better view. My investigation was rewarded by the discovery of a second snow-gully, which ran up and terminated against a buttress a hundred feet below, and, after falling for some distance in a direction parallel to the ridge we were on, turned sharply at right-angles and was lost to sight. As far as could be seen, it offered not only a practicable but an easy line of descent, which I pointed out to François, who suggested the possibility of the lower portion being precipitous. We agreed, however, that all the indications went to show this to be extremely unlikely, and that no more promising way out of our difficulties offered itself.

The view before us had been, during the last hour, one of surpassing beauty; while we slowly descended a projecting buttress, our position gave us a raking view of the peaks of the main chain, and we were at the same time at a sufficient elevation to overlook the whole of the southern sub-Caucasian district. The Radscha lay at our feet, a labyrinth of green ridges and dark forest-clad ravines, through the centre of which the Rion finds a devious way to Kutais and the Mingrelian lowlands. The waters of the river flashed in the sunshine, and pointed out the deep cleft through which it passes before reaching Oni. On the one side rose the imposing mass of the Schoda; on the other was a cluster of snowy peaks, remarkable for their elegant pyramidal outlines, situated to the south-east of Gurschavi, and separating some of the headwaters of the Ardon and the Rion. Far away in the west a high glacier-crowned chain arrested the attention; reference to the map showed that it must be the Leila mountains, which form the southern boundary of Suanetia. The mists, which some hours earlier had been sufficiently numerous to cause anxiety, had now melted away, and left the blue sky unclouded. It was a perfect summer's evening, and the sloping rays of the sun, already sinking rapidly towards his rest, flooded and transfigured the wide landscape with a golden glory, which overcame the indifference to the charms of nature too often brought on by fatigue, and roused us to make constant appeals to one another to admire some freshly-discovered beauty.

A very short scramble down the crags brought us to the head of the gully, where we were delayed by an incident we regarded only as ludicrous at the time, but which caused us a provoking loss. In stepping off the rocks on to the snow, Tucker suddenly subsided into a deep hole, the existence of which was concealed by a thin and treacherous crust. He went down at least ten feet below the surface, and the chasm was so narrow that considerable exertion was required to haul him out again. It was not discovered till some time afterwards that an excellent telescope had been wrenched off his shoulders in the struggle. It was an illustration of

the proverb, 'Misfortunes never come single,' and we could
only console ourselves with the obvious reflection that it
was impossible for us to go on losing a drinking-cup and
telescope every day. The snow in the upper part of the
gully was rather hard; but by keeping close to the side,
and digging steps with our heels, we got along capitally,
and soon reached the corner, whence we saw a straight
unbroken trough leading down to the base of the cliffs, and
on to a snowfield which stretched down to green pasturages. We felt, for the first time, that the fight was over,
and the victory won, in so far that we were secure of
sleeping below the forest level on the southern side of the
chain, of which there had been, up to this moment, considerable doubt. We were as pleased at the prospect
of a bivouac on the turf, in the place of a night spent
in kicking our heels against frozen rocks, or, worse still,
in the bosom of a crevasse, as an Alpine traveller is at
unexpectedly discovering a good hotel where he only
looked for a poor châlet. We slid merrily down the
snow-gully, in the track of a gigantic snowball which
was now reposing at the bottom. Having observed from
above, that by traversing the slopes to the right, we
should, without the need of any further ascent, cross a
low ridge dividing two hollows, and enter the one most
likely to lead directly towards the valley, we heroically
withstood our disposition to go straight down to the
grass, and kept for some time at a level, at the cost of
half-an-hour's rather tiresome walking.

At last the head of the hollow, still covered with fast-
melting snow, was reached; here, as in many other places,
the red colour of the surface attracted our notice. Where
the snow no longer lay, its meltings made the turf a
perfect playground for watercourses, through which we
splashed on, anxious to gain the forest before nightfall.

The first sign of life was a troop of horses; a little lower a distinct path appeared, which we gladly accepted as our guide until it brought us to a brow some height above the stream, and then turned away down a slope to the left. It was already growing dusk, and we had just entered the highest copse of birches; water was, of course, a necessary adjunct to our halting-place, and hesitating to leave the stream, still close at hand, we determined to go down through the copse and sleep at its foot, beside the water. The chief objection to our camping-ground proved to be the absence of even a square foot of level soil. After treading down the long grass, it was necessary to break off branches and lay them on the lower side of the spot selected by each for his bed, to prevent the sleeper rolling away down the slope. Having lighted a fire, we ransacked our bags, laid together what little provision there was left, and set aside one roll for the morning; the next thing was to divide the rest into portions; each man got a slice of bread about two inches square, and half the limb of a chicken. After this frugal supper had been disposed of, we covered ourselves as far as possible with our mackintoshes, and lay down to court sleep, but we had not long dozed off, when several big drops of rain effectually roused us. A thunder-shower had blown up, and the dark clouds which obscured the moon held out very unpleasant threats of a ducking. Luckily, they passed off without any serious fall of rain, and having exchanged mutual grumbles, we again drew up our mackintoshes over our faces, and relapsed into uneasy slumbers, or reflections on the work done and the sights seen during the past two days.

The object of our double passage of the mountains had been to discover what lay behind the great snowy wall which bounds every northward view from the Upper Rion, and to learn something of the breadth and character of this portion of the Caucasian watershed. The result of our expedition was satisfactory, for it enabled us to form a tolerably correct idea of the general character of the chain westwards from the Mamisson Pass to the sources of the Rion. Tau Durdisula may be taken as a point of division; from it to the Koschtantau group, the central ridge is too narrow to support any vast snowfields or first-class glaciers, while to the east of it the chain branches into a network of ridges, the spaces between which are filled by vast névé-reservoirs, of which the Karagam glacier is only one of the outlets. No attempt to distinguish these ridges, or to name the numerous peaks which rise out of them, has been made by the authors of the Five Verst Map; but the portion of the chain between Tau Durdisula and the Mamisson Pass, from the number and height of its summits and the size of its glaciers, forms undoubtedly one of the most remarkable mountain-groups of the Caucasus.

July 12th.—Our quarters were not so luxurious that we cared to remain in them longer than necessary, and we rose at daybreak. Our breakfast was workhouse fare, stale bread and water, and very little of the former. It did not take long to dispose of, and we were glad to set out and shake off, by a brisk walk down to the valley, the chill and stiffness produced by our night's lodging and previous hard work. We soon lost sight of the track of the previous evening, and ran down the steep and, at first, only partially-wooded slopes in search of another. Before very long we lighted on a broad path, which, after skirting the hillside for some little distance, descended by a succession of steep zigzags to the stream, which, rising in the glaciers of Adai Khokh, joins the Glola-Squali between Gurschavi and Glola. When we got fairly into the forest, the variety of the foliage and the beauty of the wild flowers gave us a constant interest. Dwarf honeysuckles, campanulas, and wistarias were abundant, and the tiger-lilies shot up in clusters, each spike bearing from two to seven blossoms. Near the stream the sombre foliage of the pines added, by contrast, to the effect of the deciduous forest. Shortly before crossing the Glola-Squali to the Mamisson road, we met a party of peasants apparently going to look after their flocks, and armed, as usual, with daggers.

We arrived at Glola in four hours from our bivouac, and going at once to the house where we had before found lodging, attempted, by the aid of a few Russian words and a great deal of pantomime, to explain our wants to the people. Necessity is a wonderful sharpener of the wits, and we all got on famously by the language of signs; but there was no question as to who was the leading pantomimist of the company, and we thought Moore perfect in his grand performances of milking the cow and mimicking a hen's cackle, in order to procure us milk and eggs, until our friend surpassed himself by 'riding a cockhorse' on an ice-axe, and prancing about, in order to intimate our wish for horses to carry us to Gebi. Pantomime, on the part of the villagers, explained that the horses were all on the mountain, but should be fetched. We waited for them so long that after we had finished our meal, Moore and I lost patience and walked off, leaving François with Tucker, whose heels, although they had carried him pretty well across the pass, were still in a tender condition. When the horses came at last, the mistress of the house presented Tucker with two loaves, made of a better quality of flour than the common bread, and sent her son in charge of the animals.

The day was fine, and the heat in the valley was great. We retraced our footsteps of the previous Thursday until opposite Chiora, where, instead of fording the river, we kept along the path which continues to follow its right bank. The valley, although of considerable width, is almost entirely filled by the stony bed of the Rion, and the path, forced to wind over the spurs of the southern range, is in consequence very uneven. After turning the base of a projecting and densely-wooded ridge, about halfway between Chiora and Gebi, the valley ceases to be entirely devastated by the torrent, and the path becomes level. It is shaded by thickets of alders and hazel, the stems of which were girt round by wild hops. Before reaching Gebi, four streams, issuing from as many lateral glens of the Schoda chain, had to be crossed — a matter of some difficulty, as the popular idea of a bridge in the Rion valley seems to be a thin and rough branch laid from bank to bank, to traverse which successfully requires some training in the customs of the country. The glimpses of luxuriant foliage and snowy peaks up these side-glens more than repaid us for the trouble we experienced with their torrents. The last and largest, the Latkischora, which falls in nearly opposite Gebi, has a wide and stony bed, and a considerable volume of water. The Rion itself is crossed by a wooden bridge before entering the village, which stands most picturesquely on a steep-sided promontory on the left bank of the river, and just above its junction with the Tchosura, the stream flowing out of the valley we had looked down into from the slopes above Chiora.

Gebi, unlike the other villages of the Rion valley, is provided with towers of defence similar to those which are universal in Suanetia. These towers are built of large unevenly-shaped blocks of stone, and the walls contract towards the top, which is covered with a sloping roof of wood or slate, like that often put on an unfinished church-tower at home. They add much to the picturesque effect of the place, which, from a distance, looks like a large feudal castle. Some of the houses are built in a knot, as closely as possible together; others are scattered round an open space like a village-green. In the centre of this we observed a small wooden building, round which was gathered a crowd of idlers; we rightly surmised that Paul had taken up his quarters there, and was now the centre of attraction. He was of course delighted to see us, having spent three rather melancholy days, surrounded by the inquisitive and troublesome villagers; no difficulties however had arisen, and the lodge which had been assigned to him was far more comfortable than we had any right to expect. It consisted of two rooms and a balcony; in the outer apartment there was a bench and a fireplace; the inner we constituted our bedroom. We never entirely satisfied ourselves as to the use to which the building was commonly put, but, as far as we could understand the explanation given through Paul by the villagers, it was designed as a kind of court-house, where the elders might meet, and any public business be transacted.

A tall fine-looking peasant, the headman of the place, came formally to bid us welcome, and to assure us that all our wants should be supplied. A high sheepskin hat distinguished him from the general crowd, numbering at least 150 men and boys, who, attracted by our arrival, had formed a circle outside the door to watch our proceedings. There was even a greater variety of head-gear amongst the peasants of Gebi than in the bazaar of Kutais. Some carried the 'baschlik' with the hood over the head, and the point turned upwards like a fool's-cap; a few wore the small Mingrelian bonnet, almost invisible in the middle of their heavy shocks of hair; the greater number had soft felt wide-awakes—a bell-shape was perhaps the most fashionable, but no two could be found exactly alike. Even the boys were

armed with daggers, and many of the men carried their
guns in sheepskin cases across their backs; their clothes
were for the most part soiled and ragged. Though not, on the
whole, a fine-looking race, like the Ossetes, they did not
bear in their faces any peculiarly vicious expression, beyond
an air of lazy stupidity. One or two of the men wore
Russian medals, showing their complete and voluntary
acknowledgment of the Government. We found, invariably, that in proportion as the natives are brought into
contact with their rulers, they improve in manners and
civilisation, and that the districts which the Russians
have left to take care of themselves are those in which
the old customs of petty warfare, robbery, and murder still
prevail.

Our dress, our accoutrements, and our luggage proved
inexhaustible sources of amusement to the large circle of
which we were constantly the centre. What caused the
greatest excitement was the sight of our pocket-handkerchiefs, and our manner of using them. Upon the first
occasion of blowing our noses, a roar of admiration burst
forth, and afterwards the slightest sign of a repetition of
the performance sufficed to raise a murmur of excitement
amongst the expectant crowd.

We had always intended to halt at least a day at
Gebi, and if possible to make it our headquarters for
some excursions amongst the mountains round the sources
of the Rion, where in most maps the name of Pasa-Mta
(said to be derived from Phasis-Mta?) is printed across
the main chain, in a way to indicate the existence of a
noteworthy peak. Tucker and I agreed that a day's *dolce
far niente* would be very pleasant, but Moore, whose
energy was still unspent, hankered after a mountain, and
settled, if the night was fine, to start at 2 A.M. with
François, and climb the Schoda (11,128 feet), the bold

summit of which rises above the lower ridges on the south of Gebi, and must, from its isolated position, command a perfect panorama of the main chain. Our plans for the morrow being thus fixed, we postponed the settlement of our further arrangements, and retook possession of our mattrass with great satisfaction.

July 13*th.*—The weather changed during the night, and Moore was prevented from starting for his proposed expedition. In the morning it rained heavily, and our day was spent chiefly in cooking a sheep we had purchased, and discussing the means of getting across the wild country at the sources of the Zenes-Squali to Jibiani, the highest hamlet in Suanetia, close to the glaciers of the Ingur. The height of this place (7,064 feet) suggested to us the idea of a sort of Pontresina, whence we should be able to make a series of excursions into the great mass of mountains marked on its north in the Five Verst Map. Elated by the successful accomplishment of the two glacier-passes, we planned, about this time, various magnificent expeditions, which weather and other hindrances ultimately defeated. We were desirous of coming to such an arrangement with our porters as might enable us to camp for a day or two at the sources of the Rion, and see what excursions could be made there. It was, however, so difficult to make the peasants understand our intentions, and to prove to them that if they sat and smoked all day, while we climbed a hill, they were not entitled to the same pay as if they were carrying our luggage over a stiff pass, that we gave up the attempt in despair; and finally arranged to engage seven men as porters, at 1 rouble and 20 copecks a day apiece, as far as Jibiani, a journey which they assured us was generally made by hunters in three days. We enquired about a pass named in the Five Verst Map, and laid down as leading up the glen of the Tchosum, and over the main chain, into the valley of the Urach; it was described as very much of the same character as the Gurdzieveesk Pass.*

The heavy rain, which continued to fall all day, caused the Rion to rise very rapidly, and to threaten with destruction the bridge below the village, the centre pile of which, a clumsily-constructed wooden breakwater, was exposed to the full force of the current. The danger of the whole structure roused the people from their usual laziness, and delivered us for a time from the constant crowd of lookers-on. The whole population trooped down to the bank, and carried stones, to fill up the interior of the framework which supported the centre of the bridge, in order to give it weight to resist the violent attacks of the stream. Their efforts were successful for the moment, and the weather clearing up late in the afternoon, the river gradually subsided.

Having cooked the necessary supply of meat, purchased some luxuries—such as sugar, and muddy grape-juice, here called wine—and, as we thought, concluded our

---

* These two passes are laid down by Klaproth, in the map appended to his 'Voyage au Mont Caucase,' published in 1823, and he describes at some length his journey across them in the year 1809. Starting from Mazdok, on the north side of the chain, he ascended the Urach valley, and crossed the Gebi-Ga Pass, to which he gives the name of Tsiti-Klong, but which is no doubt the pass of the Five Verst Map, above referred to. He descended the Rion valley as far as Oni, but was deterred from going farther by the disturbed state of the country, and therefore retraced his steps, regaining the Urach valley by a pass which can be no other than the Gurdzieveesk. In both cases the details of the actual passage of the chain are very meagre and unsatisfactory, but many particulars are given of the valleys on either side which could scarcely have been acquired by hearsay. The part of the story most hard to believe is that horses were got over both passes; so far as the Gurdzieveesk is concerned, we should certainly have declared it impracticable even for the steeds of the Caucasus.

Dubois de Montperreux denies the truth of Klaproth's statement on the authority of the inhabitants of Mazdok; but they were little likely to know anything about the matter, and I do not think their opinion is of much value.

arrangements for porters, we announced our intention of setting out next morning, and dismissed our visitors.

July 14th.—The weather was still showery and unsettled. Our first question with the villagers was on a claim for higher payment for our food and lodging, which was at first laid at nearly double the sum we offered, but was very soon brought down, by firm resistance on our part, to a petition for an extra rouble. The next difficulty raised was one less easy to settle satisfactorily. The seven porters for whom we had agreed struck, on the ground that our luggage was the load of ten men. Anxious to smooth matters, we conceded this point, and allowed them to fetch three of their friends, whereon the whole team struck again for higher pay. This we absolutely refused, declaring that, if further difficulties were made, we would ride down to Oni, and report their behaviour to the Commandant. The ten, finding that we could be as obstinate as themselves, gave in, after a long and irritating wrangle, and agreed to come at the pay previously promised. At last all the packs were separated, and each man's burden tied up into a form convenient for transport. After watching the ten defile before us, and seeing that nothing was left behind, we followed. Having crossed the bridge, all the men sat down, and held a protracted council with some friends who joined them, as to which path they should take, while we fumed with useless impatience.

The ordinary path up the valley follows the left side of the Rion, on which there is a good deal of cultivation for some distance above Gebi; but news having been brought in of the destruction of the upper bridge, our porters had at once crossed to the right bank, and were now discussing how they might best make their way along it. When the council was at last over, we were led by a narrow footpath, which mounted steeply through the forest. After climbing several hundred feet, we left it, and plunged into a dense underwood of azalea-bushes, now nearly out of blossom, through which we gradually fought our way back to the level of the Rion. The boulders of the river-bed afforded a less fatiguing path, and did not hinder our advance so much as the tangled thickets into which we were often forced to enter. Torrents, emerging from lateral glens on the south, barred our way, and at first we expended much ingenuity in attempts to cross them dryshod, by extemporising bridges with a fallen tree, or attempting impossible jumps. As each of us in turn failed and got wet, we gave up the struggle against our too numerous foes, and quietly waded through. After 3½ hours of very slow and tiring progress, aggravated by the sight of meadows and a fair path on the opposite bank of the river, we came to the broken bridge, and joined the usual track up the valley. There was now a pause of some duration in our struggle with untamed nature. It was, however, quite impossible to make up for lost time by a spurt, as our train of porters absolutely refused to be hurried, and treated our remonstrances with utter, although good-humoured, contempt.

The scenery of this portion of the valley does not equal in grandeur that of the eastern branch, but the woodland effects are very beautiful, and quite unlike anything in the Alps. Dense forests of deciduous trees, amongst which the beech and the maple are conspicuous, clothe the lower mountain-sides, which conceal all view of the snowy chain. Pines gradually disappear, and none are found near the head of the valley. On a high bank on the right of the Rion, opposite its junction with the Zopkhetura, a tributary which nearly doubles its volume, are some fields, the highest cultivated land of the inhabitants of Gebi. Rude wooden huts have been constructed by the peasants, as shelters for the night when they come up either to sow, or to gather in their crops. The valley now makes a sudden bend to the north, and several small streams fall into it from the surrounding mountains. The path becomes very uneven, winding up and down on the steep broken slopes on the right side of the river, which flows rapidly in a broad stony channel. Before long the track was altogether lost, and we followed out the ideas of the leading porter as to the best line of march—now forcing our way through the forest, now scrambling over the boulders of the river-bed. Several strong torrents had to be waded, and the heavy rain which began to fall completed our wetting, and made us look forward with some dread to camping-out.

The valley contracted almost to a gorge before it opened out slightly, and left space on the right bank for a meadow and two log-huts, which mark a summer station of the herds, known to the natives, according to Herr Radde, by the name of Sassagonelli. The huts were in a very dirty and dilapidated condition, and we decided at once, despite the wet, to pitch our tent, and leave such accommodation as they offered to the men. It is not a pleasant thing to put up even so small and easily-managed a tent as ours was in pouring rain. We had brought away with us from Gebi a winebagful of the liquid called wine in these parts, and we now had some of it mulled, in which form it was by no means nasty, notwithstanding a strong flavour of gutta-percha. We had taken over eight hours to reach Sassa-gonelli from Gebi, and it grew dusk soon after our arrival; so soon therefore as we had supped, we tied up the tent-door, wrapped ourselves round in our rugs, and made things as snug as possible for the night. Our men found at least shelter inside the hut, which some of the porters shared with them. As a rule, the inhabitants of this country care little for any further protection from the elements than their big sheepskin 'bourcas,' which can be arranged so as entirely to envelope the figure, and may very likely have given rise to the fable that the Caucasians are in the habit, when on the march, of carrying with them small tents, and taking shelter in them from the rainstorms, for which these mountains are justly celebrated.

July 14th.—The morning was fair, and the clouds were blowing off the surrounding summits when we emerged from our tent. The view up the valley was closed by the snowy mass of the Edenis-Mta, and a small glacier which descends from its flanks. The meaning of this name is the Mountain of Paradise, and a tradition, similar to that told of the 'Grand Paradis' in the Graian Alps, is related by the inhabitants concerning it. The track we now followed abandoned the deep-cut channel of the Rion, and climbed the very steep grass slopes of the range on the west. Even the birch was soon left below, and we found ourselves, after a long ascent, on a wide sloping pasturage enamelled with alpine flowers, except in the hollows where the snow still lay unmelted. The roots of the Rion valley were now at our feet; above them the main chain rose in a steep wall of rock, over which the two glaciers in which the river has its source poured in narrow and not very imposing icefalls. To the west the view was limited by the ridge we had to cross, but behind us, over a gap in the lower hills, rose a serrated icy crest, the summits of which, although not on a scale of grandeur comparable to Adai Khokh and its neighbours, would be considered fine mountains anywhere in the Alps.

Moore had been altogether upset by the hot brew of gutta-percha-flavoured grape-juice in which we had indulged on the previous night, and had great difficulty in making any progress; so Tucker and I pushed on by ourselves, leaving François to help our friend, and Paul to keep an eye on the team of porters. We walked briskly over the wide Alp, anxious if possible to reach the ridge (called by Radde the Goribolo Höhe) before the clouds had again shrouded the mountain-tops. In this we were unsuccessful. The pasturages were of great extent, and were linked to the chain separating the Rion and Zenes-Squali by a long flat-topped ridge, the ascent from which to the actual pass was very considerable. One of the porters had pointed out to us the spot we were to make for, a rocky eminence considerably to the right of the lowest point in the ridge. On reaching it we found that the mists had already enveloped all the western chain, and that our hopes of learning something of Koschtantan and its neighbours were disappointed. We had, however, a good view of the western end of the Rion basin, and of the picturesquely-shaped summits of the Schoda chain. At our feet on the west was a short glen, running down to the wooded ravine of the Zenes-Squali; a spur parallel to that on which we stood separates the two sources of that river. A faintly-marked zigzag, by which the peasants of the Rion valley reach the upper snowfields, and pass over them to the pasturages round the headwaters of the Tcherek, on the north side of the mountains, could be traced climbing the steep slopes of the main chain on the west of the Rion sources. The pass is probably free from serious difficulty, as cattle are sometimes 'lifted' over it, but it must lead across a wide expanse of snow and ice.

It will be seen, on any of the modern maps of the Caucasian provinces, that it is at this point of the chain that the name Pass-Mta is printed, in characters which seem to indicate the position of a peak only inferior in height and importance to Elbruz and Kazbek. We could see nothing of the kind, and Radde, who climbed as far as the pasturage, and was lucky in a clear day, asserts that the Pass-Mta of the inhabitants of Gebi is nothing more than a rocky buttress projecting from the main chain. Our predecessor states this clearly in the following passage:—' One is very much surprised, after having looked out so long on the journey for the Pass-Mta, to find in it nothing imposing or out of the common. It is far inferior in height not only to the Edenis-Mta, but also to the Lapuri (a summit lying farther to the north-west), and the flattened dome to which it rises scarcely attains the height of the snow-line. Its distinguishing characteristic is that it pushes forward from the main chain, here represented by the Lapuri and Edenis-Mta, and thus encloses the source of the Rion on one side, while the ridge called Goribolo shuts it in on the other.'

The solution of the apparent inconsistency between the wide reputation of the mountain and its real insignificance is not, I think, difficult to discover; to us, at least, the following explanation seems sufficient and satisfactory. The name of Pass-Mta has been applied, by the people on the south side of the chain, to the mountain they cross in going over to the Tcherek. The traveller on the hills which gird the lowlands of Mingrelia on the south, sees, when looking at the opposite chain of the Caucasus, a great snowy mass midway between Kazbek and Elbruz. He does not know its name, and can find no one to tell him, but by the map he makes out that it must be somewhere near the source of the Rion: the only mountain well known there is, he is told, the Pass-Mta, so, putting two and two together, he settles that the great wall of ice which has attracted his attention, and which is in reality the southern face of the Koschtantau group, must be the Pass-Mta of the people of Gebi.* This portion of the chain is the one which seems most completely to have puzzled geographers, and many books and maps fall into the serious error of representing the Zenes-Squali as rising entirely on the southern side of a spur of the main chain. They thus deceive a traveller, by giving the idea that only one ridge separates the sources of the Rion and Ingur, whereas it is in reality necessary to cross no less than three in going from one to the other.

We had been full half an hour on the top before the porters came up, escorted by Paul; Moore and François were not far behind, and we all made our midday meal together. When it was time to think of pursuing our journey, we enquired what course was usually taken in descending to the Zenes-Squali: the porters pointed out a long and manifestly absurd circuit, involving a considerable further ascent along the ridge on our right. There was no difficulty in going down the steep sholy rocks and snow-filled gullies immediately below us, into the head of the valley; but when we intimated our intention of doing so, the men gave us to understand that if we liked to risk our lives, they did not mean to peril their own, and that nothing should induce them to follow us.

Having fixed as our meeting-point the junction of the stream in the glen below us with the eastern Zenes-Squali, we abandoned our train to the consequences of their folly, and set off down the rocks, which were perfectly easy to anyone of mountaineering habits. A short scramble enabled us to get into a snow-filled trough, down which we slid rapidly, until the foot of the declivity was reached, and the gully came to an end amongst stones and uneven

---

* I have been confirmed in this theory, since I wrote the above, by seeing, in the Atlas to Dubois de Montperreux' 'Caucase,' a profile of the Caucasian chain, in which the outline of Tau Tötönal and the Jibiani peaks is clearly given, and the name Pass-Mta is applied to them.

ground, cut into deep furrows by the melting of the winter snows. At first the vegetation, amongst which we again found ourselves, took the form of stunted bushes, the tangled branches of which might occasionally trip us up, but offered no material impediment to our progress. As the trees grew thicker, and no trace of path appeared, we were glad to take advantage of the partially dry bed of a torrent, which was narrow enough to permit of our jumping from side to side as occasion required. When it joined the main stream, and no space was left between the foaming waters and the steep bank, we were obliged to enter the wood. First, we forced our way through a dense thicket, where we had to push aside the upper branches with our arms, whilst we scrambled as best we could in and out of deep rivulet-beds, and over or under the trunks of partially-fallen or prostrate trees. On emerging into a glade, we did not find our condition much bettered by the change. Although at a distance, and when seen from above, the smooth and flowery surface had suggested no difficulties, we now found that it was composed of a dense growth of umbelliferous plants, growing to an average height of six feet above the ground.* We were at first at a loss whether to admire the extraordinary luxuriance of the cloak which nature has spread over the soil in this mountain region, or to grumble at the toil it cost us to make each step in advance; but the latter sentiment soon gained the mastery.

Moore, unwell as he was, could not maintain the exertion

* According to Herr Radde (who is an eminent botanist), this phenomenon of the vegetable world is thus produced:—' The frosts of autumn kill down the summer's growth, and leave it rotting on the ground; the rich soil formed by its decay is covered by the winter snows, often to a depth of thirty feet. As spring advances, the water of the melting snow percolates the ground, and when it is at last laid bare to the warm rays of a Caucasian sun, the herbs spring from the saturated soil, as from a hotbed.' Whatever may be the cause, it is certain that the Zeuro-Squali would gain a prize for weeds anywhere.

without repose, and we left him to follow at his leisure in our
trail, which once made was far too broad to be missed.
Treading down ruthlessly under our feet alike the dense
masses of hemlock and the tall spikes of gorgeous tiger-
lilies, we slowly drew near to the junction of the two glens.
In the thickest of the wood we came upon a track which
seemed somewhat too broad for that of a man; it led us to a
hollow trunk, the home of a bear; but the brown gentleman

Source of the eastern Zeun-Uquall.

was out for his afternoon's stroll, and we had not the pleasure of making his acquaintance, although we valiantly
took the covers off our ice-axes, and got ready our revolvers,
in case of a chance encounter with a cub. We were never
lucky enough to see a bear, except in captivity, while in
the Caucasus: indeed, we scarcely saw any wild animals,
much to our surprise, as we had been told that bears,

bouquetin, and chamois abounded on the higher mountains. It must be remembered that we never went out of our way to look for game; of its existence there can be no doubt, and this very glen is a favourite hunting-ground with the inhabitants of Laschketi, the highest village in the valley, who come here in winter on snow-shoes. They form a party, consisting of as many as forty or fifty hunters, surround a large tract of country, and drive the game together. In this way thirty-one bouquetin were killed in one day in the winter of 1863-64.

We halted where the valleys met on a large level meadow, of course covered with a crop of tall-stemmed, broad-leaved herbage, on the banks of the eastern branch of the Zenes-Squali. The valley is closed by a rocky cirque, the centre of which is occupied by the Lapuri glacier, terminating abruptly on the edge of a cliff, down which the stream makes its way in a bold leap. After waiting for more than two hours without seeing or hearing anything of the porters, whose figures had been visible on the skyline long after we had reached the bottom of the glen, we reflected that we should be unable to push any further that evening, and that it would be as well to make what preparations we could for our bivouac. A spring of clear water, which burst out of the stony channel of the Zenes-Squali, served to fix the position of our camp, and we proceeded to cut down the herbage, dig up stones with our ice-axes, and level the inequalities of the soil. We finished our work, and still no porters arrived; at last, after we had waited for them four hours, our shouts were answered, and we distinguished the train rambling leisurely along on the opposite bank of the river. We called to them to recross the stream, to do which they were obliged to extemporise a bridge by throwing a fallen log across it. No rain had fallen during the day, and our camp was consequently more enjoyable. Paul, with the aid of our little kitchen, prepared us a capital dinner of soup, cold mutton, broiled ham, and tea, which even Moore—who had by this time rejoined us, and recovered from his sickness—was able to enjoy. During the night we were pestered by swarms of mosquitoes, and small but very venomous black flies, which, despite all our endeavours, found their way into the tent, and most effectually murdered sleep.

July 15*th*.—The valley was filled with clouds, which threatened rain before the day was much older. We packed up our tent, and, after the usual delays, got off about 7.15 A.M. At once crossing the river, we struck the bed of a small stream descending from a hollow in the range on the west side of the valley. The stones in the channel were of no great size, and we made comparatively rapid progress, ascending gradually until the woods on either side thinned, and we found ourselves in a recess surrounded by steep but not lofty ridges of a loose shaly rock. The deeply-indented gap, through which we must pass to reach the valley of the western Zenes-Squali, was now clearly visible in front; the ascent to it was at first up an exceedingly steep grass-slope utterly pathless, and for the last 150 feet by a narrow trough. How Herr Radde's guides can by any possibility have succeeded in getting horses over this pass we could never understand, although the feats performed by Caucasian steeds are truly marvellous, whether on snow or rock.

When, in two hours from our camp, we reached the summit of the Nöschka Pass (8,460 feet), we found that there was a snow-filled trough on the farther side exactly similar to that by which we had ascended. The view must at all times be limited by the higher ranges close at hand, and now even the summits of these were concealed by clouds. The porters again pretended to think the direct descent too steep, but, with the results of the day before as an argument in our favour, we were not disposed to let them have their own way, and insisted on their taking advantage of the route nature had provided in the snow-gully at our feet. Having started the whole troop—who descended with an air of the greatest trepidation, and screamed with fright when one of us above dislodged a small stone which fell amongst them—we followed ourselves by a rapid glissade, but, warned by previous experience not to leave the porters, we accommodated our pace to theirs, which was in consequence somewhat improved.

A long wooded slope led down to the western Zenes-Squali. The herbage was as rank and the woods as thick as on the day before, but we avoided some of the fatigue by allowing the whole train of twelve men to march before us, and taking advantage of the trail thus formed. We got on at a very fair pace down to the left bank of the stream, into which a pretty waterfall tumbled from the opposite hillside. The skirt of the Maschquar glacier, from which it takes its rise, was visible under the clouds at the head of the valley. The whole afternoon, from 12 till 6.15 P.M., was spent in forcing our way down the wildest valley we met with in the whole course of our wanderings. There was no trace of path, so, following the custom of the country, we clambered for some distance over the boulders in the channel of the river, and when this was impossible, forced our way through the virgin forest which lines its banks. Dense thickets, prostrate logs, and swamps into which we sank deep at every step, were the leading features of the walk, while the same luxuriant vegetation which we had before encountered was everywhere remarkable.

Heavy rain now began, and continued to fall for the rest

of the day. Our guides seemed confident in their knowledge of the right direction through this wilderness, and tramped on with praiseworthy perseverance, diverting the tedium of the march, sometimes by raising a wild monotonous chaunt,* led by one man, with a refrain taken up in succession by his companions—sometimes by excursions in quest of the stalks of a huge umbelliferous plant, for which their appetite seemed insatiable. Each man must have cut and peeled for himself several pounds of this juicy but tasteless vegetable food in the course of the day. When it became necessary to cross the river, the water was too deep and violent to be forded, but a young tree was soon felled, and laid across to enable us to pass. The valley, the configuration of which is most incorrectly represented in the Five Verst Map, now broadened out, and a cirque crowned by snowy peaks and some small glaciers opened on the right.

A densely-wooded spur projected from the main chain, turning the course of the valley we were following more directly south, and separating it from the glen of the Scena. The scenery here is probably very striking in fine weather. We passed a hunter's lair sheltered under a bank, and soon afterwards noticed the ruins of a tower rising out of the dense forest, and affording a proof that these solitudes have not always been so deserted as they are now. On our right, deep channels were cut through the friable soil by the glacier-streams, the passage of which cost us a good deal of time and trouble. We had now attained a height of from 500 to 800 feet above the river, and a bend in its course enabled us to look down to its junction with the Scena, the point fixed on as the probable limit of our day's walk. The slopes we were traversing

* Herr Radde has been at the pains to collect and translate many of these songs, which seem to possess more meaning and merit than would be imagined by a person hearing, for the first time, the succession of gutturals and uncouth exclamations of which they consist.

were exceedingly steep, and it was an immense relief when our porters happily hit on a faint hunter's trail, which, though frequently lost, was always recovered after a slight delay. The channel of the river beneath us was narrowed into a gorge, and the opposite mountain-side was even steeper than that we were laboriously traversing. As we drew near to the angle of the mountain, which projects over the confluence of the Scena and the western Zenes-Squali, pines mingled with the deciduous trees, and lower down their gigantic cones of sombre foliage clothed, from top to bottom, the sides of the tremendous ravine into which the valley contracted.

The scene, which revealed itself, bit by bit, through the breaks in a dense veil of mist, was one of the most savage of its kind imaginable, and totally unlike anything I had ever seen, except in some of the mountain landscapes of Gustave Doré. Meantime the rain fell in merciless torrents, which even the thickest pine-branches could only partially keep out. It seemed as though we should never reach the entrance of the Scena valley, but at last the corner of the mountain was turned, and we began to descend; the forest grew thicker, and a few hundred feet above the stream, we found a group of pines so dense that a patch of ground beneath them was still fairly dry, and promised to afford our men a better resting-place than they could have hoped for. The first thing to do was to set up the tent as quickly as possible, no pleasant or easy task, when the ropes and strings were all in a soppy condition. Once inside, we tried to put on dry clothes, the waterproof saddlebags having gallantly withstood the rain, and preserved their contents from wet. Our tent was always small, and the sloping sides, which in fine weather could be stretched taut, were apt when wet to flop heavily against our faces whenever we attempted to sit up, in a way calculated to test the temper of even a Mark Tapley. This inconvenience much hampered the proceedings of the two outsiders, one of whom expressed in no measured terms his disgust at the situation, and his wish that he was enjoying the creature-comforts of Pätigorsk. The man in the middle saw things in a more cheerful light, and the spirits of all were a little raised by the

Our Camp-fire in the Forest.

arrival of Paul with some broiled ham and tea. The porters had made a roaring fire in the centre of the dry plot, and the glimpse of the group of picturesque peasants clustered round the blaze, and the dark background of pines, revealed to us as our tent-door was thrown open, was enough to repay us for all the disagreeables of the day. Wrapping ourselves round in the driest folds of our rugs, which shared in the general humidity, we composed ourselves for the night, in hopes that the rainstorm of the day had been too heavy to last.

July 17th.—The morning was cloudy, but not actually wet. Our porters assured us that we should arrive early in the afternoon at Jibiani, and, in the still unsettled state of the weather, we looked forward with pleasure to the prospect of again sleeping under a roof. We had now to mount the valley of the Scena, the most western of the three principal sources of the Zenes-Squali. Our course lay at first almost due west, but after a time the valley bent round to the north, and we began to ascend rapidly by the side of the stream, which foams at the bottom of a deep and narrow cleft. For the first half-hour we were in the forest, which was composed of noble pines, though the trees were not equal in size to those on the further side of the river. A narrow path, always ill-defined and in places scarcely traceable, led across meadows of the rankest herbage, gay with subalpine flowers, lilies, lupins, and vetches, which showered down heavy drops on us as we passed. The ruins of a village, or some huts, could be distinguished amongst the trees on the opposite side of the valley. To our surprise the path grew more distinct the further we went, and gradually assumed the character of a sledge-track. It led through thickets of underwood, and over much marshy ground, the source of numerous springs, which hurry to reinforce the torrent, here leaping noisily over the granite boulders it has brought down with it from the central chain.

When we reached the level ground at the actual head of the valley, we found that the bridge over the Scena, which the peasant who acted as guide assured us had existed when he last made this journey, was now no more, and that we must make the best of our way through the water. The stream proved fordable, but the strength and speed of the current were so considerable, that some care was needful to avoid missing one's footing amongst the boulders. The scenery at the source of the Scena must be very imposing in clear weather. The usual clouds cut off all view of the summits, but did not conceal the twin icefalls of the Korüldä glacier, which, unlike the other glaciers of the Zenes-Squali, survives its fall, and re-makes itself at the bottom of the glen. It is surrounded by abrupt snow-streaked cliffs of great height. The track, now broad and distinct, turned sharply up the steep western hillside, and raised us to the verge of a wide upland pasturage, surrounded by comparatively low ridges. The slopes were bare of trees, but covered with rhododendron-bushes, and with a beautiful herbage, short in comparison to that found lower down in the valley, but capable of feeding immense herds of cows and goats. It was the sort of scene which in the Alps would have been enlivened by numerous châlets, but here there was nothing of the sort, and François, who longed to transport the whole hillside within reach of Chamonix, was loud in his lamentations over the shortcomings of a population who could allow such natural riches to run to waste.

The stream which waters these pasturages bears the name of Lastilagel. It has three sources, and there is more than one way of crossing from its basin to the headwaters of the Ingur. We descended somewhat, in order to cross the main stream, which flows in a deeply-cut channel, and having made our midday halt on its banks, climbed a long and steep zigzag, from the top of which the path bore away at a level, on the left of the most southern source of the Lastilagel. The snow

had apparently melted very recently, and the ground
was saturated with moisture. The stream we had followed
is nourished by the springs of an upper level of pasturages,
rising to a broad saddle which forms the Naksagar Pass
(8,813 feet)—the watershed between the Ingur and the
Zenes-Squali. The pass itself is so broad and flat at the top,
that it is difficult to tell the exact moment when the summit
is reached. Round grass-covered hills shut in the view on all
sides; the sledge-track goes downhill, at first very gently,
afterwards more rapidly, but there is no point where the
descent can be called steep. The stream, which rises on the
west side of the pass, and joins the Ingur at Jibiani, is
called the Quirischi. The path, becoming broader and
more beaten as it draws nearer the village, clings to the
slopes on the right bank of the torrent, which is joined by
another flowing through a short glen from the steep and
jagged flanks of the Ugua.

Signs of an inhabited country now followed one another
in rapid succession. Large herds of heifers were feeding
on the slopes, the projecting knolls were crowned with
stonemen, and we passed presently a hut near which was
a cluster of women and boys, wilder and more unkempt-
looking specimens than any we had yet seen. A tall tower,
a portion of a now ruined castle (said to have been built
by Queen Thamara), appeared perched on a commanding
knoll on the left bank of the stream, and gave us the first
warning of our approach to Jibiani. Our Gebi porters,
instead of seeming anxious to finish their job, took every
possible occasion to loiter on the road, and we vainly endeavoured to incite the slow unwilling train to a final spurt.
Our entrance to Jibiani, and the commencement of our
Suanetian experiences, will be best placed at the beginning
of a new chapter.

CHAPTER X.

SUANETIA.

Free Suanetia, Past and Present—Herr Radde's Experiences—Physical Features—Fortified Villages—Jibiani—Pious Savages—A Surprise—Glaciers of the Ingur—Petty Theft—Threats of Robbery—Alarms and Excursions—A Stormy Parting—The Horseman's Home—The Ruined Tower—A Glorious Icefall—Adisch—Sylvan Scenery—The Mushalalis—Suni-Ups and Downs—Midday Halt—Latal—A Suanetian Farmhouse—Murder no Crime—Tau Tötönal—A Sensation Scene—The Caucasian Matterhorn—Pari at last—Hospitable Cossacks.

SUANETIA is the general name bestowed by geographers on the upper valley of the Ingur, and is derived from the inhabitants, who from very ancient times have been called the Suani, or Suanetians. This people is not, however, entirely confined to the valley of the Ingur, as many of the higher villages on the Zenes-Squali are occupied by the same race. Their inhabitants are now distinguished from their neighbours round the sources of the Ingur, as the Dadian's Suanetians, from having been subject to a native prince who bore the title of Dadian. His authority, or that of other members of the same family, extends over the western portion of the Ingur basin; but the groups of hamlets, which cluster thickly in the network of glens containing the sources of the stream, are at the present time independent, and are known as Free Suanetia.

Since Russia has succeeded in converting her long nominal suzerainty over the Caucasus into real dominion, the native princes have naturally been treated by her simply as landed proprietors with certain manorial rights. Constantin Dadisch-Kilian, the Suanetian prince, resident at Pari, was about eight years ago suspected of some intrigue, and was in consequence summoned, by the Governor of Mingrelia, to meet him at Kutais. He obeyed the summons, and was told that he must leave his home and live for the future in Russia. High words ensued; the Russian officer was firm, the Prince grew violent, and finally, drawing his dagger, stabbed and killed the Governor. He escaped for the moment, but was ultimately taken and shot at Kutais. His former residence, Pari, was selected as the Russian military post in Suanetia, and for a short time the whole district was kept under control by a considerable force. The expense and difficulty experienced in carrying out the unprofitable task of preserving order in this mountain fastness appear to have disgusted the Government, which probably thought that if the free Suanetians were left to fight out their quarrels, the race would, like Kilkenny cats, soon be self-exterminated. Whatever the motive, the troops were withdrawn, and ten Cossacks, stationed at Pari, are the entire executive force at the disposal of the chief of the district, and the upper or western valleys are, for all practical purposes, independent, and at full liberty to follow their own wicked ways of theft and murder, to their hearts' content.

This is, I believe, a correct description of the present state of the country. Its past history is obscure and complicated, and I cannot pretend to have made any very deep researches concerning it. Suanetia seems usually to have been united with Mingrelia, but at times to have been treated as a province of the Imeritian kingdom. The internal disorder to which Imeritia was the prey, and the weakness of its rulers, aided the Suanetians in establishing their independence.

At the end of the fourteenth century, they made a successful foray into the Radscha, and burned Kutais; worsted in the field, they were again compelled to submit to Imeritia, and a prince was imposed on them, whose usual residence was on the Zenes-Squali. During the fifteenth century, we again find the Suanetians at war with their southern neighbours. After ten years' hard fighting, they were forced to surrender the Upper Rion district, as the penalty for the murder of an Imeritian prince. Traces of their former connection with Suanetia may still be recognised in the style of building and the towers of Chiora and Gebi, and their inhabitants are still looked upon by the population of the Radscha as a foreign race. By degrees the people about the Ingur sources established their independence, but members of the princely family, known by the title of Dadisch-Kilian, were, at the time of the establishment of the Russian dominion, still regarded as the feudal chiefs of the lower villages on the Ingur, and, as such, took the oath of allegiance to the Czar.

Ethnologists seem to have come to the conclusion that the Suanetians are a branch of the Georgian family, and a study of their language has convinced Herr Radde that it has much in common with the Imeritian and Mingrelian dialects. When it is remembered that in the eleven upper communities on the Ingur, after the successful assertion of their independence, a fugitive from the lower country could obtain not only immunity from punishment for past offences, but also personal liberty, and freedom from princely exactions, it will not be thought wonderful that the population of this district at the present day bears marks of a mixed origin.

Thus much for the history of Suanetia. As to the character of the people, I shall quote Herr Radde, the latest traveller in this country, and the only one, I believe, who, aided by all the facilities the Russian authorities could give him, and by his knowledge of the native dialects, has set himself seriously to work to study the customs and manners of this sequestered mountain-tribe. This gentleman arrived at Jibiani in company with a native priest, who served as his introducer to the villagers. He found them engaged in hostilities with the neighbouring hamlet of Murkmur, and men wounded in the skirmishes, which were of constant occurrence, were brought in from time to time during his stay. The Herr was here robbed of a horse, which was only recovered after much trouble. To avoid the scene of battle, instead of descending the valley, he made his way across the mountains, and slept in the open air, in order to pass through the village of Adisch by night, on account of the ill-name borne by its inhabitants. At Pari, the former residence of the native princes, and present post of Russian Cossacks, he stayed for some days, collecting information as to the language, ballads, and customs of the country. The result of his experiences and researches he sums up in the following words:—'Amongst the Suanetians intelligent faces are seldom found. In their countenances insolence and rudeness are prominent, and hoary-headed obstinacy is often united to the stupidity of savage animal life. Amongst these people, individuals are frequently met with who have committed ten or more murders, which their standard of morality not only permits, but in many cases commands. They are of a taciturn disposition, and their manner when endeavouring to impose upon strangers is most disagreeable.'

I may add to this the opinion of Malte Brun, who says of the Suanetians:—'Nothing can equal their want of cleanliness, their rapacity, and their skill in making weapons. We may consider the Pthirophagi, or eaters of vermin, who according to Strabo inhabited this country, as the progenitors of the Suanes.' No character can be more accurate, only that at the present day the relations of the vermin and the population have been reversed.

The nature of the country has no doubt had a great share in forming the savage and wild character of its inhabitants. A large basin, forty miles long by about fifteen broad, is shut in on all sides by glacier-crowned ridges, and the only access to it from the outer world is by means of a narrow, and at times impassable, ravine, or over lofty mountain-passes. The main chain of the Caucasus forms its boundary on the north, and this reaches its greatest elevation and true central point in the huge glacier-seamed, peak-surmounted wall which towers over the sources of the Ingur. Tau Tötönal (or Tetnuld) must be over 16,000 feet, and the summits of the serrated range, which stretches from it to the east for several miles, do not average less than 15,000 feet in height. Three glaciers, the Nuamquam, the Goroscho, and the Adisch, pour down from this wall into three separate glens. They descend to a level of about 7,000 feet, which may be taken as the lowest point reached by glaciers on the southern side of the Caucasus. The chain between Tau Tötönal and Uschba (called also Besotch-Mta by Radde) makes a semicircular sweep to the north, and at least two considerable glaciers, the Gatun Tau and the Thuber, descend from it into the Mushalaliz branch of the main valley. Uschba itself is a gigantic promontory, standing out between the glens of two of the northern tributaries of the Ingur; like so many others of the great peaks, it does not seem to be on the watershed, but it is the only one I know that is on the southern side.

A long lateral ridge, forming the western boundary of the Nakra valley, through which one of the best-known passes leads to the northern side of the mountains, runs out at right-angles to the central chain, and forms the limit of Suanetia on the west. Its spurs are separated from those of the Leila mountains, which intervene on the south between Suanetia and the lower country, only by the deeply-cut and densely-wooded gorge through which the Ingur makes its escape. The Leila chain is of considerable height; several of its peaks exceed 12,000 feet, and the glaciers on its northern flanks are by no means despicable. Its formidable barrier runs in an unbroken line along the south bank of the Ingur, and is connected at the source of that river with the main ridge of the Caucasus, by a grass-covered range (crossed by us in passing from the Zenes-Squali to Jibiani), which thus completes the circle of mountain barriers with which nature has fortified this region. The topography of the interior of the Upper Ingur basin is exceedingly complicated, and can only be understood by careful study of a map. The stream of the Ingur flows generally along the foot of the Leila range; the country to the north, between it and the main chain, is divided by spurs, none of which attain the snow-level, into ravines and meadow-basins, through which flow tributary streams coming from the glaciers above. These are of a greater size, and descend lower into the valleys, than anywhere else on the southern side of the Caucasus.

The foregoing description will have prepared my readers for the character of the people whom we are about to encounter, and, if read with a map, may assist those who care to follow our wanderings through the intricacies of the Ingur sources. We had not had time to study attentively Herr Radde's volume before leaving Tiflis, and had no reason to anticipate a worse reception or greater difficulties than we had previously met with; so that, wearied out with the dawdling ways and monotonous chants of our Gebi men, we had looked forward with some pleasure to dismissing them, and making a fresh start. We hoped to spend several days at Jibiani, and to make it our headquarters for the exploration of the surrounding mountains. The Five Verst Map showed us that behind the watershed of the main chain, situated on a northern spur, stood two great peaks, Koschtantau and Dychtau, respectively 17,096 and 16,925 feet in height—both therefore higher than Kazbek; and our object was to gain the watershed, at some point whence we might enjoy a view of these giants, and of the glaciers surrounding them. It was therefore with feelings of pleasure, unmixed with any apprehension, that we hailed our first glimpse of Jibiani, Tschubiani, and Murkmur, a community known collectively by the name of Uschkul.

Most of the villages in Suanetia are in clusters of two to four, and go by a collective name, distinct from the individual appellation of each knot of houses. Adisch is, I think, a solitary exception to this rule. Jibiani and Tschubiani are built on the projecting brow above the junction of the Quirischi with the infant Ingur, which has here run but a few miles from its cradle in the glaciers of Schkari and Nuamquam, at the base of the great chain. Murkmur is a little lower down, and on the opposite or right bank of the united torrents. The appearance presented by these hamlets was most strange and picturesque. The meadows at our feet were dotted by an array of stone-built towers, irregularly grouped—some of them white, but the majority of various shades of dinginess. In the three villages, all of which were in sight at the same time, there cannot have been less than sixty towers. The only comparison which will give an idea of the appearance from a distance of these fortified Suanetian villages (for they are all alike), is to picture a group of square-sided armless windmills, closely crowded together, and surrounded by low stone-built barns with sloping roofs. The situation of Jibiani is not striking, when the peaks of the Nuamquam are veiled in clouds. The slopes above the village are rounded and bare, or only partially clothed with low brushwood, and the traveller fancies himself carried back to the valley of the Ardon or Terek.

Our porters volunteered to introduce us to the inhabitants, and to aid us in obtaining some kind of lodging. On first entering the place we did not find many of the house-owners at home. The first barn that was offered us we declined, on account of its gloomy and dirty appearance; a second was then shown, which was a shade cleaner, although scarcely less gloomy, owing to the extraordinary smallness of the loopholes which served as windows. It had, however, the advantage of a smooth plot of grass outside the door, where we could sit in the sunshine, when there was any. Our parting with the Gebi porters was of a very friendly character. They were in a hurry to start on their return, so we paid them at once, giving each man a trifle over the contract price, which, after the numerous differences we had had on the road, was more than they expected. Though lazy and stupid, they were free from more active vices, and were, in fact, far more of fools than knaves. Having purchased some provision for their return march, of the high price of which they loudly complained, the ten set off the same evening, unwilling apparently to trust themselves a minute longer than they could help to the friendly disposition of the Jibiani populace.

A crowd of villagers had by this time collected on the green outside our quarters, and formed a circle round us. We were struck at once with the wholly savage aspect of the assemblage, especially of the children, who pressed to the front to stare at us. The men, and

even boys, were all armed with daggers; many also had
pistols attached to their belts, or guns, in sheepskin covers
slung across their shoulders. Their clothes were far
shabbier and more tattered than those of the peasants of
the Rion valley; the ordinary Caucasian type of costume
was still distinguishable, but the coats were often sleeve-
less, and the headpiece was nothing more than a bit of
rag tied into the form of a turban. Some of the men wore
sheepskin caps turned inside out, a peculiar arrangement,

A Native of Jibiani.

which at the same time shaded their eyes, and added to
the uncouth ferocity of their appearance. The women
were uniformly ugly, and their dress presented no peculiar
character to attract attention; it was simply a shapeless
bundle of rags. The children were wild-looking raga-
muffins, with matted locks, and ran about half-naked, clad
in one tattered garment of old cloth or sacking; some of

the girls had the most savage faces, more like brute animals than human beings.

We told Paul he must set to work to get us some dinner, and for this purpose it was necessary to purchase food of the villagers, as we had eaten up the sheep killed at Gebi. We were dismayed at being refused milk or cheese—butter is unknown in this part of the country—on the very unexpected ground of its being a fast-day of the Church. Jibiani was scarcely the place where one would have looked to find the outward forms of religion scrupulously observed, and the rule of fasting seemed to be peculiar, as we were allowed to purchase fowls and eggs. There was no fireplace or chimney in our barn, so, to avoid filling the place with smoke, Paul endeavoured to do his cooking in an open shed close by. Each little purchase, such as eggs and firewood, had to be paid for separately, and at once; no change could be obtained, and although we were, fortunately, fairly supplied with small ten and twenty-copeck pieces, it was often difficult to make up the exact sums called for. As a rule, in the Central Caussasus, all the natives, though preferring silver, will take the Russian paper-money, only the notes must be new; if in the slightest degree torn, they are, in nine cases out of ten, absolutely refused.

Paul was much hampered in his movements by the crowd of stupidly curious men and inquisitive children, and we agreed that in future cooking must be done inside the barn, even at the expense of a little discomfort from smoke. The furniture of our quarters consisted of a long wooden bench, and a layer of hay in one corner, the thickness of which was, at our request, doubled by a fresh importation. There was no bolt to our door, and as long as daylight lasted we were more or less troubled by visitors, who dawdled in and out, and stood by the half-hour, gazing at our proceedings with an air of absolute stupidity which was provoking to witness. When we got rid of them for the night, we drew the bench across the door, piled our baggage in the corner close to us, and with our revolvers under our heads dropped off to sleep.

July 18th.—François, who was the first to go out in the morning, came back with the intelligence that there was 'quelque chose à voir.' On our following him out of the barn, and looking towards the head of the valley, where on the previous afternoon nothing but clouds had been visible, our eyes were greeted by an enormously high mass of rock seamed with snow and ice. Over breakfast we held a consultation as to our arrangements, and agreed that, at all events, the day must be spent at Jibiani, and that we might walk up towards the sources of the Ingur, and gain a view of the glaciers from which it rises, while Paul and François were employed in cleaning and mending the parts of our equipment which had been injured in our journey from Gebi, and in bargaining for and cooking a sheep. On a knoll above the village stands the church of Jibiani, which, like all those in Suanetia, is a low square building without tower or belfry. Ruddle met with some opposition when he proposed to enter it, and we did not attempt to do so. According to him, the interior contains a collection of the horns of chamois and bouquetin, two crosses on either side of the altar, and some remains of rude frescoes on the walls.

The path, after mounting for some little distance on the left bank of the stream, crossed it by a bridge, from which there was a picturesque view down the cleft through which the water finds a channel. On the way we passed herds of cattle, all bullocks, and families of lean pigs, wandering about the hillside. The head of the valley is occupied by a wide bare pasturage, above which the central chain rises in a gigantic wall, supporting two glaciers, the Schkari and the Nuinnquam. An hour's walk above the village was sufficient to give us a perfect view of this great mountain 'cirque.' The prospect from a scenic point of view was superb, but offered very little encouragement to a sober-minded mountaineer. Opposite us the range was crowned by a massive rock-peak, and the lines of icefall and precipice on both sides of it seemed equally inaccessible. If anyone ever gets over to the north side of the chain from this point, it will be by a route as unpromising at first sight as, and probably more difficult in the passage than, the face of Monte Rosa above Macugnaga; indeed, the Italian side of Monte Rosa is the only mountain-wall in the Alps on a scale to vie with this part of the Caucasian chain. Finding the clouds were already covering the summits, and that we should gain nothing by pushing our researches further, we lay down on the turf for some time, and gave ourselves up to the enjoyment of the double luxury of fine scenery, and freedom from persecution by inquisitive natives.

The reflection that we had all our arrangements to make for our journey to Puri, at the further end of the valley, and that our men had better not be left alone with the villagers longer than necessary, somewhat hastened our return. We found Francois, as usual on a rest-day, immersed in the cares of the laundry; Paul was in a state of just irritation at the proceedings of the people, but had made considerable advances towards an arrangement which seemed likely to render our further journey easy. A man had come to him, and offered to provide, by the next morning, two horses to carry our luggage. This worthy's name sounded like Islam, and he recommended himself as the inhabitant of a lower and less barbarous village, as having been for some time in the service of a native prince, and as being now in receipt of a pension for services rendered to the Russian Government. We were only too glad to secure his assistance, and willingly ratified the bargain entered into by Paul.

During our absence, the tent, our mattress, and other articles, which were still damp after the rainstorms of the Zenos-Squali, had been put out to dry in the sunshine. The villagers took advantage of the numerous objects which divided the attention of our men to commit sundry petty thefts: some English string, a couple of spoons, a stray volume of Tennyson, and a tent-pole, made up the list of our losses; the latter was the most important, as, from its peculiar construction, it would have been impossible to replace it.

During the afternoon, the behaviour of the crowd on the grassplot round our barn became more and more unpleasant; familiarity was evidently producing its proverbial result, and we began to wonder what would be the upshot, and whether the success of petty theft would encourage attempts at open robbery. For some time we were amused in watching the athletic sports of the juvenile portion of the population, who were enjoying themselves on a piece of level ground on the opposite side of the sunken lane which led up to our barn. The popular game seemed to be for one boy to seize another's head-gear, and retain possession of it, by flight or struggles, as long as possible. Girls as well as boys took part in the amusement, which was of a very violent and noisy description, and was at times enlivened by a general scrimmage, which reminded me of a 'rouge' in an Eton football match. Unfortunately, the game was too exhausting to be long continued, and when the players joined the group, already sufficiently rude and troublesome, which surrounded us, we found it necessary to carry in all our possessions, and to retire ourselves into the shelter of our barn, unless we wished to incur further loss, and to submit to a close overhauling of our own persons, and such things as we carried about them. When the populace understood that there was nothing more to be seen outside, they came to the natural conclusion that they had better follow us in, and we soon found ourselves sitting in a knot in front of our possessions, and closely pressed upon by a growing crowd—some of the people simply sucking their thumbs and staring in stupid astonishment, while others, more lively, pointed in our faces the finger of covetousness, or of scorn, as the case might be. These persecutions grew too troublesome to be borne without a protest, so, wanting to eat our supper, we called the owner of the barn, and told him that we desired to be left alone, and that the crowd must and should turn out. Finding that we might repeat this sentiment as often as we liked without result, we took active measures, walked the people out before us, and shut the doors. After some talking and jeering outside they were violently kicked open. We again shut them, but, expecting the offence would be repeated, I waited close by, and sallying out unexpectedly, caught a boy in the act; the culprit was summarily collared and shaken, whereupon he made feeble demonstrations with his pistol, but took care for the future to keep in the background. Two or three men, from time to time, took opportunities of intruding themselves, but at last, having ordered the horseman to be ready for an early start, we succeeded in shutting ourselves up for the night. Before doing so, however, Moore went outside, fired off the five barrels of his revolver in rapid succession, and then ostentatiously reloaded it—a demonstration which produced for the time all its intended effect, for, although sounds of talking and wrangling were audible, we were free from further annoyance.

July 19th.—After the turn things had taken on the previous evening, we made up our minds that we could scarcely expect to get away without a dispute, and we held ourselves in readiness for what might occur. We were up early, and found, to our disgust, that it was raining heavily. Presently the horseman appeared, and announced that he could not get a second horse, and that we must hire porters to carry part of our baggage. Unwilling to place our power of effecting a start at the mercy of Jibiani men, as we must have done had we consented to this proposition, we determined to carry the extra saddlebags on our own shoulders. It was now suggested that we might get back the tent-pole by a small payment; the old story of the London dog-stealer was reproduced, and we were asked to believe that it had been found by a native of the next hamlet, Tschubiani, who would be happy to restore it, for a consideration. We consented to place the sum demanded in the hands of the horseman, the only one of the crowd over whom we had any hold, and, after some parley and delay, the missing stick was returned. Meanwhile we were assailed on all sides by clamours for money: one man wanted a ridiculous sum for some loaves he had brought us, another asked their weight in Russian paper for his eggs, a third had a large bill for firewood, and the master of the barn required an extortionate price for our lodging. Some of these demands we resisted, others we partly conceded—at the same time finishing as quickly as possible the packing of our saddlebags, and taking care to keep them under our eyes.

The conduct of our horseman caused us much uneasiness, as he refused to put our luggage on the horse, pretending that the villagers would not allow him to do so until we had yielded to their claims. At last, chiefly by our own exertions, the horse was loaded, and then, having paid everyone who had any fair claim, we agreed to make a decided effort to start. One of us was to lead the horse, for it was evident its master could not be relied on; the others were to carry saddlebags, and keep together as much as possible. Lifting our luggage on our shoulders, we prepared to leave the barn in a body, but our two men foolishly loitered, to make sure that nothing was left behind. The natives took advantage of the blunder, and immediately shut them in: looking round, we saw the state of the case, and ran a tilt, with our ice-axes, at the wooden doors, which were rudely constructed, divided in the middle, and opened inwards; the blow sent them flying back at once. François, who was close by inside, endeavoured to come out, when a peasant put himself in the way; but I suddenly brought the cold barrel of my revolver into contact with the scoundrel's cheek, on which he retreated hastily. François escaped, and Paul was allowed to follow him. Once more united, and forming a kind of square, with the horse in the centre, our saddlebags on our shoulders, and our revolvers in our hands, we descended into the hollow lane which led out of the village. Some of the inhabitants, yelling and jabbering, jumped down in front to bar the way; others brandished swords, daggers, and pistols on either wall; a few ran off, making signs that they would fetch their guns; while the women, screaming and endeavouring to restrain the fury of their relations, added by their wild cries and gestures to the confusion of the scene.

Whether their interference was due to any kindly feeling towards us, or to a fear lest our revolvers should make victims of their friends, we never knew. The crisis was really serious, and a peaceful solution seemed almost hopeless, when a trifling demand, screamed out by a man on the right-hand wall, suggested to us an imitation of our predecessor Jason's policy in the same country. We scattered our dragon's-teeth, in the shape of two or three small copeck-pieces, among the group, and our foes began to scramble and squabble; their attention being for a moment diverted, we pushed on as rapidly as possible, and before they had recovered their surprise at our sudden move, were clear of the village. A portion of the crowd came in pursuit, but two of us faced round in the narrow path, and brought them to a halt until the horse had gained a slight start, when we followed it. We passed hurriedly through Tschubiani, where most of the inhabitants seemed to be out, or amongst the Jibiani crowd. The owner of the horse had rendered us no assistance, and was now loitering somewhere out of sight; the villagers, who followed us, motioned us to halt, but we kept straight on, and having crossed the bridge, passed underneath the houses of Murkmer, and along the bank of the stream. We were now in open country, and might consider ourselves fairly out of the clutches of the men of Jibiani. Paul told us, that when he was released from the barn, the villagers said to him, 'If it was not for those wonderful pistols of yours, we would have tied you all up, and taken everything you had,' and there is no doubt that our revolvers alone saved us from open robbery. The knowledge that you have fifteen barrels at your disposal has a moral effect even on the most barbarous race. The difficulty lies in enforcing the impression while keeping clear of actual fighting. Had a shot been fired, we must inevitably have lost our luggage, and, considering the odds against us, might have had great difficulty in effecting our own escape.

The horseman now came up, of course professing entire

ignorance and innocence as to the whole proceedings. With him were two natives of Davkar, a lower village, who had volunteered to carry our saddlebags; but the sum they asked was so ridiculous, that we had declined their services. Now they would have been very glad to relieve us of our loads for a quarter of the payment previously demanded.

Near Murkmer there is a good deal of cultivated ground, but the valley soon narrows into a defile, and the well-beaten path attains a great height on its northern side. The loose soil of the mountain-slopes has been worn by the spring-torrents into deep gullies, out of, or round, which we were frequently forced to make long ascents or circuits. The scenery gradually improved, and, although no snowy ranges were in sight, the clumps of fir and other trees which dotted the sides of the gorge, and the frequent bends in the deep bed of the stream, combined to form a varied and interesting series of landscapes. The rain of the early morning had ceased, and the sun now shone out hotly between the clouds, lighting up the long reaches of the Ingur, which wound along the bottom of the green gorge. The second group of villages, known collectively as Kal, now appeared before us, built on opposite hillsides, above the junction of a northern tributary with the glen of the Ingur. The towers were not quite so numerous as at Uschkul, but were still sufficiently so to give each knot of habitations the appearance of a castle rather than of a village.

On the further side of the valley we noticed a solitary building, which is called on the Five Verst Map a monastery, but we could obtain no certain information about it through Paul, and it was too far out of the way to visit. The nearest hamlet, Davkár, was situated at the foot of the hill on the banks of the river, and we had to make a long descent to reach it. Here our horseman lived, and we hoped by his aid to obtain a second horse to carry on our goods. Having eaten our lunch in a barn similar to our late lodging, we commenced negotiations for our further journey; but though there are plenty of horses in the country, they had all been sent to the upper pastures, and there was not one to be found. Two men, however, were ready to carry the load between them, and we should have had little difficulty in settling terms but for the interference of the horseman. He caused a hitch in the arrangement, and, when we told him to stand aside, snatched his old flint-and-steel pistol from his belt, and brandished it in our faces. Finding that his conduct provoked more laughter than fear, and not exactly knowing what to do with his weapon, he walked off in a huff. There were only about a dozen people in the hamlet, so there was no risk of a row; but we were in considerable difficulty as to how to get forward, and were obliged to condescend to cajole Islam back into a good humour. This Moore successfully effected, and we concluded an agreement that we should be taken in three days to Pari, and should employ two porters until a second horse was met with.

We were surprised to hear that the direct path down the glen of the Ingur was impassable for horses, and that we must mount the valley of the Kalde-Tshalai, opening behind the village, to its head, and then cross a pass to Adisch; but as this route promised to fulfil our purpose of seeing as much as possible of the country, we made no objection to its adoption. The natives proposed to start at once, and spend the night near the head of the valley, where they knew of a shelter. We willingly agreed, and, about 0.30 P.M., again set out, to reclimb the steep path we had descended to reach Davkar, until a grassy brow was gained, high above the junction of the valleys. Here the paths forked, and we followed the one which mounted beside the

Kalde-Tshalai. The scenery of the lateral glen, which runs due north towards the base of the snowy chain, is very grand, the torrent flowing in a deep channel between precipitous walls of rock. At one spot a bridge leaps boldly from side to side; at another, two streams fall in showers of spray into the bottom of the cleft.

Our men did not take us across the bridge to Agran, a village on the opposite slope, probably to avoid any chance of a disagreeable encounter; for they asked Paul not to mention where they came from if we met any villagers, as the relations of Duvkar and Agran were not friendly. We kept up along an ill-marked track until, above the gorge, the remains of a winter avalanche bridged the torrent, and enabled us to join the more beaten path on the opposite side of the valley. The glen widens out near its head, and the pasturages are strewn with granite boulders, transported by ice or water from the central chain. A small ruined tower appeared before us, and was pointed out as our resting-place for the night, and a recess in the hillside on the east marked the course by which a pass practicable for horses leads directly to Jibiani. The large Göröscho glacier pushed its terminal moraine far down into the head of the valley, and we could see through the clouds that the mountain-wall rose precipitously above it. The tower housed our men; we preferred to pitch our tent under the shelter of a low stone wall by which it was surrounded.

July 20th.—The morning was fine, and only a few fleecy vapours hung upon the range at the head of the valley. The cliffs above the Göröscho glacier are very similar in character to those which overhang its next neighbour on the east, the Nuamquam, but the shapes of the peaks which crown them are even more varied and picturesque. The lower portion of the glacier is level, but its upper snows descend in torn and tangled networks of towers and crevasses, which offer little temptation to an assault. Nowhere in the Alps have I seen any barrier which approaches the apparent impracticability of this portion of the central ridge of the Caucasus; between Tau Tötönal and the sources of the Ingur its magnificence can scarcely be overrated. The ease of our start, and the absence of any crowd of greedy peasants, was a pleasant change from the annoyances of the previous day. Our course was clear enough; a double track, which showed that the Suanetians drag sledges over the pass, ascended, in well-turned zigzags, the flowery slopes of the ridge, known by the unpronounceable name of Dschkjtimer, which separates the Kalde and Adisch valleys. Like most mountain walks, the ascent is divided into three stages—a steep climb, then a gentle rise over shelving pasturages, followed by a short pull up to the final ridge.

It took us two hours to reach the top of the pass from the lower. During the last few minutes of the ascent, an apparently lofty snow-peak showed just enough of its head, over the bank we were climbing, to stimulate our curiosity, but in no way prepared us for the magnificent scene which burst into view from the summit. The first thing which fixed our attention was the icefall of the Adisch glacier.* Unlike the glaciers supplying the two eastern sources of the Ingur, which are fed only by the snow lodged on shelves of the cliffs that surround them, the Adisch glacier is the outflow of large reservoirs of frozen snow, invisible from below, and lying at the back of the line of precipitous peaks we had been gazing up at with so much awe and admiration for the last two days.

* Radde mentions a second name, Gstuntau glacier. Is not Gstuntau a corruption of Krschtantau, and do not the snowfields which feed the icefall surround the base of Krschtantau? These are questions for an explorer. The Russian engineers gave up this part of the chain as a bad job, and the Five Verst Map is quite unintelligible.

Over a break in the battlements of this mountain-wall, the ice pours down into the bottom of the Adisch valley, in a fall which, for its height, breadth, and purity, exceeded anything we had seen elsewhere, either in the Alps or the Caucasus. We estimated the height of the frozen cascade at 4,000 feet, or little more than that of the Karagam glacier, up which we had forced our way; but the fall now before us was far more broken, and, in our judgment, absolutely impracticable. From side to side stretched deep-blue chasms, the space between them filled up by a very maze of tottering pinnacles and moated towers. The whole surface was of dazzling whiteness, similar to that of the Rosenlaui glacier, before, disgusted by being treated as a grotto by troops of tourists, it withdrew to the upper world. At the foot of the fall the glacier re-makes itself, and spreads out, with a crimpled but otherwise unbroken surface, into a fanlike tail, the symmetry of which is slightly marred by a projecting hillside. By some strange mistake, the Five Verst Map marks a known pass straight up the centre of the icefall!

Two mountains, worthy of their post, stand like giant sentinels to guard either side of this crystal staircase, let down to common earth from the 'shining tablelands,' untrodden as yet by human foot, which lie in the heart of the Koschtantau group. On the west is Tau Tötönal, an elegant snow pyramid, resting on a broad rocky base; on the east is a rock-peak of somewhat inferior height, but of bolder form. Behind us the Göröscho glacier, at the foot of which we had slept, and the battlemented wall, which stretches away to the eastern source of the Ingur, were still visible. In the glen at our feet, which ran nearly south-west, we could see the towers of Adisch, and we were high enough to command a wide view over the densely-forested ridges of Western Suanetia, to the mountain ranges that encompass it. Had it not been for the clouds, which persisted in haunting us, we must have seen Uschba, and it would be worth the while of any future traveller to examine carefully the horizon in this direction, and make sure whether the dome of Elbruz does not appear above the main chain.

We had followed in the ascent, and now overtook two peasants, a man and woman, who were on their way to Adisch. The man was inclined to be sociable, and was greatly delighted by being allowed to look at the view through our field-glasses. In return we examined his gun, the barrel of which was very long and elaborately ornamented. It was arranged that the woman should descend to Adisch by a short cut, and tell the people there to begin baking some bread for us, in order that we might not have so long to wait for our midday meal. A capital path bearing towards, and consequently affording a constant view of, the glacier, led down into the valley. We followed it across a mountain-side, spotted with the large cream-coloured blossoms of Caucasian rhododendron, which gave place lower down to birch-bushes. We found the torrent too strong to be easily forded, but the friendly native, to whom we had talked on the top, had hurried down before us, and caught one of the horses grazing near at hand, on which each of us in turn rode across behind him. The fall of the glen was very gradual, but a bend to the west soon hid the glacier, and there was for some distance little to remark in the features of the surrounding scenery. The first habitation we came upon was a solitary tower, from which a similar one was visible on the crest of the ridge we had crossed; and it struck us as extremely probable that the ruined building we had camped beside was one of a chain of towers, which had served in olden times as beacons, or fire-telegraphs, from valley to valley. On the opposite side of the stream scattered trees made their appearance, and our attention was attracted to a deep hollow, or cut, in the ground, such as is called a 'graben' in the German Alps, where, owing to the friable nature of the soil, the rains have washed away the surface and laid bare the skeleton of the mountain. Large herds of horses and oxen were feeding on the grass-slopes on the right bank of the torrent. Adisch, which had been hidden for some time, came suddenly into view round a corner, finely situated on the sloping hillside, at some height above the bottom of the valley, beside a torrent descending from the snowy spurs of Tau Tötönal, a glimpse of which is the only hint of the nearness of a mighty mountain-chain.

We had outwalked our men and the luggage, and sat down to wait for them on a knoll opposite the village, which consisted, as usual, of a cluster of square stone houses, interspersed and surrounded by towers, many of which were in ruins. By the time Paul came up we were surrounded by an excited circle of juveniles, to whom our equipments were as marvellous and entertaining as a conjuror's box. Our boots were perhaps the greatest source of amusement, and the children were never tired of attempting to count the number of nails in them. Entering the village, we were led through it to a house at the farther end, where we were invited to sit down on some logs of wood, under the shelter of a projecting balcony, which was very convenient, as a heavy shower had just commenced. Paul went into the house to see how the baking was getting on, whilst we did our best to entertain and cultivate friendly relations with the crowd which, as a matter of course, gathered round us.

We thought the Adischers, as a race, more intelligent in their looks than our late hosts of Jibiani, and 'their manner in dealing with strangers' was certainly less disagreeable. They did not, however, inspire us with the amount of confidence requisite to induce us to use their village as a base for the attack of Tau Tötönal, and to leave our luggage at their mercy during our absence. From the character we afterwards found them to deserve, our caution was fortunate. We were obliged occasionally to ask the crowd to leave us a little breathing room, but they quite took the point of the suggestion, that the wider the circle the more would be able to see, and both parties were perfectly good-humoured. Paul's appearance with the first batch of loaves was greeted with enthusiasm, for the rations served out at breakfast had been scanty, and we were all ravenous. The flat unleavened cakes of the Caucasus are very palatable when hot, and have the advantage of being seldom sour, like the detestable black bread common in Russia. On the present occasion we made short work of the first baking, and were perfectly ready for a second supply of a superior character, with a layer of warm cheese in the centre, very nice and indigestible.

When the time for payment came there was of course a difficulty, or would have been one, had we not preferred to pay three times their value for the loaves we had consumed, rather than engage in another dispute. It did not seem to us a case where, in the interest of future travellers, we were bound to resist extortion; and we preferred laying ourselves open to the ordinary charge against Englishmen of raising prices wherever they go, to running the risk of being stuck by the dagger of an indignant Adischer for the sum of sixpence. Despite, however, our peace-at-any-price policy, we were followed out of the village by one man, with an absurd demand, to which we refused to listen.

The valley of the Adisch-Tshalai below the village is contracted to a mere gorge, the sides of which are rugged, broken, and picturesquely wooded with firs, pines, birch, and ash. The track which leads from the upper glen to the lower valleys makes a very long and steep ascent, in order to avoid the circuits that would otherwise be necessary to cross lateral ravines. A characteristic of the paths running parallel to the course of the Ingur, from the upper to the lower end of Suanetia, is that they are always uphill; at least, the ascents are so numerous and long, that at the end of the day, the impression left on the mind of the traveller by the intervening descents is almost effaced. The track, passable for sledges, along which we were now strolling, in the wake of our baggage-horse, rises steadily towards a solitary tree which stands up as a beacon on a projecting brow, marking the limit of the forest, and the point at which a sudden sweep must be made to the right, to avoid a series of deeply-cut water-channels, which fall into the gorge of the Adisch-Tskalai. From hence we could see, in the distance, the hamlet of Suni, one of the community of Tzurim, the resting-place for the night chosen by our horseman and porters. It was perched on a meadow-terrace, high above the junction of the Adisch-Tskalai and the Ingur, both of which flow through deep wooded gorges. In this direction the distant view was closed by the snowy heads of the Leila range.

Still maintaining our height, we wound above the heads of the ravines, until we found ourselves on the watershed between the northern district of Mushalaliz, the largest and most thickly-populated of all the Suanetian valleys, and the glens of the Adisch-Tskalai and Ingur, on the south. The ridge separating the two basins sank at our feet into a densely-wooded brow, broken away into precipitous ravines on the south, but falling more gently towards Mushalaliz on the opposite side. Several miles further west the ridge again rose into gracefully-shaped eminences, before it finally sank down to the bed of the Ingur at Latal, where the Mushalaliz torrent joins the main stream. The distant summits of the Leila, with their foreground of wooded gorges and village-dotted slopes, formed a landscape which anywhere else would have absorbed our attention; but we were now more anxious minutely to examine the details of the main chain, which swept round the northern side of the Mushalaliz valley, a broad green expanse relieved by the white towers of many villages. Two large ice-streams filled deep trenches in the mountains, and terminated amongst the lofty cliffs, which close abruptly the head of the valley. The lower portion of the western glacier presents a flat dirty surface, and is fed by two large icefalls, the origin of which was lost in cloud; the second slides out a long twisting tongue of ice, which descends below the level of the forest.

We now followed the top of the wooded brow. The combination of an exquisite woodland foreground with varied and magnificent distant views rendered this portion of our day's journey the most lovely walk we had ever taken. It is quite impossible to convey in words any idea of the beauty of the landscape, or the grandeur of scale which placed the scenery beyond comparison with any of the show-sights of Switzerland. Woods of ash, hazel, and fir alternated with coppices of laurel, white rhododendron, and yellow azaleas, the scent of which perfumed the air. Tall tiger-lilies, one of the characteristic flowers of the Caucasus, shot up their tawny spikes through the rich herbage, while dark-blue lupins and hollyhocks challenged their supremacy over the humbler flowers—campanulas, bluebells, and cowslips—which carpeted the ground. Every break in the wood afforded a glimpse, now over the pine-fringed gorges to the white-crested Leila mountains, now down upon the green Mushalalix, with its sparkling stream and castellated hamlets, and across it to the peaks, precipices, and glaciers of the central chain. We wandered on, feeling as if we had broken in on enchanted ground, and that it was all too beautiful to be real. The nature of the path did not disturb the even tenor of our thoughts; its makers, with remarkable ingenuity, had carried it first on one, then on the other, side of the brow, and it was for a long time almost level and free from stones. At length it turned down the northern side of the ridge, as if to descend to Mushalaliz; but, desirous of gaining Suni,* in the other valley, we plunged, under the guidance of our porters, into the thickets, and soon hit another track remounting to the left.

Just below the watershed a clear little tarn nestled among the trees. We crossed several wooded spurs before we reached the verge of the meadows on which Suni is built. The views, looking back towards Tau Tötönal and up the valley of the Adisch-Tshalai, were wonderfully fine, their effect being heightened by the rapid sweep of afternoon rain-clouds across the sky, and the bright gleams of sunshine which shone out between them. Just before emerging from the wood the horse slipped on a miry bank, where the rains had carried away the path, and fell several feet into a sort of slime-pit. Happily, he neither damaged himself nor the saddlebags, beyond covering both with mud. The fields around Suni are more like Alpine meadows than is usual in this country; they are carefully irrigated by a system of water-channels, and are fenced off from the surrounding pasturages. There are fewer towers here than in the upper villages, and the population is in appearance more

---

* Spelt thus on the map, but generally pronounced Suréni.

like that of the Rion valley. The character they bear is not quite so bad as that of their neighbours of the upper glens, but they seem to lead the same violent quarrelsome life, full of petty squabbles, in which it is perhaps difficult to draw the line between war and murder.

We had eagerly caught at the proposal that we should seek lodgings with a priest, but his house was shut up and deserted, and we were taken on to a neighbouring cottage. The first room offered to us was large, gloomy, and stable-like; when our eyes grew accustomed to the darkness, we found that the furniture consisted of two or three curiously-carved armchairs, of the broad and shallow shape common in the Caucasus, and a hay-bed covered with some cloths of very decided uncleanliness. Mysterious sounds issued from one corner of the apartment, occupied by a row of wooden hutches, also a good deal carved. They proved to be occupied by an old sow, whose dwelling we were to have the honour of sharing. Thinking we might go further and fare better, we asked what was above the pigstye, and found a hay-barn, clean, and, if draughty, at least free from the stifling atmosphere which had driven us from below. Here we took up our quarters and spent the night, sleeping on the hay.

July 21st.—The first matter which required our attention was the reorganisation of our baggage-train. The two Davkar porters, who had carried heavy loads, and walked fairly well, did not care to go further from home without an increase in the rate of pay, and the master of the house seemed desirous to supply a second horse and come with us. As a matter of course, when the arrangement was, as we thought, concluded, the fellow struck to see if he could not get more; but we finally succeeded in settling the business, paid off the porters, and started afresh with our new companion. The weather was still the same—hot gleams of sunshine alternating with sharp showers—but to-day the sunshine predominated. The distance in a straight line from Sani to the next group of villages is not great, but we were two hours traversing it, partly owing to the immense circuits made by the path round lateral ravines; it is compelled to keep at a high level on the rugged hillside, as the bottom of the glen is only wide enough for the Ingur, which flows for miles through a gorge clothed by thick pine-forests. The chief cause, however, of our slow progress during this and the succeeding day was the absurd behaviour of our two horsemen, who dawdled along at their horses' heads at a pace of scarcely two miles an hour, and resented the remonstrances made by Paul to the extent of drawing their pistols on him.

The position of the hamlets constituting El is picturesque; they stand at different heights on a sunny slope, which falls into the gorge of the Ingur from the wooded crest of the ridge dividing it from the Mushalaliz valley. They are surrounded by fields, and the products of the soil—hemp, Indian corn, and various kinds of grain—showed that we were gradually approaching a milder climate. We descended (I believe unnecessarily) in order to pass through a lower hamlet, at which one of our horsemen had a message to leave, and then faced a steep and apparently interminable climb, through a beautiful forest, thick enough to shut out all but occasional glimpses into the bed of the Ingur. The only new feature in the foliage was the prevalence of pines,[*] which are seldom found near the heads of the Caucasian valleys. Numerous sledge-tracks branched off up the hill, and fear of missing the way obliged us to keep close company with our sluggish horsemen.

[*] Pinus sylvestris, Abies Nordmanniana, and Abies orientalis are, according to Radde, found in the Suanetian forests.

Y

At last, when we had all begun to grumble at the toil and trouble of descending a Suanetian valley by a succession of climbs of 2,000 feet each, we came upon a glade in the wood, evidently a favourite halting-ground with the peasants of the neighbouring villages. A fringe of pines and birches surrounded the level plot of smooth greensward and screened off the hot sun, the perfume of azaleas filled the air, and the eyes rested on the noble picture made by the deep pine-clad gorge beneath, and the central mass of the Leila chain directly opposite us. 'Three silent pinnacles of aged snow' sent down long snake-like glaciers towards the Ingur, one of them being a remarkable specimen of what is technically known as a 'glacier remanié.' A wall of rock lying across its path scotches but cannot kill the glacier-snake; as the ice slides steadily downwards, masses fall over the cliff in the form of avalanches, and form a fresh glacier below, which creeps down some distance further towards the forests. A thick girdle of pines clothed the lower slopes, which are broken by rocky spurs dividing the beds of the several glacier-fed torrents. For once we were glad to see the horsemen halt and unload their animals. A fire was soon lighted with moss and pine-chips, and Paul and François busied themselves cooking 'kabobs,' while we reclined on the turf, regarding the mountain-tops, more, I fear, in the spirit of lotos-eaters than in that befitting members of the Alpine Club. Undoubtedly the difficulties of everyday travel in the Caucasus exhaust much of that energy which finds vent in Switzerland in scaling the highest peaks, but even in the Alps there are moments when it is pleasant

> 'To watch the long bright river drawing slowly
> His waters from the purple hill.'

without any thoughts of scaling the silent pinnacles in the distance.

When the time came to pursue our journey, we bade a

lingering farewell to the lovely glade, and surrendered
ourselves again to the annoyances of everyday life, and
the penance of accommodating our movements to the
pace of our horsemen. After winding now up, now down,
through the forest, we came suddenly on open hayfields,
in the midst of which rose a group of towers, most of them
in ruins. Several of the houses were still inhabited, and
a group of peasants—all, as a matter of course, wearing
daggers—were at work haymaking. They soon gathered
round us, and inspected with interest our revolvers and
field-glasses; so far as pantomime could carry it, our
intercourse was of the most friendly character. Our tail,
which as usual we had left behind, came up, and the
horsemen seemed rather to look down upon our new
acquaintances, who, I suppose, had not committed murders
or robberies enough to entitle them to rank among the
upper classes of Suanetia.

Thus far we had been all day climbing round the
irregularities of the steep slope which forms the north
side of the valley of the Ingur; now the path turned over
its summit, and began slowly to descend towards the
torrent issuing from the basin of Mushalaliz, which we
had looked down upon the previous afternoon. The last
view of the Leila summits, before we lost them behind
the ridge, was exceedingly beautiful. We soon found
ourselves in a fold of the hills, surrounded by grassy
eminences, which cut off the view of the valley, but were
not high enough to shut out the snowy buttresses of
Uschba and the summits near it. Clouds hid the actual
peaks, and prevented us from distinguishing their forms
or making any estimate of their height. We traversed a
succession of meadows alive with haymakers, and set in
frames of hazel and birch copses, through which we had
occasionally to force our way, keeping close company with

a purling stream, the image of an English brook. This delightful scenery lasted for a considerable space, as far as the verge of the descent to the Mushalaliz torrent. Here we rested for some minutes under the shade of a beech-grove, before running down the steep path, which bore obliquely along the hillside, covered as usual with a thick mantle of greenery; the bay, the laburnum, and the wild honeysuckle, now mingled with the shrubs of the higher regions—the rhododendron and azalea—forming a dense underwood on either side of the way, which was overshadowed by beeches and hazels.

A crazy but most picturesque bridge spans the narrow cleft in which flows the stream issuing from the Mushalaliz. The Caucasian cattle have stronger nerves than those of the Zillerthal, in the Tyrol, where a boarding is put up on one side of any lofty bridge, lest a cow should be alarmed, or take a suicidal fancy to leap into the foaming torrent. Here not only is there no railing, but numerous holes are left in the wooden framework of the floor, and it is wonderful that the animals do not often break their legs. A slight rise on the further side brought us to the level of the fertile strip of ground lying between the Ingur and the northern slopes, on which are situated the hamlets of the Latal community. The houses are scattered among the fields, and look less like fortresses or dungeons than those of the upper valley; while the fields are surrounded with neatly-woven fences and tall trees, amongst which we saw, for the first time, walnuts growing in clumps by the side of the path. The height is only 4,500 feet, and the produce of the fields and the abundance of fruit-trees bear witness to a milder climate. Tobacco, Indian corn, millet, peas, and beans are extensively cultivated, and the grass crops, which the peasants were now busy mowing, seemed very rich.

The people in the fields were a wild-looking race, and the women had the rough inquisitiveness of savages. François' personal appearance seemed to take their fancy, but though flattered at their admiration, he was not disposed to return it. We none of us lost our hearts to the female portion of the population on the south of the Caucasus; the vaunted and undoubted beauty of the Georgian race cannot withstand the exposure to weather, and the field-work at an early age, which is the lot of the women of the mountain tribes, and the traveller who wants to see the houris with whom popular fancy peoples the country had better stay at Kutais or Tiflis. We passed through several clusters of houses, and, having left behind us a knoll crowned by one of the small square chapels characteristic of this region, were led by our horsemen to an isolated farmhouse. The owner was out haymaking, and his wife, a hideous old shrew, would not open the door to so large a party of strangers without leave from her lord. After a long delay, during which we had nothing to do but to watch the clumsy, heavily-laden carts, drawn by oxen, bringing in the hay, an old peasant arrived, and we were admitted within the gates.

We found a regular farmyard, stocked with pigs and poultry, and guarded by a dog. The building consisted of two or three rooms on the ground-floor, and a large barn, full of new hay, which we appropriated to our use. A projecting balcony afforded us a sunny spot on which to spread our mattress, while waiting the result of the always lengthy preparations for dinner. So peaceful and pastoral was the scene before our eyes, that it was difficult to realise how many deeds of warfare and bloodshed had taken place here, even within the last few years. The inhabitants of Latal were formerly engaged in constant struggles against the Dadisch-Kilians, whose authority

extended over the neighbouring communities of Gegeri
and Betscho on the west and north. Five years ago they
were at strife with the nearest village of the Lendjer
group, in the Mushalaliz valley. Herr Radde thus describes his reception at Latal, and the character he heard
of its inhabitants:—

'We found shelter in the courtyard of an old Suanetian
castle, with a friendly priest who came from Imeritia, and
had spent a year and a half as a missionary amongst
the Suanetians. He had brought his wife with him, and
built himself a small cottage in the inner court, which
was surrounded by a high wall of defence. The account
he gave of the progress of God's word among the Suanetians was very disheartening. They are deaf, he said, to
all instruction; only by kindness can they in some degree
be drawn towards it. They dread being subjected to a
conscription, and distrust all opportunities offered of
bettering themselves. Despite the frequent church-ser-
vices, generally held here on Saturdays and Sundays, they
remain strangers to church principles. They show no
desire to allow their children to learn the Georgian language, although the Government, partly by the appointed
priest, partly by a school lately established at Pari, has
provided them the means of doing so. In Latal, as in
most of the villages of upper independent Suanetia, it is
difficult to find a man who has not committed one or
more murders; for instance, two brothers, who lived
near the priest, were well known to have killed seven
or eight Suanetians. They were two hearty old men,
with fearfully savage countenances. At night every man
drives his cattle into the courtyard, and carefully secures
the great wooden doors of the outside wall.'

We witnessed an illustration of the practice last alluded
to, in the return of the cows shortly before sunset, and

took care to secure draughts of fresh milk as soon as possible. The arrival of strangers did not seem to surprise the peasants here so much as in the upper villages, and we enjoyed comparative privacy; although we heard afterwards, from Paul, that our real character was a subject of deep discussion between our horsemen and our host's family. We were not exactly like Russians, and bore no resemblance to the only other class of visitors with whom they were acquainted—the inhabitants of the north side of the chain, who are said to cross here by the Thuber glacier, either on business to dispose of their merchandise—iron, salt, and sheepskin cloaks—or attracted by the apples and pears which abound in this neighbourhood. Altogether they were fairly puzzled, until our Davkar horseman hit on a happy solution of the mystery, and pronounced us to be, beyond doubt, wandering Jews, on the ground of our not observing church-fasts.

As the evening closed in, the clouds melted away from the summits, and, to our surprise, at the head of the Mushalaliz, Tau Tötönal shone out, a silvery spear, poised at an amazing height in the air, the point of which flushed rosy as the sun sank in the west. Immediately opposite us stood the glacier-crowned, forest-girt Leila mountains, one of their summits curiously resembling in shape Monte Rosa from the Gornergrat. It was one of those heavenly evenings which come once or twice in a summer, when the whole atmosphere seems steeped in roses and purples.

An hour after we had been lost in admiration of the sunset scene we were all writhing in slow torments in the hayloft, devoured by hungry insects, and half suffocated by smoke, which rose through the floor from the room underneath.

July 22nd.—The morning was cloudless, and the great white peak of Tau Tötönal looked superb against the blue sky. When the time came for starting, the old farmer, who had been our host, claimed an exorbitant recompense. We offered him half as much again as was due, but he threw down the rouble-notes on the floor with contempt. Taking no notice, we continued to load the horses, and, after a good deal of bluster, he came to his senses. We amused ourselves at parting by drawing out a British passport, a document certainly seen for the first time in

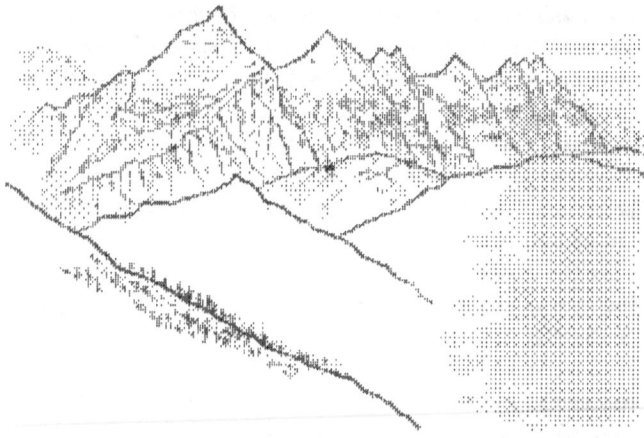

Tau Tetnuld from above Latal.

Suanetia, and bidding Paul tell him that he little thought whom he had entertained unawares. The lion and the unicorn worked the effect intended, and our host forgot his threats of not allowing us to leave till his demands were satisfied to the full, and bade us farewell with an air of mingled fear and relief.

The path down the Ingur has to cross another steep ridge to reach the next basin, that of Betscho, watered by a torrent flowing due south from the central chain. The climb

up the hillside was hot and severe, past another hamlet of
the Latal group, clinging to a slope so steep that it looked
as if the houses must soon slip down and join their neighbours on the flat ground below. Above it the path winds
along the slopes, through young woods of beech, oak, hazel,
and aspen, until the summit of the spur separating Latal
from Betscho is attained. Just before reaching it we sat

Uschba from above Latal.

down to admire the outline of Tau Tötönal and the
southern face of the mountains above Jibiani, which had
come into sight over the lower ranges. Nearer, and
scarcely less beautiful, the range of the Leila formed a
second picture, which elsewhere would have riveted the
attention. A shout from Moore, however, hurried me
on to where he was sitting with François, 100 yards

in advance, apparently gazing in a state of astonishment, that for the moment rendered them unable to express their feelings, at some castle in the air.

On reaching my companions I was at no loss to discover the cause of their emotion. Due north, above the low wood of the adjoining hillside, shot up two towers of rock, one slightly in advance of the other, forming, as regards height, steepness, and outline, beyond all comparison the most wonderful mountain mass we had ever beheld. Tier above tier of precipices rose straight up from the valley, culminating in two tremendous towers, separated by a deep depression. The twin summits resembled one another in form, and appeared to be long roof-like ridges, falling away in slopes of mingled rock and ice of terrific steepness. The idea of climbing either of them seemed too insane to be so much as suggested, and even the lower spurs of the mountain above the meadows of Betscho are so tremendous that it looked as if a stone dropped from the top of either of the peaks would scarcely stop rolling before it reached the valley. There was no mistake about it, the Caucasian Matterhorn was found at last, only here we had one Matterhorn piled on another, and then multiplied by two. It was a sensation scene of Nature's own devising. The name of the mountain was unknown to us at the time; it was unmarked on the map, and our horsemen were sulky and uncommunicative. We learnt afterwards that it is generally called Uschba—' usch,' in Georgian, means rain or storm, so that this seems to be a parallel instance to the Swiss Wetterhorn. The two peaks also resemble one another in being great promontories projecting from the main chain and immediately overhanging the valleys at their feet, so that they are likely to be the heights round which clouds first gather.

Our natural course was to descend as soon as possible to

the Betscho-Tshalai, and follow the main valley of the Ingur. Our horsemen, who were more insufferably indolent and impudent than ever, refused to take a track leading apparently in the right direction, and conducted us instead by a path up the Betscho glen. The way was through a wood of copper-beech and aspen, where the branches formed a frame to the majestic Uschba, whose base we were approaching. The men had told us we should be in Pari by the middle of the day, and trusting to their account we had not, as usual, brought provisions with us. Having already been out three hours, and appearing by the map to be little nearer our destination than at starting, we sent Paul into the nearest village to see if he could get any bread; he succeeded in purchasing one loaf, which had a medicinal taste, as if flavoured with castor-oil, and was only made eatable by extreme hunger.

The villages of Betscho are castellated; one of them, the Mazer of the Five Verst Map, is of large size, and looks very picturesque from a distance. Turning at length down the glen, we retraced our steps at a lower level along the grassy banks of the Betscho-Tshalai,* until nearly opposite a hamlet named Doli, on the right bank of the stream, to which we crossed by a good bridge. As long as we were in the Betscho glen, the forest was still dense, and the foliage varied; but when, after a tedious ascent, we rounded the brow that overlooks the junction of the streams, and turned along the hillsides above the Ingur, the character of the scenery underwent a sudden change. At the western extremity of Suanetia the Ingur flows in a deep defile, between the projecting bases of the main chain and the Leila mountains. All the villages are built at a height of from 2,000 to 3,000 feet above the stream, on the northern

* Called Dodra by the people of Pari.

slopes, which are composed of a very friable slaty rock, in which every torrent has cut itself a deep channel. The constantly-recurring necessity of making the circuits of these ravines renders the walk from Doli to Pari extremely laborious.

The afternoon was hot, and the succession of hay and cornfields, dotted with poor-looking stone houses, through which our road lay, was a bad exchange for the timber to which we had become accustomed. The path climbed higher and higher, till at last it crossed a bare slate-bank, which broke the uniformity of the hillside, and gained a knoll, whence we looked across a wide and tolerably level terrace, watered by a torrent and supporting several villages. Pari, however, was not amongst them, and the meadows had to be crossed and a second brow gained before the resting-place to which we had been looking forward for the last few days came into sight. Between us and it was another deep ravine. Two of the party tried a short cut, with the usual result, and reached the torrent only after a stiff scramble through thickets which were very nearly impassable. There was no bridge, and some waded through the water, while others rode over on one of the horses. Close to the torrent a ferruginous mineral spring, slightly effervescent and very palatable, bursts from the ground.

A last ascent brought us up again to the level of Pari, and we reached the hamlet — for it is nothing more — tired and hungry, about 4.30 P.M. Its position is surpassed in beauty by many of the villages we had lately rested in. The Ingur flows out of sight at the bottom of a deep gorge, and the hillside on which Pari stands is capped by rocky summits little exceeding the snow-level. The horizon is formed on the west by a long spur of the main chain, which runs out at right-angles from it, and forms on this side the boundary of Suanetia; between its lower slopes and the final spurs of the Leila, the Ingur finds its way out to the low country through a deep and densely-wooded gorge. The first building in the hamlet was a whitewashed house, built in the manner of ordinary civilised dwellings, with a little garden at the back. A hundred yards farther we came to an open space surrounded on three sides by buildings, and on the fourth by the ruinous and blackened walls of the late residence of Constantin Dadisch-Kilian, the native prince who, as has been before related, some eight years ago murdered the Governor of Mingrelia at Kutais. The village is composed chiefly of mean-looking cottages, without the tall towers of defence characteristic of Free Suanetia. We told our horseman to take us to the quarters of the Russian Cossacks who, since the crime of Constantin and the confiscation of his property, have been stationed here to keep a watch over the district. The full force consists only of ten men, and is of course too small to preserve order except in the immediate neighbourhood, where the power which they merely represent has been once felt, and is now better appreciated.

Herr Radde's account of the policy adopted by the Russian Government towards the wild mountaineers of Suanetia is very curious. 'It is remarkable,' writes the Herr, 'how here too the Government deals on a policy of mildness with the wildest mountain-tribes. When it is borne in mind that Free Suanetia contains a thousand armed men, there can be no doubt to what issue energetic measures would lead. It is only by patient persuasion, and an avoidance of all misunderstandings, that the government can be carried on in outlying districts. With such obstacles to contend with, the results must necessarily be slow, but they are more certain than those an energetic military discipline would produce. The upper valley of the Ingur is incapable of maintaining a great military force, and the natural difficulties which here oppose the transport of food and ammunition would render its maintenance exceedingly costly. For what good end would a military force keep in subjection the neglected but conquered Suanetians? Murder would not be prevented, agriculture or cattle-breeding encouraged. The perseverance and tenacity with which the Government adheres to its principles must in time tame these wild Suanetians. They will by degrees adopt the Georgian faith and language, and accustom themselves to a more peaceable existence.'

How much of the mildness of their rulers the people owe to the military difficulties mentioned above, and how much to the real belief of Russian officials in a conciliatory policy, it would perhaps be unkind to enquire. Paul told us that at one time the Government had collected a small house-tax from the Suanetians, but for the last year or two they have been relieved from even this slight mark of subjection. Arrived at the Cossacks' quarters, we rejected the officiousness of our Davkar horseman, who now strove to render himself important, and sent Paul to represent who we were and what we wanted. The chief of the outpost was away, but his place was filled by a very civil fellow, who at once found us lodging in the now confiscated dwelling-house of one of the native princes.

Our rooms were on the first-floor, which was approached by an outside staircase. The skill in wood-carving common to so many mountain populations was conspicuous here, in the elaborate ornamentation of the roof of the principal apartment, and of the spacious balcony which was our favourite lounge. Our lodgings were quite bare, but a fireplace gave our men the necessary facilities for cooking.

Nothing could exceed the kindness and readiness to meet our wants shown by the Cossacks, and we rejoiced in our deliverance for a time from the constant struggle with the barbarous natives. When our Davkar horseman came to be paid, he had the impudence to ask for a present, a demand which we promptly crushed by threatening to inform the Cossacks that he had drawn a pistol on us. No more was heard of one of the greatest scoundrels we ever had to deal with, and he left the village soon afterwards. We slept as usual on our mattress, but with a pleasant sense of security which had been wanting since our arrival at Jibiani.

July 23rd.—The day was spent in idleness, while collecting information and preparing provisions for our further journey. We had now reached a point due south of Elbruz; it was therefore necessary to turn north, and cross the main chain, in order to reach the foot of the great mountain. We had been unable, either at Tiflis or Kutais, to obtain any information as to the passes leading from Suanetia to the northern valleys, and were naturally anxious to learn the character of the difficulties we might expect in completing the only link now wanting in our mountain route from Kazbek to Elbruz. The Cossacks told us that there were two passes used by the people, and practicable for cattle, though too rough for horses. The one they recommended us to take leads from the head of a glen called Nakra, which joins the Ingur valley below Pari, and we were told that by this pass Uruspieh, the chief village of the Upper Baksan, could be reached in three days. The usual need of porters arose, but the chief Cossack undertook to find some for us, and shortly introduced two pleasant-looking men, with whom we made an agreement that they, with five friends, should carry our baggage to Uruspieh for six roubles (seventeen shil-

lings) apiece. A man was sent up to the pasturages to
buy us a sheep, for the price of which our provider the
Cossack apologised, alleging as its cause the rapacity of
the villagers, which made living at Pari exceedingly dear.
There was no shop in the Suanetian capital, where wine,
sugar, and butter are unknown luxuries, but we laid in a
store of what provisions we could set hands on. Our
hopes of honey, a dainty for which the district is celebrated, and which it exports to the neighbouring provinces, though long deferred, were finally fulfilled. During
the day we devoted much thought to the subject of dinner.
Moore's hungry eye having observed an old pig with a
large young family, he prevailed on Paul to procure the
slaughter of two of the innocents, and we looked forward
to them as a welcome relief to a long course of tough
mutton and tougher fowls. Notwithstanding the unusual
appliances at Paul's disposal, the dish was not a success.

My friends varied their amusements by taking a Russian
steam-bath, in a little building, specially constructed for
the purpose, opposite our house. What tortures they
underwent I never clearly learnt, but they came out
looking half-boiled and as red as lobsters. A considerable
fire is necessary to heat the building, and as large logs
are scarce and dear at Pari, owing to the imperfect means
of transport, although the forests are barely three hours
distant, all the Cossacks got a steam afterwards, and were
glad thus to enjoy their sole luxury at our expense.

CHAPTER XI.

FROM PARI TO PÄTIGORSK, AND ASCENT OF ELBRUZ.

A Captive Bear—Moore Harangues the Porters—Camp in the Forest—
A Plague of Flies—Lazy Porters—A Nook in the Mountains—Cattle
Lifting—Across the Chain in a Snowstorm—A Stormy Debate—A Log
Hut—Bakesan Valley—Uruspieh—The Guest House—Villany Rewarded—
Minghi-Tau—An Idle Day—An Enlightened Prince—Passes to the Karat-
chai—Tartar Mountaineers—A Night with the Shepherds—A Steep Climb
—Camp on the Rocks—Great Cold—On the Snowfield—In a Crevasse—
Frigid Despair—A Crisis—Perseverance Rewarded—The Summit—Pano-
rama—The Return—Enthusiastic Reception—The Lower Bakesan—A Long
Ride—A Tcherkess Village—Grassy Downs—Zonitzki—Pätigorsk.

July 24*th*.—It is not such an easy matter in the Caucasus as in Switzerland to start in the morning at the hour fixed overnight. Our new troop of porters had to be gathered together from their respective abodes, and each article of the luggage lifted, in order to test its weight, before the business of arranging the burdens could proceed. More than half was placed on the back of a ridiculously small donkey, a meek-looking specimen of his race, with long ears given to flap uncertainly backwards and forwards. This animal was to accompany us for some hours, to relieve our porters of a portion of their burden.

On parting we presented the chief Cossack with an English knife, with which he was very much pleased, and we had consequently to submit to a repetition of the hugging and kissing business so popular abroad, and particularly in Russia, but which is not appreciated by the reserved and unsympathetic Anglo-Saxon. Bidding farewell to the kindly Cossacks, we took a path which connects Puri with the few villages to the west—the last in Suanetia, above the defile of the Ingur. These, like the capital, are built on terraces, high above the river, and separated from one another by deep and broad ravines. A description of the details of our morning's walk would be nearly as wearisome as was the reality; there was little change in the character of the views, and our time was chiefly spent in going down and up steep zigzags. So great were the circuits we were obliged to make, and so dawdling were our porters, that it was time for a midday halt before we reached the last hamlet built on the southern slope of the spur that projects on the eastern side of the entrance to the Nakra valley. We sat down under the shade of a large tree, above the houses, from which the villagers soon crowded round us; with them came a small bear, led by a troop of boys, who cuffed and dragged about poor Bruin very mercilessly. A short and gentle ascent over rich meadows brought us to the top of the spur, whence there was a fine view. At a great depth beneath us lay the glen of the Nakra, backed by snow-streaked ranges, and clothed with the finest pine-forests we had yet seen. The single trees were magnificent, and, standing some distance apart, rose above the other foliage in distinct dark-green cones. The scene strongly resembled some represented in photographs of the Himalayas, and was exceedingly striking. Turning southward, we had before us the deep cleft in the hills which forms the gate of Suanetia, and affords an outlet to the river and the road to Sugdidi, a small Abkasian town, halfway between Kutais and Soukhoum-Kalé.

Our Pari porters showed so strong a disposition to walk in the ways of the men of Davkar—that is, to dawdle on for ten minutes, and then sit down and chat for fifteen—that we took the opportunity of their suggesting that the sum fixed on as their pay was insufficient, to read them a lecture. Sitting in a row, we summoned them before us, and told them, through Paul, that we wished to know, once for all, whether they meant to carry out the agreement they had made with us. If they were dissatisfied, and wished to be off it, we were willing to return to Pari, but, in that case, we should pay them nothing. The porters one and all declared that they wished to come on, and were quite satisfied with the terms. Having been then treated to a spirited harangue from Moore, on the 'whole duty of man' in connection with mountain-walking, and the advantages they would reap from behaving well, the troop resumed their march. For some distance the path, which, though broken in places, was very distinct, skirted the hillside; but at last, by a steep and sudden descent through a grand forest of pines and beeches, the bottom of the Nakra valley was gained. Beneath the close branches of these trees underwood seldom flourishes, and the ground was carpeted with a soft moss, like that of an English glade.

On reaching the banks of the torrent, we turned up the valley by a beaten track, which, but for the occasional obstruction of a fallen tree, would have been practicable for horses—now through the thick wood, now across glades where the rich herbage recalled to our recollection the woods of the Zenes-Squali. One of the porters brought us the branch of a small shrub, with the explanatory remark 'tchai,' and we recognised the tea-plant. We had already been informed at Pari of its existence in the neighbourhood, and the Cossacks there told us that some of their predecessors had turned the leaves to practical account, and acquired sufficient skill to manufacture out of them a very tolerable beverage. We caught occasional vistas of the foaming torrent of the Nakra, which falls in a prolonged rapid, dashed into sheets of white foam, over the granite boulders it has brought down with it from the central chain. The blue sky was bright overhead, and the sun's rays very powerful; we longed for a photographic camera to turn them to account, and fix some of the pictures of wood and water that constantly met our eyes.

At an early hour in the afternoon, our porters surprised us, by sitting down on the banks of the torrent, at a spot where the roar of the water was so loud that it was difficult to carry on any conversation, and proposing we should remain there for the night. We refused to listen to the proposition, and prevented its renewal by walking on, at a pace which the men were unable or unwilling to keep up with. About 5 P.M. we found a position suitable for a camp, on the further side of some clear springs, which were surrounded by dense herbage, rich in flowering plants. Here our tent was soon set up and Paul had a roaring fire of logs to cook by. After a good dinner we flattered ourselves we should spend a quiet night; but at sunset the odious hum of the mosquito commenced, and we were attacked by the venomous little insects, as well as by swarms of a small black fly, no bigger than a pin's head, but armed with a sharp sting. Our Puri men now told Paul that the people of the last village we had halted at purposed following us and stealing our goods during the night, and that we had better divide our baggage amongst them to guard. Feeling sure that, if any robbery was attempted, it would be with the connivance of our attendants, we ordered all the baggage to be piled against the opening of the tent, and told them that our fifteen barrels would be fired without warning if we heard anyone stirring near it. Although no robbers appeared, the night did not pass wholly without alarm. The silence of the forest was broken by a loud crash, like a rattle of musketry, which we found, in the morning, had been caused by the fall of a large tree within fifty yards of our camp.

July 25th.—Our rest was disturbed, and before the early

sunshine lit up the pine-forests on the opposite slopes, we were up, and woefully detailing to one another the sufferings of the night. François' eyes were almost closed by the bites of the small flies, and all the party were more or less disfigured. As soon as the porters had finished tying up their loads, and could be persuaded to put them on their backs, we started. On leaving Puri we had understood that we should cross the pass on the second day; but our men now declared it would be impossible to do more than reach the southern foot of the last ascent before nightfall. Having passed through a belt of pines, we entered a region of scantier and less lofty vegetation, and saw, for the first time, the range of cliffs which, to all appearance, closes the head of the glen. The hillsides were now steep, and slopes of débris, brought down by avalanches from the hollows in the mountains overhead, extended down to the torrent. Numerous and clear fountains sprang out of the ground, and, dammed by the surrounding blocks, formed crystal pools, the banks of which were adorned by clusters of lilies of the valley. Two pretty cascades dashed down the rocks on either side of us, but lacked body of water to rank as firstrate waterfalls. The last birches grew in the hollow at the foot of the cliffs which had barred our view all the morning.

We now saw that the Five Verst Map was entirely wrong in its representation of this part of the chain. The officers employed on the survey evidently contented themselves with a distant view of the Nakra valley, and, deceived by appearances, have in consequence represented its head as a symmetrical horseshoe basin. In reality the sources of the stream lie in a recess of the mountains unseen from below, and therefore ignored on the map. The way to it lay up a steep slope on the left bank of the torrent, which is joined by a good-sized tributary, pouring over the cliffs on the opposite side, from a glacier only a portion of which was visible. In this direction our porters asserted a pass to exist, leading to the villages on the headwaters of the Kuban, in the Karatchai district.

It was only midday, but our lazy troop wanted to halt for the afternoon, alleging that it was impossible to cross the pass before evening, and that if they went any further they should be frozen during the night. With much persuasion we prevailed on them to follow us for two-and-a-half hours more. A sharp ascent, marked with the last traces of a path on this side of the pass, brought us to the level of an upper valley, for some distance bare of herbage, and covered with the snow and rocks of spring-avalanches. The direction of this trough is for about two miles due east, when it splits into two glens, running respectively north and south, of which the former is the most considerable. Having crossed the stream by a snow-bridge, we came to a grass-slope, broken by projecting boulders, just at the junction of the glens. It was difficult to find a plot of turf for the tent, and we were obliged to dislodge the porters from a noble bivouac they had appropriated to themselves under a huge boulder, where alone the ground was level. When the tent was put up the space proved ample for all.

The view from our 'gîte,' which was entirely surrounded by snowy mountains, was very grand. Deep beneath us lay the lower Nakra valley, the range on its western flank crowned by a fine ice-coated peak, occupying the position of the Tau Borkushel of the map. The stream we had lately crossed had run but a short course since leaving its cradle—a glacier flowing round the base of a very remarkable mountain, the perpendicular cliffs of which were overhung by an ice-cornice of enormous thickness. The head of the glen, on the north, was also closed by a glacier;

the route, however, does not lead over it, but makes for a gap between two rocky eminences in the chain on its east. The afternoon sun beat full against the face of the rock under which our tent was pitched, and the heat inside the canvas was great. I rashly sought coolness in a shady nook in the rocks, and thereby caught a chill, which, on the top of a previous slight indisposition, made me for a time very unwell. Wrapping myself up as warmly as possible in my plaid, I took at intervals small doses of chlorodyne, a medicine which throughout our journey we found of the greatest service.

Towards evening we observed four men, armed with guns and swords, and driving eleven cows, descending from the direction of the pass at a hurried pace. Paul enquired of our men where they came from, and was told that they were natives of Lashrash, who, according to custom, had been on a cattle-lifting expedition over the pass, and were now returning with their unlawfully-gotten booty, stolen from one of the herds belonging to the Tartars of the Upper Baksan. Our porters naturally felt uneasy as to the reception they, as the countrymen of the thieves, would meet with on the north side, and we had great difficulty in persuading them to proceed any further. We succeeded, however, in convincing all but one hoary-headed old rascal that the fact of their being with us would be a conclusive proof of their innocence. The man in question knew, apparently, that his character would not bear examination, and, conscious probably of some recent misdemeanour, begged to be allowed to return—adding that his two sons, who were also with us, would share his load between them. To this family arrangement we of course made no objection. The sunset hues were gorgeous, but too vivid to promise a continuance of the fine weather of the last two days.

July 26th.—The weather had changed during the night, clouds were creeping up from the south, and the morning promised to grow worse rather than better. Pain and constant sickness had driven away sleep, and I felt much more fit to lie in bed than to cross a pass; but it was absolutely impossible to remain where we were, and the idea of returning to Pari, now a long day's walk in the rear, was insupportable, besides offering few advantages over the only alternative course—that of reaching the watershed before bad weather came on. We therefore started, and although at first I could scarcely crawl along, even with the help of Tucker and François, necessity proved a wonderful spur; my strength gradually returned, and each step gained towards the top of the pass was an encouragement to further progress.

Above our bivouac the head of the valley was paved with snow; the surface had not frozen in the night, and its softness added considerably to my troubles. In an hour's time we reached the point where the glen is quitted, and the traveller desirous to cross the chain must climb the steep banks of grass on his right. These lead him to the edge of a considerable snowfield, which fills a recess in the rocky ridges on the southern side of the pass. The gap in the crest before us was marked by two stonemen, and we had a further guide in the deep trail of the cattle that had passed the previous afternoon. The clouds now swept up round us, the wind howled dismally, and, as we drew near the top, heavy snow began to fall. The change in the weather was a great disappointment, for we had been looking forward, ever since leaving Pari, to the view of Elbruz during the descent, and all hopes of seeing it were now at an end.

When we reached the top, there was nothing for it but to sit down with our backs to the snowstorm, and munch a crust of bread before descending. A few flowering plants were growing on the rocks, at a height of over 10,500 feet. On the north side is a small glacier, steep enough in places to admit of glissades, but so thickly covered with snow that few crevasses were visible. The falling sleet soon turned into heavy rain, which followed us for the rest of the day. The stream which flows from this glacier is very soon lost to sight under another and very extraordinary ice-stream, which seems to be fed entirely by the avalanches falling from the cliffs of a great mountain, now partially hidden by clouds, but afterwards well known to us by the name of Tungzorun. Like Uschba, unmarked in the Five Verst Map, it is probably the second in height of the mountains of the main chain west of the Koschtantau group.

The glen on this side of the pass runs directly north, and the path, marked out at first by occasional stonemen, soon becomes very distinct. We kept along the western slopes at some height above the glacier, which is covered by débris, and has a dirty appearance; its tongue curls over a steep brow, and at its lower end meets the highest firs. Beneath us, under the clouds, we could see the green Baksan valley; but opposite, where Elbruz should have displayed his full height, nothing met the view but a sea of mist, from which two glacier-tongues protruded only at intervals. A descent, rapid at the last, brought us to the banks of the Baksan. Its upper valley is a trough, with a level floor about a quarter of a mile in width, through which the three streams flowing respectively from the glaciers of Elbruz and Tungzorun, and from the larger ice-stream which fills the head of the valley, run parallel for some distance before joining. It is covered with a thick forest of firs, which in this district entirely take the place of pines. Our men turned several hundred yards out of the way, to visit a very rudely-constructed and now deserted log-hut, where, in partial shelter from the incessant rainstorm, we ate some food.

Not altogether to our surprise, the porters refused to re-shoulder their packs, and demanded immediate payment, declaring that they had fulfilled their contract to take us over to Baksan, and persistently ignoring the express stipulation we had made with them, that Baksan was to be taken as meaning the chief village in the upper valley. One of the men, who had been disagreeable in his manner throughout, now became very violent, and made pretence of drawing his dagger. Moore took the leading part in the diplomacy on our side. His policy was to ignore the ruffian, and refuse to have dealings with any but the two men with whom we had first made the agreement at Pari, who were far the best of the party, and little disposed to join in the violence of some of their companions. The rage of the chief ruffian at being ignored was ludicrous, but, after a long and wearisome wrangle, the malcontents gave in, and the train again got into motion—the virtue of the well-disposed men being confirmed by the promise of an extra rouble on our arrival at Uruspieh, if no further questions occurred.

Although we had no more rows, the delays to which we were subjected were frequent and vexatious. It is a trial of temper to sit on a log, wet through, out of sorts, and in a pouring rain, while half your attendants hurry off without apparent purpose, and the remainder refuse to stir until their companions return. We learnt, afterwards, that the men were looking for the shepherds, whom they knew to be somewhere in the neighbourhood with their flocks. Having been living for the last three days on the provisions they carried with them from home—which consisted only of coarse flour, baked every evening on hot stones by a wood-fire—they naturally wished to have a good meal, and were now anxious to purchase a lamb for their supper. We could not at the time understand their motive, which for some unknown reason they were unwilling to explain, and we naturally grew very wroth at the constantly-recurring halts. After a long walk, in a deluge of rain, through dripping fir-forests, we reached a log-hut, well-built, and fortunately quite watertight. We were glad enough to find a resting-place where we could get off our wet clothes, and warm ourselves round a roaring fire. Our rugs we generally managed to keep dry, by rolling them up inside the mattrass, so we had something besides the ground to lie on. Milk and cheese were procured from the shepherds, and after an attempt to make a brew of arrowroot, the results of which were not wholly satisfactory, we rolled ourselves up in one corner of the hut, and enjoyed a tolerable night's rest.

During the evening we had some amusing conversation with our men. We found that the names 'England' and 'English' conveyed no idea to their minds, and that the only peoples of which they had any knowledge were Russians, Turks, and 'Franghi' (foreigners). After this it was rather startling to be suddenly asked, what we considered the best form of government? Moore shirked the question by replying that certainly that form of government could not be considered good under which the people of one valley could carry off cattle belonging to the inhabitants of another—an answer which, when interpreted by Paul, seemed to tickle our friends amazingly, and to be considered fully adequate.

July 27th.—The morning was fine, and we started hopefully. We were led by a circuitous track along the northern hillside, owing to our men's ignorance of the direct path down the valley. The forest comes to an end suddenly, and is succeeded by green meadows, amongst which stand several groups of buildings, answering to Alpine châlets. They were long, low, and irregular-shaped huts, built of very massive unsmoothed fir-logs, with flat grass-grown roofs. The first we came to were uninhabited. The Baksan valley, after running for some distance north-east, bends northwards, and then again resumes its former direction. At the elbow it is joined by two tributary glens—one on the right, running up towards the main chain; another a mile lower on the left, which leads towards the foot of Elbruz. Opposite the first opening there was an exceedingly striking view of a cluster of snowy peaks, remarkable for their fantastic forms and close grouping.

A farmhouse, apparently the highest permanent habitation in the valley, is situated opposite the mouth of this glen; in a small enclosure at the back, potatoes and other kinds of vegetables seemed to flourish. We kept along the left bank of the Baksan, and presently crossed the powerful torrent which flows from the eastern icefields of Elbruz, but the gap out of which it flows is not wide enough to admit of any view of the great mountain. We succeeded in finding an old man at one of the huts built beside the torrent, and in obtaining a bowl of fresh milk —a rare luxury in the Caucasus, where it is generally turned sour directly, and is then very unpalatable. The lower portion of the walk from this point to Uruspieh is undeniably dull, the principal feature being a bold rock-wall on the right-hand side of the valley, which rises just above the snow-level. The mountain-sides are no longer wooded, and have an arid burnt-up appearance. An enormous barrier, abutting on the northern chain, very similar in form to the Kirchet above Meyringen, blocks the valley, and the path has to climb over it. From the brow, Uruspieh is seen for the first time, still separated from the traveller by a long stretch of level ground. The road, now passable for narrow carts, crosses the Baksan at the base of the mound, and traverses a succession of meadows on the right bank, recrossing only just before it enters the village of Uruspieh. The character of the houses is entirely different from the Suanetian fortresses, and far less picturesque; built on a gentle slope, the low flat-roofed buildings are scarcely distinguishable at a distance, and offer no external attractions on nearer approach. A strong torrent, flowing out of a ravine in the northern hillside, cuts the village in half; to the south another lateral valley opens towards the main chain, and some snowy summits are visible at its head.

The view of the Baksan valley is closed by the icy mass of Tungzorun, which from here rather resembles in form the Zermatt Breithorn. These distant vistas rescue Uruspieh from the charge of positive ugliness, which will certainly be brought against it by those visitors from whom clouds veil everything but the brown barren slopes immediately surrounding the village. A large building, just beyond the bridge, was the abode of the princes of the Uruspieh family, who have given their name to the place. A group was gathered round the door; the men were dressed in the tall sheepskin hats and long coats of the country, and our porters' equipments seemed shabby when brought into contrast with their silver-mounted daggers and handsome cartridge-pouches.

We were naturally most anxious as to what the character of the people would turn out, as upon it depended whether we should be able to attack Elbruz at once, or whether we must descend to Pätigorsk, and make the mountain the object of a separate expedition from thence. Happily our hopes, founded on the favourable report given by the Cossacks at Pari of the people of Buksan and the Karatchai, were not doomed to be disappointed. Some villagers came forward, and at once conducted us to a clean-looking cottage, which proved to be a regular guest-house. It contained two rooms, the inner one provided with a wooden divan in one corner. The walls and roof were constructed of the most massive fir-trunks, and the ruddy hue of the timber, combined with the scrupulous cleanliness of the floor, gave a snug appearance to our quarters. The princes, it was intimated, would soon pay us a visit, and in the meantime we hastened to settle with the Pari men, who seemed anything but at ease, and anxious to set off home again as soon as possible.

As paymaster of the forces, I had told out the necessary quantity of notes, separating them into the proper shares for each man, and, with Paul's aid, was in the act of distributing the money, when there was a stir at the further end of the room, and the princes entered. In the consequent confusion, Paul allowed one porter to secure two shares, and of course, when the turn of the last man came, there was nothing left for him. I felt certain I had handed over the proper amount of notes, but the porters all protested that each had only his own share. The matter was suddenly settled by a villager stepping forward, and, to our great amusement and delight, pointing out the noisy ruffian, who had given us so much trouble on the road, as the recipient of the double portion. The money was at once taken from him, and he seemed too doubtful as to his position to venture on any resistance, although he indulged in a display of indignation, and pretended to be ready to be searched. Finding, however, that he was an object of universal laughter, even to his companions, he speedily retreated, and we heard no more of him. Villany having thus met with its deserts, we rewarded the comparative virtue of the two men who had held aloof in the dispute at the head of the valley with an extra rouble, and the whole troop departed, after much hand-shaking.

All our attention was now due to the princes, to whom we apologised for the disturbed state in which they had found us. Our hosts were three brothers, tall fine-looking men, with open and kindly countenances, and dressed in the full Caucasian costume. The younger brother, Hamzet, had been for some time in the Russian service, and spoke Russian fluently. The interview commenced with the usual enquiry as to our nationality: instead of the stolid ignorance exhibited by the Mingrelians at the mention of the English name, Hamzet's face at once brightened up, and he exclaimed, 'Anglicany, karasho (good), Williams Pasha, Kars, karasho.' It was quite like coming back into the world again, from some region where everyone had been asleep for 500 years, to find men acquainted with the events of the Crimean War.

We recounted our ascent of Kazbek, and our journey across the country (at both of which great surprise was expressed), and then explained our wish to attempt Elbruz. The princes, while admitting Kazbek to be the more precipitous mountain, expressed great doubts of our reaching the top of Elbruz, adducing the very good argument that no one had ever done so. They promised to send for the peasants who had accompanied former Russian travellers bent on exploring the mountain, and said we should be taken at least as far as anyone had been before us.

Ararat was not unknown to the princes, and they were aware of the legend by which the Caucasian mountain is connected with the Armenian. According to local tradition, the Ark grazed on the top of Elbruz before finally resting on Ararat. The correct appreciation of the relative heights of the two mountains might fairly be used as an argument for the truth of the story. If it meets with general acceptance, we are ready cheerfully to waive any claims to the honour of the first ascent of Elbruz in favour of the crew of the Ark, or, as François happily phrased it, 'la famille Noah.'

Our hosts were acquainted with the name of Elbruz, but it had to be translated to the circle of villagers, who only knew the mountain as Minghi-Tau. The introduction having been thus happily effected, our new friendship was cemented by the timely arrival of a trayful of tea and cakes, which were placed on a low three-legged stool, which served as a table.

The princes requested us to ask for whatever we wanted, and offered to supply us with food as well as lodging. Unwilling to put them to unnecessary trouble on the one hand, and also preferring Paul's cookery to the hunches of boiled mutton which form the staple dish of a Caucasian cuisine, we asked only that our servant might be aided to procure what was wanted. In this way we were able to pay for the large stock of provisions necessary for a campaign of at least four days against Elbruz. Paul found all sorts of luxuries, including butter, potatoes, and sugar—the latter coming from the princes' household—to all of which we had long been strangers. Having drunk nothing but tea and water since leaving Kazbek, with the exception of the muddy wine of Glola, we welcomed enthusiastically a villager who brought us some very fair native beer. This is probably the liquid referred to by old Klaproth, who mentions that the beer made by the Karatchai and Baksan people 'is nearly equal to London porter,' although, in that case, either the London brewers must have improved since his time, or the native manufacture deteriorated. We were supplied at night with the unwonted luxury of pillows and sheets, and were thus able to sleep out of our clothes for the first time since leaving Kazbek. It is a curious fact, the reason of which we failed to comprehend, that while in the Mahommedan districts cushions in abundance are generally found, they seem utterly unknown in the nominally Christian parts of the country. The reason is obscure, but the fact remains that, whether at Christian villages or Russian post-stations, the traveller must carry his own mattrass, or be content to lie on boards.

July 28th.—The day was given up to eating and doing nothing, which we succeeded in enjoying thoroughly. Relays of tea and cakes filled up the intervals between heavier meals, and the spare time left at our disposal was spent in sunning ourselves at the door of the cottage, or in conversation with our hosts, who introduced us to a visitor, a Suanetian prince of the Dadisch-Kilian family, allied to them by marriage. He was probably one of the rulers of Betscho, the branch of the Ingur valley lying at the base of Uschba, as Radde mentions their connection and frequent intercourse with the tribes on the north side of the chain. The Suanetian was haughtily aristocratic in his personal appearance and manners, and his presence seemed rather a restraint on everybody else. He was tall, with regular features, but a very unintellectual expression of countenance, and a supercilious dandified air, which would have done credit to a man more accustomed to civilised life.

The native princes were far better-informed men than any we had yet met in the mountains. Only two days' journey from Pätigorsk and Kislovodsk, Uruspieh is frequently visited by Russian travellers or officials, and even the rambling photographer has carried his camera thus far. The last visitors had been two Frenchmen in search of rare woods, who had come here to see what they could find. We heard of them elsewhere, and ultimately saw their purchases on the quay at Poti, on the point of being shipped. The people are thus brought into contact with the European world, but its rumours echo faintly in this remote corner of the continent. The princes themselves are men of taste: one is a good musician; the other, whose mind seemed to be of a practical turn, has gained some information during his Russian service beyond that of a purely military character. Accustomed at home to nothing but the perishable cream-cheeses which are alone made in the Caucasus, he had been struck by the 'Gruyère' eaten in Russia, and had set to work to imitate it with very tolerable success. Before our departure we saw also a number of improved carts, which had been constructed under his directions, to replace the clumsy machines formerly in use. The facts connected with our country most deeply impressed on his mind were, that it had produced a great dramatic author named Shakespeare, and that Englishmen lived entirely on beefsteaks and porter; he was profuse in his apologies at being unable to supply us with our national food, and offered to send up to the pasturage and have a bullock slaughtered, a proposal the execution of which was only prevented by our declaring ourselves quite contented with the sheep we had just bought.

We endeavoured to extract as much information as possible as to the customs and mode of life of the people, but it is very difficult to talk on any but the simplest subjects through an uneducated interpreter. The sum of what we gathered was, that the natives of this and the upper valleys next to the east consider themselves a distinct race from the Tcherkesses, who dwell on the verge of the steppes and in the mountains to the westward. The people here claim to be the old inhabitants, and to have been dispossessed of their ancient supremacy when the hordes of Tcherkesses from the Crimea inundated the country. Their language is Tartar, and their religion, as far as they have any, is Mahommedan; the princes seemed, however, to be very broad and tolerant in their views. The imperial sway of Russia does not press hardly on these mountaineers, who pay only a light house-tax, are exempt from conscription, and are too remote to be exposed to those petty restraints which a once-free people often find the hardest to bear. Their local government has been generally described as feudal; it seemed to us that patriarchal would be the more fitting word. The princes are the recognised heads of the community; they live in a house four times the size of any other in the village, they are richest in flocks and herds, and on them falls the duty of entertaining strangers; but their word is not law, and they can only persuade, not compel, their poorer neighbours to carry out their wishes.

We acquired some geographical information as to the neighbouring mountains. There are two routes into Suanetia—the one by which we had come, through the Nakra valley; and another leading up the glen, due south of Uruspieh, and crossing, as far as we could understand, to the Betscho district. This last, though higher than that which we had crossed, was said to be practicable for horses. The traveller desirous of reaching Utechkulan,[*] the principal village in the Karatchai district, has the choice of skirting the northern or southern flanks of Elbruz. If prepared to undertake on foot a glacier-pass, he will go up to the sources of the Baksan, and traverse the range connecting Elbruz with the watershed, to the Upper Kuban. If he prefers a less toilsome

[*] Quite unconnected with Uschkul, the collective name of the highest group of hamlets in Suanetia, of which Jibiani is one.

journey, he will ride over two northerly spurs, descending between them to cross the valley of the Malka. By either route Utschkulan can be reached on the third day. The Malka is the stream which rises in the northern glacier of Elbruz, and it is from its head that most of the Russian explorers have viewed the mountain, and that the first and most famous attempt to reach its top was made, by the expedition under the command of General Emmanuel, in 1829. There was a report in the village that some Russian officers had lately been seen on the Malka, and we felt some alarm, lest the news of our success on Kazbek had stirred up the officials to endeavour to anticipate us by a prior assault on Elbruz. We never heard anything further of our supposed rivals, and if there was any truth in the story, it referred, I believe, only to a pleasure-party who had come up from Kislovodsk to look at the mountain.*

The prince promised that the necessary attendants for our expedition should be ready early in the morning, and also that they would supply us with large loaves, better fit for carrying than the small crumbly cakes usually eaten in the villages. The terms asked by the men who were to act as porters were two roubles apiece for each day, to which we made no objection. In the evening we were amused by the athletic sports of the youngsters who were gathered outside our door. Two boys began wrestling, and were incited to the most valorous struggles by the promise of a twenty-copeck piece to the winner. The

---

* Any mountaineers who visit the Caucasus are likely to go to Uruspieh, and I may therefore, while on the subject of the routes leading to it, suggest an expedition which, in point of interest and fine scenery, would, I am sure, repay a mountaineer, and is very unlikely to prove impracticable, or even difficult. It is to ascend the valley opening due south of the village, and, turning to the right from its head, effect a pass over the glaciers into the glen, the torrent of which joins the Baksan halfway between Uruspieh and its source.

villagers were constantly passing and repassing in front of our door, and we had ample opportunities of studying their characteristics. The men were a fine race, with a very high type of countenance. The women we saw were prematurely old and wrinkled, with the exception of the quite young girls, who were many of them pretty little things, with close-fitting caps hung round with coins, and long elf-locks streaming out from beneath. The men can-

Woman of Urospieh.

not possess all the beauty of the race; but as this is a Mahommedan country, the young wives and marriageable maidens are probably kept more or less in seclusion.

July 29th.—As usual in the morning, although we got up and breakfasted early, the porters did not appear till two hours later, and then only dropped in one by one. The bread too had not been baked, and it was not till 8.30 A.M.

that all our preparations were completed. We had with
us five natives, who put the greater part of their loads,
for the present, on the backs of two horses, which were
to go with us as far as the highest pasturage. Our
companions were equipped with poles, armed with tremendous iron spikes about two feet long, gradually tapering to a point, and a species of 'crampon,' to attach to
the heel in climbing ice or slippery turf. They soon
proved themselves far better walkers than any we had yet
had to do with, and we retraced our steps up the valley
at a very tolerable pace. Our plan had been to turn up
the glen leading to the eastern glacier of Elbruz,
which, by the map, is manifestly the most direct route to
the mountain, and, owing to our difficulty in conversing
with the porters, it was not till the point where we proposed turning off was reached that we found their intentions differed. They declared that we must go up the
main Baksan valley to its head, and then turn to the right,
in order to reach the south-eastern Elbruz glacier. The
objections to the route we proposed were diverse; there
were no shepherds in that direction, there was no path up
the glen, and it made such a circuit that it would take
three days to reach the foot of the mountain. The first
two reasons were plausible; the third was ridiculous, and
entirely contradicted by the views we subsequently had in
the course of the ascent.

As soon as we understood the points of the case, we
acquiesced in our men's wishes, and continued in our old
tracks up the valley, occasionally profiting by their local
knowledge to make short cuts through the wood. Close
to the hut where we had spent the night after crossing the
range, wild strawberries grew in great profusion, but, generally speaking, they do not abound in this country. A bend
in the direction of the valley hides its head from the

traveller until he has rounded a projection of the northern mountain-side, the base of which the stream hugs so closely, that the path is obliged to wind along the slopes overlooking the thick fir-wood beneath. Here we met some hunters driving two donkeys, each laden with a fine bouquetin, recently killed on the edge of the glacier. The head of one carried a noble pair of horns, but the second was a comparatively young animal. Having at length turned the corner, we saw before us the source of the Duksan, a large glacier filling the head of the valley. At a deserted hut we halted for a consultation, and our men gave us the choice of turning up a glen opening on our right, or going still further up the valley. We decided on taking the former, as being the most direct course.

The climb into the glen was rather rapid; above us, on the left, rose a striking mass of columnar basalt strangely contorted, and of a deep ruddy hue. The long grass was full of snakes, which, as a rule, are rarely found in the Caucasus. One of the porters beckoned me to follow him a few feet up the slope above the path, and pointed out a flattened snow-dome, just visible over the top of the fine icefall that closed the glen, as 'Minghi-Tau.' This was our first sight of Elbruz since we landed in the Caucasus, our only previous glimpse of the mountain having been from the Black Sea steamer, when approaching Poti. Half-an-hour's walk below the end of the glacier, we found the shepherds, who had fixed their quarters in a level meadow, which we reached in nine hours from Uruspieh. In order that we might be within easy reach of capital milk, cheese, and 'kaimak' (a species of Devonshire cream)—delicacies of mountain life which had been long wanting on the south side of the chain—our tent was pitched close to the herdsmen's bivouac. The sheep, apparently disturbed by the novel erection which disfigured their resting-place, determined to get rid of it, and spent the greater part of the night in charging down on the sides of our shelter. Fortunately, an Alpine tent is not easily upset, but our slumbers were constantly broken by the uneasy consciousness of an angry animal butting within a foot of our heads.

July 30th.—The morning was fine, and the cold wind seemed likely to be the harbinger of a spell of settled weather. We did not expect a long day's work, as we were already at a height of about 8,000 feet, and were not likely to find an eligible spot for our tent above 12,000 feet. We started, however, in fair time, in order, in case of need, to be able to push on, and reconnoitre the work before us. The horses came on, for half an hour, to the foot of the glacier, which has retired considerably of late years; there are distinct ancient moraines, now overgrown with herbage, a quarter of a mile in advance of the present termination of the ice. Crossing the torrent, which was divided into several branches, we began to ascend the steep hillside on the right of the icefall, which is tolerably clean, and finely broken into towers and pinnacles. After some time a line of crags appeared to bar the way; they are, however, easily turned in one place, near a little fall which tumbles in a pretty shower of water-rockets over the almost perpendicular struts of basaltic rock. Above this, gentian-studded slopes of short turf were soon succeeded by alternate beds of snow and boulders, extending to the foot of a steep bank, from the top of which we gained a clearer insight into the configuration of the mountain-side.

We were on a rocky ridge, the summit of which, still some six hundred feet above our heads, confines the upper snowfield, which overflows towards the Baksan by two channels—one, the icefall beside which we had ascended;

and a second, farther west, and nearer the head of the valley. The porters had made a long circuit in order to avoid the steep bank we had just climbed, and were now out of sight. Moore was unwell, and had walked thus far with much difficulty; Tucker, François, and I therefore set off to climb the ridge before us, in the hopes of finding a suitable spot for a bivouac near its summit. The boulders were very big, and, although there was no difficulty in scrambling over them, it was long before we could find a plot of ground six feet square which, by any stretch of language, could be called level. When we at last succeeded, we announced the fact by a shout to our friend below, and hastened on to see what was above. The highest rocks were soon passed, and a further climb of about fifty feet brought us to the level of a great snowfield, surrounding the final cone of Elbruz, which rose immediately before us, resembling in shape an inverted tea-cup. The mountain appeared to have two summits, of nearly equal height, and both easy of access to anyone accustomed to Alpine climbing.

Thoroughly satisfied with our inspection, we returned to the spot we had chosen for our tent, and set vigorously to work to make the surface level. To effect this we dug out, with our ice-axes, nearly a foot of stony earth at the upper side, and spread it below—increasing the breadth, which was insufficient, by breaking off masses of rock on one side, and throwing them down on the other. We then completed and filled up the interstices of the natural wall of rock to windward, and, having finished our labours, sat down very contentedly to admire our handiwork, and await the long-delayed arrival of the rest of the party. At last our porters came up, and the tent was pitched. The evening view from our eyrie—the height of which was about 12,000 feet—was superb. Looking nearly south, across the trough of the Buksan to the central chain opposite, the square-headed Tungzorun rose grandly, its cliffs capped with a huge cornice of ice, and a broad stainless glacier streaming down one of its flanks. Further east, in a double-toothed giant, we recognised our startling Suanetian acquaintance Uschba, bearing, it is true, more ice on his northern side, but quite as inaccessible in appearance as from the south.

Our enjoyment of the scene was interrupted by a perfectly unforeseen disturbance. Our porters presented a demand for their first two days' pay; we reminded them that they had distinctly agreed that the settlement was to be delayed until our return to Uruspieh, but at the same time offered the money in two notes. They refused it, and required that each man should be given his exact portion; we told them we had not sufficient small notes with us, on which they announced their intention of returning home, and leaving us to get our luggage back as best we could. Such unreasonable conduct could only be met with contempt, and we answered that they might do as they pleased; that we should start soon after midnight, and should return in the afternoon, and, unless our goods were carried safely down to the shepherds' bivouac before nightfall, should pay them nothing; we added, that if any of them were willing to attempt the ascent it would give us great pleasure, and that they should have every assistance from our rope and ice-axes. On receiving this message from Paul the five all departed, as they would have had us believe, never to return; but in less than half an hour they came back, like boys who had had their sulk out, and made a half apology for their behaviour. This difficulty being satisfactorily smoothed over, the men retreated to lairs somewhat lower down the hillside, while we prepared for the night. Paul had, contrary to our advice, insisted on coming with us; he was so anxious to ascend the famous mountain, which he had lived near and heard talked of all his life, that we did not like to check his enthusiasm, especially as there seemed no reason why he should not accomplish his desire. The night promised to be cold, and we invited François to come inside the tent—which, as we had proved on Kazbek, would at a pinch accommodate four—while Paul found a sheltered couch in a trench we had dug at the head of the tent.

July 31*st*.—The cold during the night was so intense, that the water in a gutta-percha bag, which we had filled overnight and hung within the canvas, was frozen before morning into a solid sausage of ice, and in consequence, having no firewood with us, we could procure nothing to drink. At 2.10 A.M., having attached ourselves with the rope, in the knowledge that *terra firma* would soon be left behind, we set out alone, the natives not answering to our shouts. In climbing the steep snow-banks which lead to the 'grand plateau,' Paul slipped about helplessly, and Tucker had almost to drag him for some distance. When, in a quarter of an hour, we reached the edge of the great snow-plain, Elbruz loomed before us, huge and pale, but, to our surprise and disgust, partially shrouded by a black cloud. The walking was now easy, and we tramped on in solemn, not to say surly silence, our ice-axes under our arms, and our hands in our pockets. We were well protected from the severe cold by Welsh wigs, scarves, cardigans, and muffetees, though, owing to our men having mislaid my gaiters, I offered one weak point to the enemy's attack.

A few benighted people still reiterate the assertion that the true beauty of nature ceases at the snow-level, and that those who go beyond it get no reward for their pains except the satisfaction of having treated a great mountain as a greased pole. As we tramped over the snowfields of Elbruz, I could not help wishing we had some of these unbelievers with us, because, while they would have been compelled to admit the startling grandeur of the situation, the intense cold would have inflicted on them a just punishment for their past offences. The last rays of the setting moon lit up the summits of the main chain, over the gaps in which we already saw portions of the southern spurs. The icy sides of Uschba and Tungzorun reflected the pale gleam of the sky; a dark rock-peak further west stood in deep shadow. We were high enough to overlook the ridges that run out from Elbruz towards the north-east, in which direction a dark band of vapour, illuminated by fitful flashes of sheet-lightning, overhung the distant steppe. The thick black cloud was still on the mountain before us; otherwise the sky overhead was clear, and the stars shone out with preternatural brilliancy.

Near the point where the snow began to slope towards the base of the mountain, the crisp surface broke under my feet, and I disappeared, as suddenly as through a trap-door, into a concealed crevasse. Paul, who was next behind me on the rope, was horror-struck, and his first impulse was to rush to the brink to see what had become of me, a course of proceeding which had to be summarily checked by my companions. The crevasse was one of those which gradually enlarge as they descend, but the check given by the rope enabled me at once to plant my feet on a ledge on one side, and my back against the other. The position was more ludicrous than uncomfortable. I had both hands in my pockets, and my ice-axe under my arm; and owing to the tightness of the rope, and the cramped space, it was not easy to make the axe serviceable without fear of dropping it into the unknown depths below.

The snow-crust on the side of the hole I had made broke away beneath my arms when I first tried to raise myself on it, and it cost us all a long struggle before I was hauled out and landed safely.

The slopes now steepened, the cold grew more intense, and the wind almost unbearable, so that altogether the prospect was far from cheering. The morning star aroused us to a temporary enthusiasm by the strange accompaniments and brightness of its rising. Heralded by a glow of light, which made one of the party exclaim, 'There comes the sun!' it leapt forth with a sudden splendour from amidst the flashes of lightning playing in the dark cloud that lay below, shrouding the distant steppe. The shock was but momentary, and we soon relapsed into a state of icy despair, which was not diminished by the sudden desertion of Paul, who, fairly beaten by the intense cold, turned and fled down our traces. For hour after hour we went on without a halt, hoping that the sun would bring with it an increase of warmth.

A sunrise viewed from a height equal to that of the top of Mont Blanc is a scene of unearthly splendour, of which words can convey but a feeble impression. A sudden kindling of the eastern ranges first warned us to be on the watch; in a moment the snow upon which we were standing, the crags above us, indeed the whole atmosphere, were suffused with rose-pink. The cloud on the summit, which had changed from black to grey as daylight dawned, now caught the pervading flush, and suddenly melted away, like a ghost who had outstayed his time. As the hues faded, the sun's orb rose in the east, and flooded us with a stream of golden rays, which were soon merged in the clear light of day. There was no increase of warmth as yet, and, despite the improved look of the weather, it became a serious question whether we could go on. By 7.30 A.M.

we were at a height of over 16,000 feet, and had now reached the rocks which form the upper portion of the cone. Finding what shelter we could among them, we stood shivering, kicking our feet against the rock, and beating our fingers, to preserve them if possible from frostbite, while the debate, as to whether we should turn back or not, was carried on in voices almost inaudible from the chattering of our teeth. On the one hand, the wind did not abate, and the risk of frostbites was growing serious; Tucker and François had no sensation in their fingers, and my toes were similarly affected. On the other hand, the rocks were less cold to the feet, and gave some shelter from the weather. Looking back, we saw, to our surprise, two of the porters advancing rapidly in our footsteps. We had almost decided to turn when they came up to us, looking fairly comfortable in their big sheepskin cloaks, and quite unaffected by the cold. A third, however, who had started with them, had, like Paul, given in. I said, 'If a porter goes on, I will go with him.' 'If one goes, all go,' added Moore. The decision was accepted, and we again set our faces to the mountain.

From this time the cold, though severe, ceased to be painful. A long climb up easy rocks, mostly broken small, with here and there a large knob projecting from the surface, brought us to the foot of a low cliff, to surmount which a few steps were cut in an ice-couloir, the only approach to a difficulty on the mountain. Arrived on the top of what had for long been our skyline, we saw as much more rock above us. Doubts were even now felt, and expressed, as to our success. We persevered, however, making but few and short halts, until the base of some bold crags, we had taken long to reach, was passed. Almost suddenly, at the last, we found ourselves on a level with their tops, and stepped on to a broad crest, running

PANORAMA FROM THE SUMMIT.

east and west. We turned to the left, and faced the wind, for a final struggle. The ridge was easy, and, led by the porters, we marched along it in procession, with our hands in our pockets, and our ice-axes under our arms, until it culminated in a bare patch of rock surrounded by snow. This summit was at one end of a horseshoe ridge, crowned by three distinct eminences, and enclosing a snowy plateau, which, even to our unlearned eyes, irresistibly suggested an old crater. The rocks which we picked up, and carried down with us, are of a volcanic character. We walked, or rather ran, round the ridge to its extremity, crossing two considerable depressions, and visiting all three tops; under the farthest, a tower of rock, we found shelter and a quite endurable temperature. There we sat down, to examine, as far as possible, into the details of the vast panorama. The two natives pointed out the various valleys, while we endeavoured to recognise the mountains. Light clouds were driving against the western face of the peak, and a sea of mist hid the northern steppe—otherwise the view was clear. Beginning in the east, the feature of the panorama was the central chain between ourselves and Kazbek. I never saw any group of mountains which bore so well being looked down upon as the great peaks that stand over the sources of the Tcherek and Tchegem. The Pennines from Mont Blanc look puny in comparison with Koschtantau and his neighbours from Elbruz. The Caucasian groups are finer, and the peaks sharper, and there was a suggestion of unseen depth in the trenches separating them, that I never noticed so forcibly in any Alpine view.

Turning southwards, the double-toothed Uschba still asserted himself, although at last distinctly beneath us; the greater part of the summits and snowfields of the chain between us and Suanetia lay, as on a relieved map, at our feet, and we could see beyond them the snowy-crested Leila, and in the far distance the blue ranges of the Turkish frontier, between Batoum and Achaltzich. Shifting again our position, we looked over the shoulders of a bold rock-peak, the loftiest to the west of Elbruz, and endeavoured to make out the Black Sea. Whether the level grey surface which met our eyes was water, or a filmy mist hanging over its surface, it was impossible to distinguish. The mists, beating below on the slope of the mountain, hid the sources of the Kuban, but we looked immediately down upon those of the Mulka. On this side the slope of the mountain seemed to be uniform for nearly 10,000 feet; and although there is nothing in its steepness to render an ascent impossible, the climb would be very long and toilsome.

We were not hungry, and, if we had wished to drink anyone's health, we had nothing to drink it in; so we gave vent to our feelings, and surprised the porters, with 'Three times three, and one more!' in honour of the old mountain, which, by the help of wind and cold, had made so good a fight against us. We then hurried back to the first summit, on which, as it seemed somewhat the highest, François had already set himself to work, to erect a small stone-man.

At this period, some one remembered that we had forgotten all about the rarity of the air; we tried to observe it, but failed, and I think the fact that, at a height of 18,500 feet, no single man, out of a party of six, was in any way affected, helps to prove that mountain-sickness is not a necessary evil, and that it only affects those who are in bad training, or out of sorts, at the time. Such is my experience, so far as it goes, having only twice suffered from it—once in an attempt on the Dent Blanche, on the first day of a Swiss tour, and again on Ararat, when quite out of condition. We reached the top of Elbruz at 10.40, and left a few minutes after 11 A.M. We had some difficulty in reconciling the appearance of the top of the mountain, when seen from a distance, either on the north or south, with its actual shape. From Poti, or Pátigorsk, Elbruz appears to culminate in two peaks of apparently equal height, separated by a considerable hollow. The gaps between the summits we visited are not more than 150 feet deep, and we were surprised at their being so conspicuous from a distance. In walking round the horseshoe ridge, we naturally looked out to see if there was not some other summit, but none was visible; and on the west (where, if anywhere, it should have been found), the slopes appeared to break down abruptly towards the Karntchai, and there were no clouds dense enough to have concealed any eminence nearly equalling in height that upon which we stood.

The ascent from our bivouac—one of 6,500 feet, or 600 feet more than Mont Blanc from the Grands Mulets—had occupied 7½ hours, with very few halts; the return was accomplished in four hours, and might have been done much faster. The rocks were so easy, that but for the trouble of coiling up, and then again getting out the rope, we should have hurried down without it. Some little care was necessary, on the part of those in the rear, to avoid dislodging loose stones, and Moore got a nasty blow on his finger from one, the effects of which lasted for many weeks. At about one o'clock we sat down on the spot where we had held our debate in the morning, and made the first regular meal of the day. We now, too, broke the icicles off our beards, which had been thus fringed since 3 A.M. We observed from hence that the eastern glacier of Elbruz flows from the same névé as the ice-streams that descend to the sources of the

B B

Baksan, and that there was no apparent difficulty in following it into the head of the glen from which we had originally proposed to attack the mountain. The snow was still in good order, owing to the extreme cold, and we slid quickly down—the two natives, though declining to be attached to our rope, gladly accepting the suggestion that they should hold it in their hands. When ascending in our tracks, they had seen the hole made by my disappearance in the crevasse, and the lesson was not lost upon them.

A cloud which had formed in the valley now swept up, and enveloped us for half-an-hour, but we found no difficulty in steering our way through the layer of mist into bright sunshine. We arrived at the bivouac to find that Paul had already left with the baggage, and we soon followed, leisurely descending the steep slopes beside the icefall. The stream, which yesterday burst from the foot of the glacier, had changed its source, and to-day spurted in a jet from the top of a bank of ice. The heat of the afternoon had swollen its waters, and we found some difficulty in crossing them. The two natives had arrived before us, and told their story to their companions and the shepherds, who, having made up their minds that we should never be seen again, were surprised and seemingly pleased to welcome us, not only safe but successful. On our appearance in camp we had to submit to the congratulations of the country, offered in their usual form of hugging and kissing.

August 1st.—We were too stiff, after our long exposure to cold, to rest very easily, and were ready to start on our return to Uruspieh at an early hour. Our men, however, had other plans, and we found that they meant to kill and eat a sheep before leaving. Wishing to take our time on the road, we left them to follow; but I no sooner attempted to

walk than one of my ankles became painful, as if it had
been badly sprained, and I was therefore obliged to stop,
and mount one of the horses we had brought up from the
village. The pain and stiffness, no doubt resulting from
the cold, gradually wore off, and I was glad to dismount
halfway. The train of porters overtook us about an
hour out of Uruspieh, and we walked in together. I
never saw better walkers than these Tartars, not only on
a hillside, but—what is even more remarkable amongst
mountaineers—upon flat ground. They gave us a start,
and caught us up easily in the ascent of Elbruz, and now,
when Tucker, wishing to try their mettle, put on a spurt
across the meadow, they walked with apparent ease at a
pace of five miles an hour, and soon caused our friend to
repent the trial of speed he had rashly provoked. These
men are the raw material out of which Caucasian guides
will have to be made, and, if the great language difficulty
could be overcome, there is no reason why they should
not, with a little practice in ice-craft, become firstrate
companions for a traveller wanting to explore the glaciers
of this part of the chain.

We entered Uruspieh, and reached the guest-house
almost unobserved; but we had not been there many
minutes before our native companions spread the news of
our return, and a crowd of excited villagers flocked into the
room. Several minutes passed before the story was fully
understood: our burnt faces, and the partially-blinded
eyes of the two men who had accompanied us, were
visible signs that we had in truth spent many hours on the
snowfields, and the circumstantial account and description
of the summit given by the porters seemed to create
a general belief in the reality of the ascent. The scene
was most entertaining. The whole male population of
the place crowded round us to shake hands, each of our

companions found himself a centre of attraction, and the air rang with 'Allah'-seasoned phrases of exclamation and astonishment, mingled, as each newcomer entered, and required to hear the tale afresh, with constant reiterations of 'Minghi-Tau!'—a familiar name, which sounds far more grateful to my ear than the heavy-syllabled Elbruz.

We underwent a crossfire of questionings as to what we had found on the top, and had sorrowfully to confess that we had seen nothing of the gigantic cock who lives up aloft, and is said to salute the sunrise by crowing and flapping his wings, and to prevent the approach of men to the treasure he is set to guard, by attacking intruders with his beak and talons. We could not even pretend to have had an interview with the giants and genii believed to dwell in the clefts and caverns of Elbruz, concerning one of whom Haxthausen relates the following legend:—'An Abkhasian once went down into the deepest cavern of the mountain, where he found a powerful giant, who said to him, "Child of man of the upper world, who hast dared to come down here, tell me how the race of man lives in the world above? Is woman still true to man? Is the daughter still obedient to the mother?" The Abkhasian answered in the affirmative, whereat the giant gnashed his teeth, groaned, and said, "Then must I still live on here with sighs and lamentation!"' The giant lost an opportunity when I was in the crevasse; for had he then put the same questions to me, an old 'Saturday Review' might have been found in my pocket, the perusal of a famous article in which would have clearly justified him in considering his period of punishment at an end.

The princes, of course, came to talk the expedition over with us, and seemed much struck by what they heard of the use we had made of our mountaineering gear, of which they had before scarcely comprehended the purpose.

Hamzet's enthusiasm was boundless; he strolled in and out perpetually, repeating each time the magic word 'Minghi-Tau!'—till at last he achieved an astonishing linguistic feat, and showed at the same time a surprising acquaintance with the manners of Western Europe, by confidentially suggesting, 'Minghi-Tau,—London, champagne fruhstuck, karasho.' He had evidently not been in the Russian service for nothing.

We went to bed with a weight off our minds, feeling that, come now what might, the three great objects of our journey—the ascents of Kazbek and Elbruz, and the establishment of a high-level route between them—were fully accomplished. Conscious virtue now proposed to reward itself, and after a month of hard work, poor living, and no accommodation, attended at times by considerable anxiety as to the successful issue of our projects, we looked forward with pleasure to a period of enjoyment of the luxuries of civilisation at Pätigorsk, the watering-place of the Northern Caucasus.

August 2nd.—We had naturally imagined that, with the friendly aid of the princes at our back, we should have no difficulty in procuring horses to ride down the two days' journey to Pätigorsk. Such, however, was not the case; the old leaven of covetousness, which seems inherent in the Caucasian mountaineer, again came to the surface; the price asked was absurd, and the arrangements were further complicated by the necessity of making a bargain with two or three men, owing to no single peasant having sufficient horses for our whole party. The princes possessed influence, though no authority, and by their aid an arrangement was finally concluded. We were asked by several of the wealthier villagers, and received a formal application from one of the princes' servants, to know if we had any gold or silver pieces with us, that we would exchange for Russian paper. In this country every man is his own banker, and either carries his balance on his person, melted into the form of gold ornaments for his belt or dagger-sheath, or else hangs it in a row of gold coins on his wife's forehead. The day was superb, but we were too lazy to go up a hill, even for the sake of seeing Elbruz.

August 3rd.—Our start was to have been early, but, as usual, delay arose from various causes. We were anxious to acknowledge the hospitality shown to us, but the means at our disposal were limited; at last we determined to quiet our consciences, when our hosts came to see us off, by presenting a compressible drinking-cup to Ismail, the eldest brother. We were on the point of departure, when the princess, his sister, sent down a servant with a special request (translated to us apologetically by Paul), that we would leave behind for her use an article of toilet, one of the very few we possessed, which she had seen and admired. The princess's wish was of course gratified, and the object on which she had set her affections—a large bath-sponge—was yielded up to her.

Two of the princes presented us with their cartes-de-visite, taken at Pätigorsk, and accepted ours in return; then, after exchanging hearty farewells, we left Uruspieh behind us, and took the road leading down the valley. Our course lay along the banks of the Baksan for the whole day, during which (between 8.30 A.M. and 9.30 P.M.) we accomplished a distance of fifty miles—a good ride, on native saddles, for men who had not been on horseback for weeks; but the excellence of the road aided us much in getting through the day's work. The valley of the Baksan, as yet the most visited in the Caucasus, is also the dullest, and its scenery cannot, by any stretch of courtesy, be called either grand or beautiful. For some distance below Uruspieh, the valley preserves the same general character and direction; the riversides are cultivated, and farmhouses are seen every half-hour; the lower slopes of the mountains are scantily wooded with firs. A narrow gorge, from the upper end of which the traveller obtains his last and finest view of Tungzorun, leads into a wide green basin, hemmed in on the north by cliffs, the tawny hue and bold outlines of which reminded us of pictures of Sinaitic scenery. The landscape was for a time perfectly bare. On the right, in recesses of the hills, we passed two villages, from one of which a low pass leads over into the valley of the Tchegem, the next tributary of the Terek on the east. The path then enters a second defile, longer and more picturesque than the first, and rendered pleasing to the eye by abundant vegetation, which suddenly succeeds to the utter bareness of the glen above. We thought a hamlet at its lower extremity would be our journey's end, but found there was still a ride of some hours before us.

After crossing the river twice, by new and solidly-built bridges, the track leads across a wide grassy plain, surrounded by ridges which no longer deserve the name of mountains. Copious springs of the clearest water burst out of the ground, and nourished a tall and rank herbage, the home of myriads of insects, which persecuted most cruelly both our horses and ourselves. A perfect plague of horse-flies swarmed around us, and the backs of our coats were so thickly covered with the insects that the cloth was scarcely visible. At the lower end of this plain a considerable tributary joins the Baksan, on the left, and close to the junction stands a group of old tombs, concerning which the natives tell numerous legends. After wading the tributary, we had still a long stretch of cornland to traverse, before reaching Ataschkutan. As night came on, and the moon rose, the coolness, and relief from our insect tormentors, were very pleasant. At last lights appeared, and we rode along the outskirts of a large and scattered village, composed of low houses, each standing apart from its neighbour, and surrounded by its own garden.

We had now entered the country of the Tcherkesses, the most famous tribe of the Caucasus, from whom the whole mountainous region between the Black and Caspian Seas is often vaguely, and very incorrectly, called Circassia. We were to lodge at the prince's house, situated at the farther end of the village, within an enclosure surrounded by barns and outbuildings. The interior bore witness to Russian influence. For the first time since leaving Kobi, we found chairs and tables, knives and forks, and other luxuries of Western life; indeed, the room we slept in would have been perfectly European in its appearance, but for the illuminated texts of the Koran hung up against the walls. The prince was a good-looking youth, of apparently less than average intelligence; in our case, at least, he meddled only to muddle. Our horsemen, dissatisfied with the bargain the Uruspieh princes had led them into, struck for higher pay, which we refused to give, trusting, somewhat rashly, to find others without difficulty. In the morning the prince seemed unable to give us any help, and declared there were no horses unemployed, an assertion which was hardly uttered when we saw a drove of at least two hundred on the opposite bank of the river.

A peasant having offered to provide a bullock-cart and one horse to take us to Zonitzki, the nearest post-station, forty versts off, we accepted his offer as the simplest means of escape from our difficulty, and set out in this novel style. The road led for many miles over rolling hills, which, but for the luxuriant herbage with which they were covered, might have been taken for part of the South Downs of Sussex, the level steppe in front looking from a distance not unlike the sea. To the south we had occasional glimpses, through the clouds, of the snowy chain we were leaving. Our progress was delayed by the breakdown of the cart, which its owner happily succeeded in exchanging for another we found on the road. Now up and now down, we traversed for hours the same description of country, passing between meadows of gigantic weeds, amongst which the wild sunflower had the pre-eminence. Lunch took place, under the shadow of the cart, beside a muddy brook, the only water we saw for miles. Our progress throughout the day was of the slowest description; one of us rode the horse in turn; the others either walked or took lifts in the bullock-cart, which creaked slowly along at about two miles an hour. We had one good laugh to relieve the dulness of the ride. François, whose notions of horsemanship are practical but not scientific, was about to mount the horse, when he felt that the stirrup was weak. He met the difficulty without hesitating for a moment, by going round to the other side and mounting with the left leg foremost, consequently with his face to the tail. This result, however, had been foreseen, and, with a dexterity for which we were unprepared, the rider wriggled himself round in the saddle, quite unconscious of the amusement his proceedings had afforded to the party in the cart.

As we drew near the river Malka the hills sank, and the country became well cultivated, the grain principally grown being a kind of spelt. The view was very striking from the brow above the slight descent to the ford. On the river-bank was a large Tcherkess village, and before us a vast plain—golden in parts with uncut corn, dotted in others with the small ricks into which the peasants first heap it—stretched to the horizon. A group of bold hills rose like islands in the distance, in the loftiest of which we at once recognised Beschtau (4,594 feet), an isolated summit rendered famous by several of the early travellers in this country, and not far distant from our goal—Pütigorsk. Only ten versts now intervened between us and Zonitzki, the green-cupola'd church of which marked its position long before we reached it. There was no difficulty in obtaining 'troikas,' though there was no regular posthouse, and the functions of head of the post were exercised by the village schoolmaster. Excited at our gradual return to civilisation, and at the discovery of a shop, we rashly ordered a bottle of wine, but failed in the attempt to swallow the vinegar which bore the name. While our carts (the too well-remembered 'paraclodnaia') were being prepared, we sat in a farmyard, where we were entertained with tea and bread by a funny old Russian woman, whose life seemed troubled by her pigs—lean and hungry beasts, that gathered round us in a circle, waiting to pick up the crumbs of our repast.

The red and purple tints of a gorgeous sunset were slowly fading away, and the symmetrical form of Beschtau stood out, as a dark mass against the lustrous sky, as we left Zonitzki. Before the light was too far gone, we caught sight of Elbruz looming indistinctly, like a huge pale shadow, on the southern horizon. As the twilight grew deeper the moon rose, and lighted us on our way across the grassy steppe. In spite of the jolting, we dozed for three hours at the bottom of the cart, and were only aroused by finding ourselves in the broad street of a village we at first thought to be Pütigorsk; it was, however, only its suburb, the Cossack 'stanitza' of Goriatchevodsk. No bridge crosses the stony bed of the Podkumok, a tributary of the Kuma, one of those streams which contradict the poet's assertion, that

    Even the weariest river
    Winds somewhere safe to sea,

by perishing miserably in the vast steppe which stretches inland from the north-western shores of the Caspian. We drove slowly through the ford, gazing with wondering eyes at what looked to us, unused to any building larger than a Tartar 'aoul,' the temples and palaces covering the opposite hillside. Our driver, lashing his horses into a final spurt, galloped up a street lined with two-storied European houses, past dozens of shop-signs, and then round a sharp corner, where the blue domes of the cathedral, surmounted by chain-hung crosses of gold, glistened in the moonlight. To our astonishment the horses turned into the courtyard of a massive building with an Ionic portico, and we found that this palatial pile was the hotel. Never probably before this night had such a queer-looking crew demanded admission of Mademoiselle Caruta, the daughter of the worthy Italian who owns the establishment. With a discrimination which did her great credit, she recognised at once our true character, showed us rooms, and suggested, in Circean tones, that we should probably like some supper. It was 11 P.M., and we had thought that our travel-worn garments would not be exposed to public gaze till the next morning, when we might in some degree have furbished them up. To our surprise the restaurant was crowded with Russian officers in full uniform (when are they not?) and, worse still, ladies in evening costumes. Dazzled by the blaze of candles and looking-glasses, and puzzled by the profusion of good things suddenly placed at our disposal, we retired hastily to the nearest table, and having ordered our food, tried to look as if we were not conscious of being dusty, travel-stained, and about the colour of Red Indians.

The contrast, characteristic of Russia, between an excess of luxury and a lack of the commonest articles of civilisation, is seen in its most exaggerated form in the Caucasian provinces. As we sat surrounded by all the luxuries of civilisation, and supplied, by assiduous waiters, with delicately-cooked dishes and brimming glasses, from which the champagne flowed gratefully down our throats, the adventures of the past four weeks seemed to us as a tale that is told, and we could scarcely believe how short a time before high sheep's brains had been regarded as a delicacy, and a pair of shooting-boots and a revolver as a luxurious pillow.

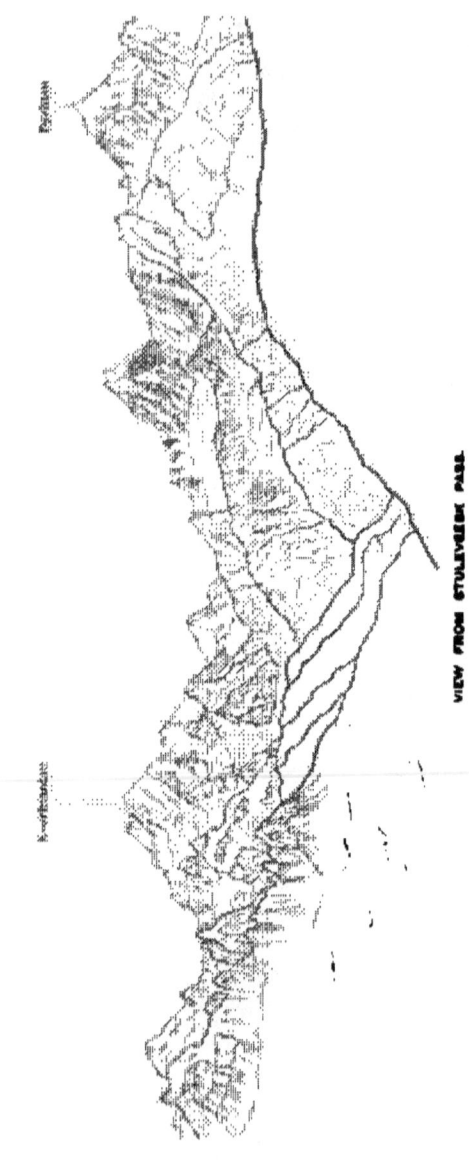

CHAPTER XII.

PÄTIGORSK AND THE TCHEREK VALLEY.

The Caucasian Spas—Their History and Development—View from Machoucha—The Patients—Essentuky—Kislovodsk—The Narzan—Hospitable Reception—A Fresh Start—A Russian Farmhouse—By the Waters of Baksan—Naltschik—The Tcherek—Camp in the Forest—A Tremendous Gorge—Balkar—A Hospitable Sheikh—The Mollah—Gloomy Weather—A Solemn Parting—Granitic Cliffs—Karaoul—A Mountain Panorama—Sources of the Tcherek—The Stulevrcek Pass—Koschtantau and Dychtau—A Noble Peak—Our Last Camp.

August 5th to 9th.—We spent five days very pleasantly in resting from our fatigues, and enjoying the good things brought within our reach in the Caucasian Capua, of which I must now give some account. The history of the mineral waters to which Pätigorsk owes its existence is curious, as illustrating the state of the country during the last century, and the gradual steps by which the Russian conquests have been extended and consolidated. In 1717 the court-physician of Peter the Great reported on the rumoured existence of mineral springs in the country of the Tcherkesses, but no Russian could then visit them. In 1780 the fortress of Constantinogorsk was established, four versts from the present site of Pätigorsk, to check the constant raids of the Tcherkesses.

Klaproth visited the sulphur-springs in 1807, and gives a vivid and amusing description of the troubled lives led by the poor patients at that time. During the day the bathers sojourned in huts built round the source, which was conducted into a clumsily-hewn basin, capable of containing six persons at a time. At night they returned to the adjacent fortress, under the escort of a strong armed force; for the country was still kept in alarm by continual raids of the Tcherkesses, who found the trade of catching and obtaining ransoms for prisoners as lucrative as the Neapolitan brigands do at the present day. As the historian of the waters naïvely remarks, it is easy to imagine that, perfect repose of mind being an essential part of the cure, the patients did not benefit by it as much as they might have done under more favourable circumstances. Still, despite all hindrances, the popularity of the springs increased, and so early as 1811, two hundred Russian families were drawn together to the spot.

In 1812, an employé at Constantinogorsk built the two first houses on the site where Pätigorsk now stands. In 1829 the transfer of the official portion of the population of Georgievsk to Stavropol gave a new impulse to the growth of Pätigorsk, which received many of the former inhabitants of Georgievsk, a town in an unhealthy situation, only occupied on account of its supposed strategic importance. In 1819 the first regular bath-house was erected. Between this date and 1830, the town as it now stands was created—partly by imperial ukases and grants, partly by the favour and influence of successive governors of the province. It was during this period that the hotel, the public gardens, the bath-buildings, and the roads in the neighbourhood, were for the most part constructed. In 1837 the Emperor Nicholas visited the Caucasus, and made an annual grant of 8,000 roubles for the maintenance and improvement of the bathing establishment.

Pätigorsk is the centre of the group of mineral springs, and the point on which the Government has concentrated its efforts to create a national bathing-place, worthy to rank with those of Western Europe. It is not, however, the only spot where mineral springs have been brought within the reach of the invalids; the 'Eaux minérales du Caucase' comprehend three other groups of sources —Geleznovodsk, ferruginous springs; Essentuky, alkaline; and Kislovodsk, acidulated carbonic. Our visit to the two latter I shall presently have occasion to describe. Geleznovodsk we did not see; it lies at some distance north of the others, and nearer the base of Beschtau; on the road to it, the colony of Karras, once inhabited by Scotch missionaries, is passed. The threatened extinction of the original stock led to the introduction of some German Lutherans, between whom and the Scotch such internal feuds arose, that the Government withdrew their support from the mission. In 1858 there was only one living representative of the original colonists, named Galloway.

Pätigorsk itself is one of the most curious phenomena of the Caucasus, and its incongruities were perhaps more apparent to us, coming upon it, as we did, fresh from the mountains. The first feature about the place that strikes one with surprise is, that, though standing far away from the last swells of the great range, in the centre of a bare and featureless plain, it yet contrives to be pretty. Its attractions are due to its position on the side of a lofty isolated hill, Machoucha by name, which has been planted of late years with wood. The Podkumok flows round the southern base of the hill, on the lower slopes of which the town is built; the hotel and best quarter are sufficiently high to command from their windows a noble panorama of the snowy chain— from Elbruz, standing out like a sentinel on the west, to the more distant summits of Dychtau and Koschtantau, on the east. The distance to Elbruz is about the same as that of

Mont Blanc from Geneva; the other mountains are from twenty to forty miles further away. The principal bath-houses, and the gardens which surround them, are situated in a sheltered hollow on the side of Machoucha. A long boulevard, shaded by a double avenue of trees, which have already reached a very tolerable size, leads up to the bath-buildings; the gardens are well laid-out, provided with numerous seats, and adorned with summer-houses, and some curious statues with Greek inscriptions found in the country. Nothing can exceed the cleanliness and comfort of the baths, those appropriated to the ladies being, with a thoughtful consideration for the weakness of the sex, even provided with large looking-glasses. On one side of a grotto, just behind the public library and reading-room, stands a brazen tablet, on which is recorded the expedition of General Emmanuel to the foot of Elbruz in 1829, the attempt and failure of the German savants to reach the top, and the supposed success of Killar in doing so. The story is, of course, written in Russian characters; we asked our companion, Dr. Smirnov, the head-physician, what it meant, and his reply was, 'Bah! c'est une bêtise.' Our own reasons for doubting Elbruz having found its Jacques Balmat in Killar, I have entered into elsewhere.*

The ridge which forms the southern boundary of the hollow in which the baths are situated is of a very extraordinary character. According to Dr. Smirnov, whose theory was certainly confirmed by the appearance of the surface, it has been entirely formed by the deposit of the sulphur-springs during past ages. The handsomest building connected with the waters is the Elizabeth Gallery, a long arcade, from beneath the arches of which a fine view of the town and the plain below is obtained. A zigzag path, shaded by thick oak copses, has been made to the top of Machoucha

---

* See Appendix I.—'The Elbruz Expedition in 1829.'

(3,258 feet), whither we climbed one cloudless morning, and enjoyed a perfect view of the great chain, from Kazbek, the crest of which was just distinguishable among the slanting rays of the newly-risen sun, to the double-headed Elbruz. The ugly little molehill, called the Yutskaia Gora, which from the town cuts off some of the lower portion of the mountain, is completely sunk, and the whole 8,000 feet of unbroken snowslope, falling towards the valley of the Malka, exposed to view. The monarch of Caucasian and European mountains brooks no rivalry; clothed in his wide-spreading ermine mantle, he stands forth a burly but not undignified sovereign, taller by the head and shoulders than any of his neighbours. The sharper peaks of Dychtau and Koschtantau are so distant that none but a trained eye is likely to appreciate their real height and beauty, and few who had not known them before would have noticed the twin summits of Uschba shooting up, keen as ever, over the intervening ranges. On the north, Beschtau was of course conspicuous; elsewhere the prospect extended over a boundless steppe, dotted by isolated mounds like the 'tells' of the Syrian desert.

A carriage-road has been lately completed round the base of Machoucha, forming a pleasant afternoon's drive for Pätigorsk society. We followed it as far as the sulphur-spring, called by Russians 'The Proval.' It is a natural grotto of the form of an inverted funnel, at the bottom of which is a deep well of sulphur-water. A Moscow merchant, who had benefited by the Pätigorsk springs, rendered the grotto accessible, by having a passage cut to it through the hillside at his own expense. Owing to its distance from the town, and the similarity of the water to others nearer at hand, it is now little used.

The hours kept by the patients are very remarkable. They dine from 12 to 4 P.M., and sup from 8 P.M. to 1 A.M.;

but despite, or rather perhaps because of, their dissipated hours, they look as sick and miserable a collection of men as one often sees. Their days are spent in drinking the waters and taking baths, or dawdling about the gardens smoking cigarettes, and listening to the strains of military music. At the time of our visit the 'Mabel Waltz' was the latest musical novelty in the Caucasus, and seemed to be very popular. The attractions of the town are not great; it consists, besides the boulevard and the villas on the hillside above it, of one long straggling street, and a shop-quarter put down on a dusty slope, with half-finished arcades ending in bare open spaces, after the untidy fashion common in Russia.

We occupied our time in roaming about the bazaar, and laying in stores for another week in the mountains, and occasionally went into the fruit-market to buy one of the huge water-melons which form the staple article of food of the people of the country. The discovery of a photographer afforded us some amusement, as the enterprising artist had been as far as Uruspieh, and had taken stereoscopic views on the road, many of which we were glad to purchase. On the whole, the attractions of Pätigorsk, to a passing traveller, are quickly exhausted; but our stay was rendered exceptionally pleasant by the kindness of Dr. Smirnov, the resident physician in charge of the bathing establishments, who bears the, to English ears, odd-sounding title of civil-general. On our calling at his house, Dr. Smirnov told us that he had already heard from St. Petersburg of our probable visit, and had expected us for some weeks.

To the bathers Pätigorsk is, I fear, sometimes slow, and the Government will scarcely succeed in their desire to attract hither any large portion of the crowd of Russians who annually visit the German Spas, until it possesses not only railway communication with Central Russia, but also the gambling-tables, which are apparently necessary, as a mental fillip, to the complete success of all water-cures. The weather during our stay was continuously cloudless; night and morning the serrated array of the Caucasus invited us to return into its recesses. Pätigorsk, owing to its position on a southern slope, is decidedly a hot place; and the constant sunshine drove many of the invalids and all the visitors to Kislovodsk, where a short course of the waters is generally prescribed after the sulphur-springs have had their effect. Dr. Smirnov proposed that we should make a day's excursion to Kislovodsk, a suggestion we were glad to adopt, more especially as all trouble was taken off our hands by the kind loan of the doctor's open carriage. Moore was, unfortunately, too unwell to accompany us; but Tucker and I set out, at 5 A.M. on the morning of the 7th, with four horses harnessed abreast, in the usual Russian fashion.

The road (I speak as a Russian) is simply a portion of the steppe where carriages ordinarily pass. It leads through a military cantonment, a row of tidy cottages surrounded by huge sunflowers, and then strikes across the plain in a south-westerly direction towards a green oasis already visible in the distance. On the north the symmetrical form of Beschtau is more than usually conspicuous; its loftiest summit is surrounded by four minor ones, so that, from every point of view, the mountain bears the same appearance, and may be compared to a Russian church with its four small cupolas clustering round the central dome. In the opposite direction, the snowy heads of Elbruz are constantly in sight, over the lower ridges that bound the plain on the south. It is seventeen verats from Pätigorsk to Essentuky, formerly a frontier-post, then a Cossack 'stanitza,' and now a bathing-place. The most has been made of an unpicturesque situation, by planting the ground round the springs, and laying out winding walks under the trees. The morning band was playing at the time of our arrival, and we met numerous patients rambling about the park, through which we ourselves strolled. The character of the landscape changes, and the road enters a shallow valley, where the Podkumok flows between low rounded hills, broken here and there by projecting masses of white rock. The country is covered with green pasturage, but entirely bare of trees. The ruins of old fortifications, still visible here and there on the flat hilltops, are records of the long period during which this was debateable ground between the Cossack and Tcherkess. We met on the way the omnibus which, for the convenience of patients, performs a daily journey between Kislovodsk and Pätigorsk, and *vice versâ*. The 'stanitza' of Kislovodsk, with its green-domed church, is left behind on the right, and the road, quitting the valley of the Podkumok, crosses a low hill, and soon descends to the baths, which have grown up round the most famous spring of the Caucasus. Kislovodsk is thirteen versts beyond Essentuky, and is situated in a narrow glen surrounded by low hills, which deprive it of any extended view; it owes its only claims to beauty to the rich vegetation with which the care of successive governors, aided by the natural fertility of the soil, has endowed it. A fine avenue of poplars leads up to the baths; the wood beyond consists chiefly of acacias.

We were driven to the 'Hôtel de la Couronne,' kept by the same manager as the hotel at Pätigorsk, where we found Dr. Smirnov, who proposed that we should at once visit the baths. The building which now covers the famous Narzan is in a style very far in advance of what one would expect to find in so remote a position. It owes much to the care of the late Prince Woronzoff, the general benefactor of Southern Russia, whose works and name are equally remembered at Tiflis, Odessa, and in the Crimea. The centre of the entrance-hall is occupied by an hexagonal basin ten feet in diameter, in approaching which a slight fizzing sound reaches the ear. This proceeds from the great spring, which bursts out of the ground with astonishing force, and is dignified by the Tcherkess name of the Narzan, or 'Giant's draught.' The whole surface of the basin is in a constant state of effervescence, owing to the escape of the carbonic acid gas, and its appearance resembles nothing so much as a gigantic goblet of very effervescent seltzer-water. A long arcade, open on the south to the sunshine, offers a promenade to the patients; the baths occupy portions of the same building, and there is a small swimming-bath, with numerous separate ones, in all of which the arrangements are of the best description. We took advantage of Dr. Smirnov's proposal that we should test the effects of the waters; he warned us to keep our heads well above the surface, a precaution necessary, from the quantity of carbonic acid evolved. We found our dip both invigorating and appetising, and returned quite prepared to do justice to the sumptuous lunch provided by the doctor, who, however, annexed one condition to the entertainment—that in the matter of drinks we should obey implicitly his prescription. This proved to be a mixture of champagne and the water of the Narzan, a preparation requiring skill, principally, in maintaining the just proportions.

The park—which, owing to its shade and coolness, makes Kislovodsk a favourite summer resort with all the officials of Cis-Caucasia, and even with those of Tiflis—had next to be visited. The first person we met was an old acquaintance, General Orlovski, the Governor of Tiflis, from whom we had parted at Kazbek posthouse; we were still more surprised to see with him Prince Ismail of Uruspieh, whose talents as a musician make his assistance valuable in the concerts which often take place here. We failed to discover precisely on what footing he stood with the Russian officers, but the impression left on our minds was, that the invitation he received amounted to a command, and that the Prince met with little superfluous courtesy from the habitués of the baths, to whose amusement he was invited to contribute.

The little stream which flows through the bottom of the glen is liable to sudden floods, and, despite the enbankments by which it is restrained, had lately broken loose, and done considerable damage. We were shown over a botanical garden, where the gardener cut and presented to us a beautiful bouquet of flowers. The walks through the woods extend, on either bank of the stream, for at least a mile above the hotel; they are nicely kept, and deliciously cool in hot weather. For those who do not require sulphur-baths, I have no doubt that Kislovodsk is a far more enjoyable summer retreat than Pätigorsk.

We had heard at Pätigorsk that General Loris-Melikov, the military governor of Cis-Caucasia, was staying at Kislovodsk, and we were anxious to call on him, to obtain such aid and advice as he could give in carrying out our plans for the next fortnight. In one of the detached cottage villas, built for the accommodation of visitors (as the 'Hôtel de la Couronne,' in fact nothing more than a handsome restaurant, contains no bedrooms) we found the General. An Armenian by birth, he is one of the numerous instances of the success attained in foreign service by that clever nation, which, like the Greek, seems capable of doing well everywhere except at home. We were received very courteously, my maps were soon spread out, and we pointed out the route we wished to take. Our plans were afterwards so fully carried out that I may here repeat, for the benefit of my readers, what was then explained to the General.

We had selected Naltschik, a small town and military post at the foot of the mountains, distant some eighty versts from Pätigorsk, as our new base of operations; thence we desired to push up the eastern arm of the Tcherek (to be distinguished from the better-known Terek, of which it is a tributary), and cross from its head, by a pass, called 'Par Stulevcesk' in the map, to the headwaters of the Uruch, the valley into which we had already looked down from the icefall of the Karagam glacier. After following this river for some distance, we proposed to turn, by a track crossing low spurs, to Ardonsk, the second station on the post-road on this side of Vladikafkaz. Our object in adding this supplementary piece to the programme with which we left England, was to gain some knowledge of the scenery and geography of the great mountain-group under the southern face of which we had rambled in Eastern Suanetia. Viewed on the map, this group appeared to resemble in shape the letter T, the top bar being represented by the watershed, from a point above Jibiani to Tau Tötönal, and the downstroke by a gigantic spur, in which are situated the second and third summits of the Caucasus—Koschtantau and Dychtau.

Had not what we had already accomplished been known, our wish to penetrate the recesses of the mountains would not have been so easily believed, and greater difficulties might probably have been suggested. The most useful piece of information we obtained was that the Stulevcesk Pass was practicable for horses—a very important fact, as it obviated the necessity of again encumbering ourselves with a troop of porters, and suggested the pleasing possibility of hiring horses at Naltschik for the whole journey to Ardonsk. The General kindly promised to write to Naltschik, and tell the Commandant there to expect us, and to take care that horses were forthcoming on our arrival. He also wrote to his subordinate at Pätigorsk to help us in hiring a carriage for the drive to Naltschik, the direct road between the two places not being furnished with post-horses. We took leave well satisfied with the result of our interview, though our minds were slightly troubled by a spectre raised by something the General had said about providing us with a 'specialist,' who would tell us where to go and where not to go, a kind of Mentor who would have found himself sadly out of place in our party. The threatened companion, however, perhaps fortunately for himself, was never assigned us.

Our carriage, which was awaiting our return to the hotel, took us quickly back to Pätigorsk. The hurry of our driver caused the only mischance of the day. Dr. Smirnov had told us that General Chodzko (who, it will be remembered, is the head of the Russian Survey which executed the Five Verst Map, and who had shown us much civility at Tiflis) was staying at Essentuky. Our coachman had been, as we believed, instructed to take us to the General's lodgings, and as he drove on, we assumed that the house must be at the Pätigorsk end of the scattered village; it was not till we were fairly beyond the place that we discovered he had no intention of stopping at all. Our ignorance of Russian made the mistake irreparable, and we much regretted thus to have lost the opportunity of talking over our experiences with one of the few Russians who have any real knowledge of the interior of the Caucasian chain.

On the following morning, the military commandant of Pätigorsk called, and proffered his assistance in any arrangements we might wish to make. Paul and the officer's servant set off together to find a carriage-master and make an agreement with him; but we did not gain much from the aid of the military, as the price asked was exorbitant, and we could obtain no abatement. The 8th was spent in replenishing our exhausted stores with such articles as the bazaar could supply; but, beyond the most commonplace necessaries, we found little in the shops except sweetmeats, which existed in every variety, from the wooden box of 'rahat-lakoum' to the gilded case of Moscow candied fruits. If the supply for sale is any index to the amount consumed, the baths must have a wonderful effect, not only in sharpening the appetites of the inhabitants of Pätigorsk for sweet things, but also in strengthening their digestions.

August 9th.—In order to avoid the heat of a drive across the steppe in the burning sunshine, we did not set out till four o'clock in the afternoon. During the day the aspect of the weather had changed, and the sky, hitherto unclouded during our stay, was hidden by dark masses of vapour, which, shortly before the time fixed for our start, discharged themselves in pouring rain. The turn of affairs was not pleasant, but we found consolation in the thought that we might have been worse off, and that, in such weather, a watertight 'tarantasse' was luxury compared to an open 'telega.' Bidding farewell once more to civilisation, we drove out into the desolate and now muddy steppe. In no other European country but Russia is the transition from the comforts and even luxuries of the towns to the barbaric lack of roads, bridges, and every necessary of intercourse, in the country, so marked. No Russian poet could have written, 'God made the country, but man made the town,' for such a sentiment would have seemed to his countrymen to savour of the grossest impiety. The country, at any rate in the steppe districts of Russia, is a wide featureless plain; the road—a mere track, converted into a sea of mud in winter, enveloped in a dust-cloud in summer —is often rendered wholly impassable by unbridged and flooded rivers. Some such reflections passed through our minds as, having left behind us five minutes before the handsome rooms of the 'Hôtel de la Couronne,' we found ourselves, after fording the Podkumok, plunging into the mud on its further bank. As far as Zonitzki, we drove along the same track as that by which we had arrived; thence we struck more to the east, and at last, after being refused admission into one farmhouse, on the very reasonable ground that the owner had that day buried his wife, we came to a halt, shortly before midnight, at another on the banks of the Malka. It was inhabited by a family of colonists, kind homely people, like most Russian peasants not connected with the postal service. The 'tarantasse' was put up in a shed in the yard, while we were introduced into the best room of the farmhouse, which was clean and tidy, but terribly close and hot, where we passed the short portion of the night still remaining.

August 10th.—At 5 A.M. we were again on the road. The sky was overcast, and we saw nothing of the mountains all day; but the absence of oppressive heat was a great comfort, and almost reconciled us to the loss of view. Crossing the Malka by a bridge, the track led us over low bare hills, until the banks of the Baksan were reached, where at a walled Cossack station, a remnant of past and more turbulent times, we learnt, to our dismay, that the bridge had just been carried away, by the floods caused by the previous night's rain; it was, however, suggested that if we could wait a couple of hours, it would probably be made passable. In a shop opposite the station, we found a room in which we sat down, while the 'samovar' was heated, and some eggs boiled. More than the appointed time elapsed, and yet no satisfactory intelligence came from the bridge; we settled, therefore, to drive down and ascertain the real state of the case, and whether it was not possible to ford the stream. We found that the central pile of the bridge, with the roadway, had been swept clean away; and as there seemed no prospect of a speedy repair of the damage, there was nothing to be done but to encounter the muddy flood. After our Juno experiences in the Araxes valley, this seemed to us, though a formidable, by no means a terrific task, and we encouraged our driver, by every means in our power, to dare the deed. He, however, did not view the matter in the same light, and entirely refused to adventure his precious life, except under the guidance of three Tcherkesses, who had just ridden up and offered their assistance,—for a consideration. After some delay, a bargain was struck, and driving some little distance up the stony river-bed, we forded successfully, though not without serious difficulty, the first branch; the second was easier, but the third was too much for our driver's pluck, and he flatly refused to go a yard further. We were, therefore, obliged to unpack all our baggage, and to ride across, carrying it on the horses of the Tcherkesses. The force of water was really formidable, and the wiry little steeds had some difficulty to maintain their footing. Safely landed on the further side, we had still to wait until sufficient horses could be found, in the adjacent village, to mount us and our men. For two weary hours, unable to leave our baggage, we sat fretting and fuming by the waters of Daksun, until Paul at last appeared with a bullock-cart for the traps, and saddle-horses for us.

A fresh delay now arose. The man of whom we hired the horses had prepared a meal for us, and we found it would be a gross breach of good manners to refuse to partake of it. We therefore entered his cottage, and, sitting down on a bench, were served with a dish of spelt-bread and toasted cheese, which was not unpalatable to hungry men. Though the Tcherkesses are Mahommedans, the women of the village took small pains to cover their faces, or avoid the eyes of strangers and infidels. They are a well-grown race, but there was nothing in any of their faces, except those of the very young girls, to attract a second look, and our host's wife, who attended on us, was a marvel of ugliness.

Having disposed of the food as quickly as propriety would permit, we jumped into the saddle, and set off at a canter across the grassy steppe, leaving the bullock-cart to follow more leisurely, under the charge of François. Once on the road, we lost no time, as our horseman was anxious to return the same night to his home, and urged us to push on. To gallop on a Tcherkess saddle is not a very easy or agreeable feat, but we were all tolerably successful, though I suffered shipwreck on one occasion, from rashly opening a map—a proceeding my horse resented with such an unexpected flourish of his heels, as to land me safely on the grass over his head, there to continue my researches at leisure. There was still another large river, flowing from the glaciers of the central chain, between us and Naltschik. This was the Tchegem, the upper valley of which is inhabited by a branch of the same Tartar race as is found at Uruspieh and Bulkar. The stream was coming down with great violence, but the frail-looking bridge had, fortunately, not as yet been carried away, although the catastrophe seemed imminent, and the structure was watched by a guard, who would only allow us to pass one by one, and at a foot-pace.

A low range of wooded hills now bounded the steppe we had been traversing all day, and we could see, at their feet, the buildings of Naltschik. As we drew nearer our goal, the hitherto barren soil was covered with scrub, and thickets of large dog-roses. The entrance to the town was a record of the old days, when it was exposed to the constant danger of attack from the mountaineers. The buildings had once been surrounded by a stockade, and the gateway at the entrance to the main street was guarded by a sentry. He enquired at once if we were the English who were expected, and despatched one of his comrades to guide us to the quarters prepared for us. We were taken to a well-built one-storied house, standing on one side of the open space, in the centre of the town; our hostess proved to be the widow of a Russian officer, who was glad to let us her front-parlour during our stay. It was a clean and cheerful room, with fairy-roses in the windows, and pictures on the walls; but there was the common Russian want of creature-comforts, and we could obtain nothing but our own mattress to lie upon.

August 11*th.*—We went off after breakfast to call on the Commandant, who had heard from General Loris-Melikov of our intended visit, and had detained some natives of the Upper Tcherek, with their horses, to accompany us on our excursion into the mountains. Our expressions of pleasure and thanks were suddenly cut short by the announcement of the price, five roubles a day (nearly fifteen shillings), we were expected to pay for each horse. As we were to take eight, it was of course out of the question that we should ratify any such arrangement; but at the moment we were too completely taken by surprise to say much; and the Commandant turned the subject, by proposing a walk in his garden, and showing us a seat, constructed to command a view of the distant mountains, the snowy summits of which are in clear weather visible over the lower wooded hills. He related to us a legend, current among the Tcherkesses, of an extraordinary treasure secreted on the top of one of these peaks, but it did not seem different from the tales of the same description common to most mountain countries.

It was evident that we should not be able to make a start early the next day, and we were forced to acquiesce in the suggestion, that an officer should call on us in the morning with the horsemen, and that the conclusion of any definite arrangement should be postponed for the present. Naltschik in itself is a neat little place, showing marks of its origin as a military cantonment, and gradually sinking, under the influence of more peaceful times, into a quiet country town, with broad streets shaded by trees, bordered by cozy-looking, green-roofed, one-storied cottages, each surrounded by its patch of garden-ground. There are several fairly-supplied shops, and in one Paul secured a ham, an article we had looked for in vain at Pätigorsk; while at another, we found, to our surprise, a bottle of very fair eau-de-cologne—a great boon to Moore and myself, who were out of sorts, and with neuralgic tendencies.

August 12*th*.—In the morning an officer called, accompanied by the two natives who, it was proposed, should provide us with horses. They were bluff hearty-looking fellows, one of whom emitted, from time to time, a most ferocious grunt, which prepossessed us, perhaps somewhat unfairly, against him. The debate was opened by an explicit refusal, on our part, to have anything to do with any previous understanding. The horsemen stuck to the terms originally suggested, and thus things seemed to have come to a deadlock. An adjournment was shortly agreed upon, and my friends accompanied the officer to the Commandant's house, where they spent two hours in protracted negotiations. It is very doubtful whether their

labours would have had any result, without the assistance
of a lady, the wife of one of the officers, and apparently
the only person in Naltschik who spoke any European
language but Russian. By her suggestion, a round sum
was offered the men for the whole journey, and at last
they agreed to take us through the mountains to Ardonsk,
with liberty to be ten days on the road, for 130 roubles.
At the rate originally suggested, we should have paid 400
roubles for exactly the same advantages. The mention of
these figures will suffice to warn future travellers, that in
availing themselves of assistance from Russian officials,
they must not leave any money arrangements in their
hands, unless they are willing to risk paying three times
the fair value for the services rendered. It was past
midday ere we shook hands over the bargain, and consequently our start was deferred until the next morning,
when our new attendants promised to have the horses
ready.

August 13th.—We set out for our ride at 6.30, after the
usual difficulty in collecting together all the animals, and
we finally left with one short of the promised number,
which the men undertook to make up on our arrival at
Balkar, the collective name of the highest group of
villages in the eastern branch of the Tcherek valley. We
were told it would take two days to reach Balkar, and
that, owing to the absence of shelter on the road, the first
night must be spent in the forest. The weather looked
unpromising; a dull grey pall clung to the hillsides, and
blotted out half the beauties of the landscape. Naltschik
is situated on a small stream issuing from the neighbouring
hills some miles west of the Tcherek, to reach the banks
of which the road crosses the shoulder of a low chain.
The ground was covered with tall coarse herbage and
thickets of small timber, mingled with wild fruit-trees,

among which we noticed the pear, the apple, the plum, and the medlar, their gnarled boughs hung with long creepers.

We came in sight of the Tcherek, where, issuing from the hills a broad rapid river, it strikes out into the steppe, and runs through marshy ground, fairly timbered in comparison to the barren tracts beyond. It was a desolate view, the surrounding country being absolutely in a state of nature, and showing no traces of the neighbourhood or care of man. Once fairly free from the hills, the river divides into a dozen branches, and, until it joins the Terek, flows through a dreary swamp, uninhabited except by wild boars. From the brow of a steep though short descent, we looked down on the hamlet of Dogüjokova, which, like most of the Tcherkess villages of the plain, consists of a long double row of one-storied cottages, surrounded by sheds and fenced-in gardens, calling to mind pictures of a South-African kraal.

We halted for lunch on the bank of a stream flowing out of the range that overlooks Naltschik. The direction of our course had now changed, and we were riding south-west in place of south-east. A few hundred yards of level land generally stretched between the Tcherek and the base of the hills, but once or twice the path was forced to climb a bold bluff, breaking down abruptly in chalk cliffs to the river. The meadows were clothed in luxuriant herbage, which, now uncut and ungrazed, was rapidly running to seed; the 'glossy purples' of the gigantic thistleheads specially attracted our admiration, and quite justified the Laureate in declaring that they can at times 'outredden all voluptuous garden-roses.' Occasionally a group of tall tombstones, each seven or eight feet high, capped by a carved turban, and chiseled with a long inscription in the language of the country, stood out above the long grass. We tried hard through Paul to get at their history, but could learn little definite, except that they were of considerable antiquity. A roadside tomb, recording the name and family of the departed, must have been an object of ambition with the former inhabitants of the country. The lower portion of the northern valleys of the Caucasus lying beyond the mountain gorges, easily accessible from the plain, and subject to sudden raids from the mountaineers, seems, despite its pastoral riches, to have been left as a debateable ground alike by Russian and Tartar. Beyond the hamlet I have mentioned, there is no dwelling-place on the Tcherek for a full day's journey, until the lowest village of Balkar is reached.

The hills on both sides of the river are rounded and monotonous, and the persistent fog robbed us of any glimpse there may be, under more favourable circumstances, of the snowy chain; so we had to be content with admiring the fine beechwoods on the opposite bank, trusting that, as we penetrated deeper into the mountains, the landscapes would become more striking. The scenery assumes a different character at the point where the western Tcherek, flowing out from amongst loftier hills, brings the tribute of the glaciers on the western flanks of Dychtau and Koschtantau to swell the main stream. The hillsides grow steeper and higher, and the range separating the Balkar and Bezeenghe valleys breaks down in a succession of most picturesquely-shaped and thickly-wooded bluffs. The western Tcherek, a strong body of glacier-water flowing in a narrow but deep channel, is spanned by a good bridge, beyond which the road, after traversing marshy meadows, is forced to climb over a projecting spur. The forest now began to change character, and there was greater variety among the trees; the sombre foliage of pines varied the lighter shades of green, and tall alders shot up amongst the beeches.

D D

After crossing for the first time the eastern Tcherek, at a most striking point, where it flows in a cleft so narrow that a man might almost have leapt across, we rode for half-an-hour through a wood, beautiful enough to demand a special word of admiration, even in this country of woodland scenery. The tall trunks between which the path wound were festooned with long streamers of creeping plants, and the lofty boughs that overarched our heads sheltered beneath them shrubs of rhododendron and azalea, growing to greater size than any we had yet seen. Crags jutted out from the green banks, affording a home for delicate ferns, and moss-cradled springs trickled down shady hollows. In an opening of the wood, we came suddenly on a round tarn fringed with grass, reflecting on its surface the surrounding cliffs and overhanging branches. The spot was so charming that we wanted to camp there, but our horsemen were obstinate, and the leader, with very decisive grunts, which there was no gainsaying, told us that we should find a much better place further on. About half-an-hour later we halted, after a ten hours' ride, under shelter of an overhanging rock, the black streaks on which showed that our camping-ground was not now used as such for the first time. Our tent was quickly pitched, and Paul set about his cookery; the horsemen unluckily discovered he was broiling some ham, and not only shunned him for the rest of the evening, but warned him not to pollute any of their saddles, bridles, or other equipments with his touch.

August 14th.—The same dull pall of cloud veiled the sky, although it hung higher on the mountain-sides than on the day before. The valley above our camping-ground was completely closed by precipitous cliffs, which seemed to form a barrier against all further progress. The path—already at some height above the Tcherek, glimpses of which could only be seen from time to time at the bottom of a deep ravine—turned abruptly upwards, and climbed rapidly through the forest. Having reached a height of at least 1,500 feet above the bed of the river, it struck boldly into the heart of the gorge, circling round ravines, and winding over the top of the perpendicular cliffs, where a fall from one's horse on the off-side would have led to a short roll, followed by a sensational header of many hundred feet. The vegetation, wherever it could find room to cling on the shelves and crannies between the precipices, was magnificent; pine and beech still predominated, though there was a sprinkling of other foliage. The way in which a single tree often crowned some projecting crag, where, destitute of any apparent source of sustenance, it yet contrived to maintain a vigorous existence, added much to the beauty of the defile. Alpine flowers now for the first time showed themselves in company with the most delicate ferns, and even the grandeur of the surrounding scenery could not altogether blind us to the presence of such old friends.

We could only appreciate the magnitude of the precipices immediately below us, when a bend in the hillside enabled us to look back on some portion of the road already traversed; those on the opposite side were even more tremendous. Halfway through the defile, its course is bent by a spur on the eastern side of the river, which juts out straight across the gap, and in fact does at one spot actually touch the opposite cliffs, leaving the water to burrow underground as best it may. The path descends on to the saddle connecting the rocky crown of this spur with the hillside from which it springs. This point, from its position, commands a view both up and down the defile, to which there is nothing similar, or in the least comparable, in the Alps. The gorge of the Tcherek is no mere crack in the lower slopes of the mountains, like those of Pfeffers and the Via Mala; it is rather a huge trench, dug down from their very summits to a depth of 5,000 feet or more. Behind us forest trees clung to every available inch of ground; looking upwards, the character of the defile was more savage. The foaming waters of the Tcherek, crossed three times by bridges, filled the bottom of the trench, the sides of which were perpendicular walls, succeeded by shelves, capped in their turn by a loftier tier of precipices. The path, a mere ladder of broken stones, brought us, by a rapid series of zigzags, to a most extraordinary spot, where the overhanging cliffs meet, and form a natural bridge over the river, which can barely be seen at the bottom of its deep bed. As we looked from this spot, the torrent to all appearance plunged directly into the bowels of the mountains, and it was impossible to discover how it found a way out of them. The savage grandeur of the scenery here attains its height, and no words will convey to others the impression it made on us.

Henceforth the cleverly-contrived rock-staircase which connects Balkar with the outside world finds room— now on one side, now on the other—to creep along the base of the cliffs at the river's edge; at last, when the careless observer would think it was hopelessly defeated, it crawls along the face of an overhanging bluff, by a gallery, partly cut into the rock, partly built out from it. This difficult passage surmounted, it leads an easier life; the mountains draw back from the river in two grand curtains of precipice, and the basin in which the hamlets of Balkar are situate gradually opens to the view. We now wound over barren and disintegrated slopes, broken occasionally by stone-capped earth-pillars, similar to those we had seen before in the Caucasus, on the Ardon, and to the well-known examples in the Val d'Herens in Switzerland.

Before reaching the first village, which is on the left side of the valley, we descended to and crossed the river. On an isolated crag above the houses—here, as at Uruspieh, flat-roofed stone cabins built against the hillside— stands a fortress, or place of refuge, which, properly defended, must have been impregnable, except to cannon or famine. The wide upland basin, now fairly entered on, is by nature bare and savage in its character, but has been rendered less so by the careful system of cultivation, which has converted every available patch of ground, not only in the valley but to a great height on the mountainsides, into a fruitful cornfield. After the neglect of natural bounties shown among the lazy tribes on the south of the chain, and also in the lower portions of the very valley we were now following up, it was strange to see how the industrious and well-to-do Moslems who dwell in these mountain fastnesses contrive to make the waste and desolate places 'laugh with corn,' thus putting to shame the slothfulness and consequent poverty of their Christian neighbours. We passed several hamlets on our left, and met numerous parties of men at work in the fields, before we recrossed the stream, and, mounting a gentle cultivated slope, entered Muchol, the village at which our horsemen wished us to stop.

We were, I believe, the first Western Europeans who had been seen in Balkar, and our sudden appearance gave rise to no small excitement. The male population surrounded us in the street; the womankind, being the property of Moslem lords, were obliged to content themselves with what they could see from the house-roofs. Their dress consists of a loose crimson robe, with a cap, from which a row of coins hangs down over the forehead. There was certainly one pretty face amongst them, and there may have been more, but no second opportunity of seeing any of the beauties occurred during our stay.

The Sheikh * himself, a tall venerable-looking old man, came forward to invite us to his house, which, like all the rest, was a low one-storied building, with a portico, supported on massive trunks, running along the whole of the front. At one end was a small room reserved for the reception of strangers, which we were invited to enter. It was, at first sight, a dark and comfortless-looking hole, but the lighting of a fire and the appearance of some bright-coloured mattrasses, which were brought for our use from the Sheikh's apartments, made us very contented—especially as our men were to be quartered in a separate house, where Paul would have abundant facilities for cooking. The 'samovar' soon appeared, accompanied by a dish of cakes, made, as the Sheikh took care to inform us, by the hands of his wives. The weather still looked so unsettled that we did not endeavour to hurry our arrangements, and were quite content with the prospect of spending a day in a place where we had met with so hospitable a reception.

Shortly after our arrival, a Mollah came in to call on us. He had given up the tunic and sheepskin of the Caucasus, and wore the turban and loose robe of an ordinary Turk. This was explained when we learnt that he had made the Mecca pilgrimage, and of course had acquired foreign manners on the journey. He seemed an intelligent man, had gleaned some confused knowledge of European politics, and knew at any rate the fact that England was reputed a staunch friend of Turkey, which made him very civil towards us. We were surprised to learn that the natives have so strong a dread of being made the subjects of religious proselytism, and being compelled to worship the relics and pictures

* At Uraspich I have spoken of Prince—here of Sheikh—maintaining a distinction which, though more nominal than real, was observed by all the people of the country, and even by Russians, with whom we talked on the subject.

of the thousand-and-one saints of the Russian Calendar, instead of the 'One God and Mohammed his Prophet,' that they would willingly, if they saw an opportunity, emigrate to some district still under the control of the Commander of the Faithful. Meantime they cling tenaciously to their old faith, carry its precepts into daily life, and observe its ceremonies. Muchol was the only place in the Caucasus where we heard the call to prayer resound night and morning through the village.

August 15th.—The weather was again gloomy, and the sun never appeared all day. The mountain-tops being hidden, it was useless to undertake any long expedition, and we contented ourselves with a short stroll up the hillside. Muchol, seen from above, has a most curious appearance; the flat grass-grown roofs of the houses, and the rough stone walls, give them more the look of a collection of burrows than of the comfortable homes of an industrious population. If the house in which our men were lodged was a fair specimen, the interiors are tolerably snug. Passing through a courtyard, we entered a large room, the walls of which were fitted with shelves, on which were ranged the brightly-painted trays in which Easterns delight, and pegs, on which hung sheepskins, swords and guns, with the other necessary equipments of a Caucasian when away from home.

All day long the Sheikh loaded us with a succession of civilities, in the very tangible form of relays of tea-cakes, and a kind of beer, peculiar apparently to these Mussulman valleys. Having finished our preparations for a sojourn of some days in the mountains, we determined, if the weather promised well, to start early next morning.

August 16th.—The clouds were more broken, and, for the first time for many days, patches of blue sky shone through them. With daybreak came the Sheikh, bringing in his wake a large supply of meat 'rissoles' smeared with honey, a finishing touch we could willingly have dispensed with. Not only were we well-feasted at the time, but, by the Sheikh's order, a number of these dainties were put into our provision-bag. On starting, when we had all mounted, a beer-jug was brought out, and a stirrup-cup presented to each of us, after which the Sheikh solemnly invoked 'Allah!' to prosper our journey. Having made what requital we could for the hospitality which had been shown us, we left Muchol, carrying away with us pleasanter recollections of its inhabitants than of those of any other village we had halted at. At Uruspieh we had, it is true, received almost equal kindness; but there the princes were imbued with a tinge of Russian manners, in contrast to which the patriarchal simplicity of Balkar was the more striking.

We left the village by the same road we had entered it, and recrossed the river to its right bank. We were some hundred yards beyond the bridge, when we saw a horseman, conspicuous by a green turban and streaming purple robe, riding after us. It was the Mollah, who, unprepared for our early start, had not been present to wish us 'Goodbye,' and now, arrayed in his best, came to repair the omission. After an exchange of Oriental salutations and farewells, including the hearty hand-shakings which are common alike to Tartars and Englishmen, our reverend friend wheeled round his steed, and followed by his servant, whom we made happy by a small present, returned home, while we pursued our journey. The path led us through a succession of cornfields, and passed two considerable villages beyond which a slight westerly bend in the direction of the valley hid what lay before us. When we had turned the corner, the character of the scenery underwent a rapid change; the cornfields and villages of the Balkar basin were left behind, and we followed the Tcherek for many miles, through a deep and trough-shaped valley, which almost deserved the name of a gorge. Tall granitic cliffs rose on either side of us too steeply to admit of any glimpse being caught of the gigantic peaks to which we knew them to serve only as foundations. The shelves and slopes were covered with dwarf firs, but the general aspect of the scenery was stern and savage. Barriers, formed of débris brought down by torrents pouring out of lateral ravines, stretched across from side to side, and made as it were steps in the valley, the level of which rises very rapidly, as it penetrates deeper into the mountains. The landscape was more Swiss in its character than anything we had lately seen, but, owing to the absence of villages or châlets, it was more savage than the generality of similar Alpine scenes. A slight turn in the course of the valley brought into view a graceful snow-peak, rising above the fork of the two glens which contain the sources of the Tcherek. We crossed a strong tributary flowing out of a cleft in the western hillside, which has its birth in a glacier (invisible from below), clinging to the cliffs of Dychtau.

Skirting the steep shelving bank of the river, we drew near the meeting of the two torrents, immediately under the spur projecting between the glens; on the right bank of the united streams, the mountains leave space for a broad and flat meadow, where herds of horses and oxen were grazing. The word 'Karaoul,' meaning (Paul said) 'guard,' printed at this spot on the Five Verst Map, had hitherto puzzled us, but we now learnt its purport. The pasturages at the head of the valley feed, in summer, numerous flocks, and it is worth the while of the community of Balkar to maintain a guard at this point, to prevent any predatory expeditions, on the part of their southern neighbours, such as we had witnessed in crossing the Nakra Pass. How the Mingrelians manage to find a way practicable for cattle across the chain it is difficult to imagine, as the easiest pass must lead over fields of snow and ice, far larger than those traversed in crossing the St. Theodule. We convinced ourselves that the ridge between this branch of the Tcherek and Suanetia is practically impassable; the robbers therefore must come either from the Rion or Zenes-Squali, which both rise on the southern side of this portion of the chain. We did not cross over to the meadow, but, after a seven hours' ride, halted under an overhanging cliff just below the junction of the streams—a spot evidently frequented by the shepherds, as low walls had been built against the rock to make the shelter more complete. Here we employed ourselves in pitching our tent, while one of the horsemen rode off to find the herdsmen, and obtain firewood and milk. The weather again looked unpromising, and we began to fear we had penetrated into the heart of the mountains to no purpose.

August 17th.—We had given orders that we should be called early, and the first sound that greeted our ears was that well-known and disheartening phrase of Swiss guides, 'Mais il y a du brouillard.' The curtain of mist, that hung only a few hundred feet over our heads, did not appear dense, so we determined to go up the nearest hill, and trust to Providence to show us something when we got to the top. The pass to the Uruch, which we followed the next day, crosses the two streams above their junction, and it is at the bridge over the first that the guardians of the flocks reside during the summer months, in a tiny stone hut. Our first intention was to ascend the gorge of the Dychsu, as the western branch is called on the map; but we were overcome by the pantomimic demonstra-

A GREAT GLACIER. 411

tions made by the guardians, to show the impossibility of this course, and were induced to climb the great hillside which rose steeply on our right. We soon reached the level of the mists, and in half-an-hour had left them far below us, and were enjoying unclouded sunshine.

Peak in the Tcherek Valley.

After a long and severe climb of 3,000 feet, up grassy slopes broken by crags and ravines, and covered in places with Caucasian rhododendrons, we gained the brow of a spur, whence we had a panorama of the ranges surrounding

as worthy to be compared with the views from such Alpine summits as the Gornergrat or Æggischhorn. Immediately at our feet, looking south, lay an immense glacier, the source of the Dychsu, fed by an accumulation of névé, filling two great basins, separated by a rocky ridge which projected from a noble ice-crowned mountain opposite. Far away to the right, above the western bay of the glacier, rose a tall peak culminating in a slender point of snow, which, though not corresponding exactly in position with the Koschtantau of the map, we assumed must be that great mountain, 17,095 feet in height. Looking eastwards, the snout of a glacier pushed round the corner of a cliff that concealed the whole of its body, and formed part of a great snowy buttress of the main chain, dividing the two heads of the Tcherek valley. The ridge over which our pass to the Uruch was to lead us closed the eastern glen, above which a noble rock-peak shot boldly into the air, sending down from its flanks a small highly-crevassed glacier.

The extraordinary feature of the view was the steepness of the chain; the peaks and the gaps between them seemed equally difficult of access, and cut off from the lower snowfields by long slopes of glistening ice and unscaleable walls of crag. The only object we had expected to see, and could not discover, was Dychtau, the companion of Koschtantau in all distant views. After a discussion whether it would be better to descend on to the glacier, or to climb higher, it was determined that I should remain and endeavour to make some outlines of the surrounding peaks, while François and my two companions went off to try and get a view of Dychtau. They so far succeeded as to catch a glimpse of his peak, the rest of the mountain being cut off by lower ridges. We loitered away some hours on the flowery pasturage, supplied with water by the meltings of the beds of snow which still lay in the hollows. Part of our time was occupied by a grand council on our future plans, and especially on the expediency of attempting the ascent of either of the great peaks. Dychtau looked on this side absolutely inaccessible, and the chances of success in any attack on the still loftier Koschtantau appeared so slight that we determined to leave him alone also. The base of the mountain was a day's march distant over the glacier, and the difficulties offered by the final peak were such as would require a strong party of step-cutters to overcome, if they were not altogether insuperable. No help could be expected from any of our native companions, and we did not wish to mar the success of our journey by undertaking an expedition leading to almost certain failure. The second and third summits in the Caucasus and Europe are, therefore, still not only unscaled but unattempted peaks. We strongly advise any mountaineer who may think of assaulting them, to go first to the Bezeenghe valley and inspect their western flanks, which may possibly prove less formidable than the defences on this side. Having decided that our best plan was to cross the Stuleveesk Pass while the fine weather lasted, we returned rapidly to our camp, where we found Paul and the horsemen getting on very well together. During our absence a hunter had brought a bouquetin, of which Paul had bought a portion, which he was busily preparing for our dinner. His companions had got a lamb from the shepherds, and were also making ready a feast after their own fashion. Paul's exertions produced a capital meal, and, cheered by the fineness of the evening, we looked forward with pleasure to the pass to be accomplished next day.

August 18th.—When the time came for arranging the baggage for a start, a very unexpected difficulty arose with our men, who declared that the horses were overloaded, and could not possibly carry the baggage over the mountain unless their masters received higher pay. While declining to accede to their demand, we pointed out that we should all walk the greater part of the way, and that the baggage might be subdivided, as our two men would be perfectly content to have one horse between them. By this means, after a vexatious amount of palaver and delay, the question was settled, and we set out up the eastern branch of the Tcherek. After passing over the level meadows, and the stony bed of a stream flowing from the glacier of an unpronounceable peak which rises grandly on the left, the path climbs a gentle ascent, whence the tops of both Dychtau and Koschtantau are for a moment visible together, and then finds itself in another plain, apparently an old lake-bed. A strong iron-spring bursts out under the hillside, and colours half the plain with a bright-red deposit.

On the opposite side of the lightly-wooded glen, a large glacier pours over the cliffs, descending from the same snowfields which feed the infant Rion on the southern side. The termination of the glacier is most peculiar and picturesque. The frozen flood descends in one great sheet, until it reaches the edge of the line of cliffs immediately over the valley, and then separates into two portions; the larger pours down in an icefall, broken into the usual minarets and towers; the other keeps a course parallel to the river, along the top of the cliffs, until it finds a curious cleft, into which it plunges, and shoots forth a long tongue of ice. Mineral waters, fine air, and mountain excursions are all ready prepared, and some day perhaps a comfortable bathing establishment may make this spot a centre for mountain explorers.

The main stream finds its way, both in and out of this basin, through deep clefts, and the path makes a rough climb along the hillside. Tucker's horse here stumbled over some rocks, and, but for the rider's quickness in getting clear of the animal, he might have had his leg crushed and been seriously hurt. Happily, he escaped with a cut knee and a severe shaking. The course of the upper glen was now nearly due east and west; the birch-wood gradually ceased altogether, and the ridge of the Stulevcesk Pass closed the view in the distance. On the hillside numerous flocks and herds were feeding, and we passed the quarters of the shepherds, who, as usual in this country, content themselves with a temporary shelter built up of a few branches covered with sheepskins. On our right a steep broken slope supported the glaciers of the main chain, far less precipitous in this portion than further west or east; it is here that the known pass to the Rion valley crosses the mountains. We met two of the guardians of the district: one, an old gentleman, who looked as if he ought to have retired from active service, irritated us by descanting to our horsemen on the difficulties before them, and the impossibility of getting horses over the snow, except in the early morning while the surface was still frozen.

In the teeth of such remonstrances, we had some difficulty in persuading our men, always on the look-out for an excuse for halting, to go on, and when we finally succeeded, the old guardian and his companion came with us, to give their aid in the perils they declared we must inevitably encounter. The head of the eastern Tcherek is a wide pasturage, into which pour streams both from the central chain and from the recesses of the northern range, the summits of which are exceedingly bold rock-peaks, lofty enough to support a considerable amount of snow and ice. The highest source issues from a mass of old avalanche débris, covering the end of a small glacier falling from the upper snowfields.

The final climb to the pass is very steep for horses; it has been rendered easier in places by the construction of a zigzag path, but near the top the snow entirely covers the surface. It was satisfactory to find that the old native's alarm was unfounded, and that, as we expected, the snow had only melted sufficiently to give the horses firm foothold.

We had hurried on, to secure a clear view from the pass before midday clouds blew up and obscured any of the summits. The actual ridge is a thin and steep comb of rock, probably about 10,000 feet in height. The views from it are superb, and we were exceptionally fortunate in seeing them almost unclouded. Looking back from our present position, we commanded the whole eastern face of the group of the Central Caucasus, as I have before designated the great cluster of granite peaks, of which Koschtantau appears to be the highest. On the left, the steep peak we had previously identified as the Nuamquam, which looks down on Jibiani on the southern side of the chain, was seen above the snowy head of another eminence, rising out of the snowfields of the great glacier we had noticed in the morning. Connected with the Nuamquam by a long snowy curtain was the massive, but yet graceful, pile of crag and ice constituting Koschtantau. The day before we had only seen one end of the magnificent mass; the fuller view we now obtained of its rocky buttresses, fretted icefalls, and high-pitched slopes, fully confirmed us in the wisdom of our decision, to leave this triumph of mountain architecture alone in its glory.

At the northern foot of the mountain, at the head of the Dychsu glacier, a comparatively low gap leads somewhere—it would be most interesting to discover whither. An explorer would probably be able to descend from it into the Bezeenghe valley. On the right of this break in the chain stand two rocky giants, the northern unquestionably the Dychtau of the map, a terrifically steep-sided peak, with a glacier flowing in a deep trench at its foot; the other a nameless summit, certainly exceeding 16,000 feet in height, and too precipitous to bear much snow or ice. While admiring, and endeavouring to carry away a recollection of, the noble outline of the group I have attempted to describe, we could not help regretting that such grand mountains should have been deprived of their legitimate honours by a mere volcanic accident like Elbruz. In Koschtantau and Dychtau the Caucasus would have had a worthy king and queen. Elbruz is at best a bloated monarch, and has little beyond size to recommend him.

Immediately to the north was the serrated chain which separates the upper valleys of the Tcherek and Uruch from the minor glens, the streams of which flow independently to the plain, between the two rivers. Looking east, the conspicuous feature of the near view was a projecting summit of the main chain, which reached a greater vertical elevation in less lateral space than most mountains. It rose immediately out of the valley in a series of precipices, separated by most disproportionately small ledges, and I believe, if the icy cap of the mountain had toppled over, there would have been nothing to stop its fragments until they reached the bed of the Uruch. The greater portion of the Adai Khokh group was hidden by the nearer ranges; beyond some of its northern spurs, two or three distant snow-peaks were visible, in one of which we easily recognised Kazbek.

The train of horses followed us quickly up the pass, and when they rejoined us on the summit, we found the men in high good-humour at the unexpected easiness of their day's work. The first descent on the eastern side was down steep frozen banks, where the horses, owing to the perfect condition of the snow, and their own experience in such work, found no difficulty, and the assistance of the timorous old gentleman was scarcely needed. On reaching the first grass, we satisfied his claims, and bade him and his companion farewell. There was very little track to guide us on this side of the mountain, and the horsemen did not seem well acquainted with the way; so that when, after following a little stream amongst rich pasturages, we emerged from its glen, on the verge of a steep descent leading into the trough of the Uruch, we found ourselves too much to the left, and the horses had to make a slight circuit to reach the valley. The source of the Uruch is in a moderate-sized glacier, chiefly noteworthy for the streaks of red snow that lay on its central portion; a pass may probably be found over it to the Rion valley.

We now ran down a rocky slope, with the precipitous peak we had seen from the pass immediately opposite. In this case nearer approach did not diminish the grand effect of its cliffs, and we agreed that it deserved to rank among the sensational features of the Caucasus. With such an object in view, the scenery could not be dull, but, putting aside the surrounding mountains, there was little to attract notice in the long flat glen through which the Uruch runs quietly for a while before it plunges downward through the ravines we were to traverse on the morrow. The torrent had made the level ground a museum for specimens of the rocks of the main chain, brought down by its parent glacier. Further on we came to a stretch of herbage, where numerous springs burst out of the ground at every step, frequently converting the path into a watercourse. At a point marked 'Kut' in the map, a stream from the northern range, the peaks of which are almost dolomitic in the fantastic boldness of their forms, makes a graceful leap over a wall of crag before joining the Urach. Here we saw traces of human habitations, and a group of peasants, in quaint-shaped wideawakes, employed in haymaking.

Having been on the road for ten hours, we thought it was time to stop, and instructed Paul to make enquiries of the natives, and find out if they were disposed to allow us to take up quarters for the night in the half-underground hut we noticed close by. The chief man among the peasants, who talked a dialect Paul had the greatest difficulty in comprehending, did not at first sight impress us favourably. His expression of countenance strongly resembled the masks worn by country bumpkins in a burlesque, and his comical, not to say idiotic, appearance was heightened by a quaint bell-shaped wideawake considerably the worse for wear. At first there seemed some question whether a welcome would be offered us, but any difficulty was soon removed, and we proceeded together through the hayfield, by the wall of which the parley had been held, to the door of the hut. Being built against a steep slope, part of it was underground, while the front projected from the hillside. The back was occupied by an extensive but gloomy stable, which was entered through a perfectly clean chamber, where our men passed the night; as the weather was fine, we preferred to pitch our tent outside. The pig-faced peasant, against whom we had at first sight conceived such an unjust prejudice, turned out a capital fellow. He brought us not only fresh milk, but a peculiar species of liquor, something between publichouse beer and sour cider, for which we expressed the greatest admiration, taking care at the same time privately to empty out the vessel containing it, on the first opportunity.

This was the fifteenth and last occasion that we slept under canvas in the Caucasus, and as Eastern travellers commonly base their ideas of tents on the spacious pavilions raised for their accommodation by Syrian dragomen, I may take this opportunity of describing the little structure for which, owing to the satisfactory manner in which it had protected us from cold, wind, and rain, we had by this time acquired quite an affection. The framework consisted of two pairs of poles, jointed for convenience like a fishing-rod, and stoutly shod with iron spikes. A single rope, forming the ridge of the tent, was passed over the fork of either pair, and attached at both ends as tightly as possible to the ground, by pegs or boulders. By this means, and by forcing apart the spikes of the poles, the canvas which was attached to them was stretched taut in every direction. I may add that the flap serving as a door could be securely tied across the entrance, and that the floor formed one piece with the sides, so that the weight of our bodies served as an additional safeguard against the risk of being blown over during the night. The internal arrangements were of a simple character. Having first fastened in a second waterproof floor, we spread out our light mattrass, bought at Erivan, laying it across the tent. The foundation of our pillows consisted of our thick boots; upon these we laid our revolvers and cartridge-pouch, crowning the edifice with a coat or mackintosh. All being then ready, we took up our respective positions; the dimensions of our snuggery were six feet square, so that no one could be restless at night without rousing his neighbours, but we found practically that each had sufficient room to sleep comfortably.

When the hour of our evening meal arrived, we ate our food seated tailor-fashion in a row on the mattrass, Paul pushing in to us, from the entrance at the other end, the various viands, tempting or the reverse, which he had prepared. The cooking was accomplished by means of a portable apparatus brought from England, combining, in a comparatively small compass, two saucepans, a fryingpan, a tripod, a Russian furnace, and a drinking-cup. Our standing dish as far as Uruspieh, at which point our store became exhausted, was soup made from the essence of beef-tea procurable in London, and Liebig's 'Extractum Carnis,' which, combined in nearly equal proportions, and with the addition of a tablet of Chollet's 'Compressed Vegetables,' produce a very palatable and supporting beverage. Dinner finished, we pushed out the plates and cups that had served our turn, cleared our pockets of heavy articles, such as aneroid barometers, compasses, and pocket-knives, that might otherwise have disturbed our rest, and, wrapped in our respective rugs, resigned ourselves to such slumber as the state of our consciences, or of our digestions, would allow.

CHAPTER XIII.

THE URUCH VALLEY AND RETURN TO TIFLIS.

Wooded Defiles—Styr Digor—A Halt—We Meet a Cossack—A Rainstorm—Zadelesk—The Gate of the Mountains—Across the Hills and Through the Forest—Tuganova—Novo-Christiansky—A Christian Welcome—A Wet Ride—Ardonsk—A Breakdown on the Steppe—Vladikafkaz—A Diligence Drive—The Dariel Gorge—Return to Tiflis—Reflections on the Caucasian Chain—Its Scenery and Inhabitants—Comparison with the Alps—Hints for Travellers.

August 19th.—The morning was cloudless, and we were fully prepared to enjoy our day's journey down the valley of the Uruch, the scenery of which, running parallel, as it does, to the central chain for many miles, must, we believed, prove in the highest degree interesting. We had to experience a fresh illustration of the perversity of the Caucasian nature before starting, and the result of the trifling concession we had made on the previous day showed the inexpedience of yielding any point in dispute with these unmanageable mountaineers. Paul and François having walked by turns over the pass, the horsemen now required that they should be content with one horse between them, while the men themselves rode. Wishing to push on as far as possible during the day, and knowing well that the animals were not overladen, we refused to start until the ridiculous proposal was withdrawn. The promise of our entertainer that, in case of need, he would find horses, enabled us to take rather a high line. After a warm debate—in which the feelings of one of our horsemen became so intense that, unable to express them by words, or even by the remarkable grunt in the emission of which he was so proficient, he gave point to his remarks by demonstrations with his dagger—they wisely succumbed, and we became as good friends as ever.

On parting, the chief of the hay-cutters requested us to take the side of his village in a disputed claim as to pasturages, which the local chiefs had been unable to settle, and was likely to come under Russian arbitration. Following the custom of the country, we promised to do anything in our power, and, after an affectionate hand-shaking, set off for our day's ride with our no longer refractory though somewhat sulky attendants. For some distance below the hut, the Uruch continues to flow through level and marshy meadows. The frequent water-channels were serious impediments to our progress, as the mud in them was often deep, and leaping is not a part of the education of a Caucasian horse. Our baggage-mule, a plucky little animal, whose performance on the pass had won our admiration, now stuck hopelessly in the mud, and had to be unloaded, an incident which recalled, to Tucker and myself, some of our Syrian misadventures. When the river does begin to descend, it does so with a will, and the path is obliged to have recourse to very steep zigzags, in order to keep pace with it.

A description in detail of the constantly-shifting features of the landscape for the next few miles would be wearisome, and give little idea of the beauty of the scenery. The path, always at a considerable height, on the left side of the valley, carried us by an alternation of terraces and rapid descents through a succession of defiles, adorned by fine fir-forests, enlivened by pretty waterfalls, and overhung on the right by noble snowy ranges. On two grassy knolls, high on either side of the stream, were groups of the half-underground huts, that here fill the place of châlets, and numbers of their occupants were haymaking in the adjoining meadows.

The dwellers on the Upper Uruch, known to the old writers on the country as 'Digors,' are, as far as we saw, the best specimens of Eastern Christianity in the country. We understood, at the time, that they were a branch of the Ossete tribe, whose territory they border on the east, but the latest authors seem to class them, with the people of Balkar and Uruspieh, as a Tartar race. A conical-shaped wooded eminence closed the view of one of the bends in the defile, and formed an admirable foreground to the peaks of the main chain. On reaching its foot, where the path abandoned the hillsides, and returned to the bank of the torrent, we found ourselves in a region of richer and more varied foliage, and constantly admired the contrast between the bright green of the lower slopes and the white shimmer of the peaks that overlooked them.

Leaving behind us the conical hill, we emerged into an open valley, and a basin surrounded by a semicircular range of precipices opened on the right. Several glaciers spread their icy skirts to the edge of the cliffs, over which streams tumbled in picturesque cascades; while, crowning all, snow-peaks, of the inaccessible order of mountain architecture which might fairly be called Caucasian, lifted their bold heads against the sky. It was a scene similar to the cirque of the Diablerets, but on a far grander scale. We rode across the rich and well-watered meadows on the left bank of the Uruch, until, under the shade of a picturesque group of chesnut and lime-trees, we halted, to wait for our horsemen. When they rejoined us, we continued our journey, and, leaving the first hamlet of the valley on our left, passed through cornfields, almost ripe for the sickle, either carefully fenced in, or only guarded by a few stakes, according to the character of the owner.

Styr Digor, where we intended to make our midday halt, is the principal place in the valley; but our horsemen, for some reason we were unable to comprehend, had made up their minds to go on to a comparatively small hamlet, an hour's ride further down. The backward view of the semicircle of ice-clad peaks and the rich pastoral foreground was very fine. At the hamlet of Moska, we found only an old woman, who had nothing to offer us but eggs, and even these she had some difficulty in collecting, as her hens had apparently mislaid them. After a shorter halt than usual, and having exchanged salutations with a priest in the orthodox clerical attire of a long sack-shaped coat and soft felt wideawake, we rode down the valley, anxious to get as far as possible on the road before nightfall. During our halt the beauty of the day had departed, and the clouds had rapidly swept up over the range, threatening rain, which began to fall slowly later in the afternoon. The path again descended to the banks of the Uruch, and the glen of its first considerable tributary, the Karagam, opened on the right. We looked out eagerly towards its head for the great glacier by the side of which we had bivouacked a month ago, when we had crossed the main chain from the valley of the Rion. The glen was longer than we anticipated, and nothing could be seen except the snout of the ice-stream on the upper portion of which we had spent so many hours.

The valley of the Uruch, having run thus far east and west, or nearly parallel to the chain, turns sharply to the north, and the scenery becomes less interesting, though the rich woods and broken crags of the continuous defile through which the river flows would attract the admiration of travellers in any region less richly gifted with mountain beauties than that we were now traversing. The road, still keeping the same course on the left bank, and now wide enough for the passage of carts, is shaded by dense thickets of hazels.

We were jogging along quietly, when we encountered a man in the ordinary Caucasian costume, but with a military air, which distinguished him from the common native. He stopped to speak to us, and, Paul being summoned to our aid, we learned that he was a Cossack in the Russian service, sent by the Commandant of Ardonsk, who had received his instructions from General Loris-Melikov to aid us in our journey. He was now on his second trip to Digor; having been previously unsuccessful in finding us, he had returned to headquarters only to be sent back again, with instructions to wait till our arrival—a piece of unlooked-for attention on the part of the Russian authorities, for which we felt the more grateful, as our horsemen had lately shown signs of renewed insubordination. The readiness to assist us, and appreciation of the real object of our journey, shown by the officials on the northern side of the Caucasus, was very gratifying, and our best thanks are due to General Loris-Melikov and his subordinates for the aid they so kindly gave us.

We soon passed a considerable village, and following for many miles a terrace-path, along steep hillsides, always green and picturesquely broken, approached the point where the Uruch turns due north, to fight its way through the limestone range which flanks the central chain. Having first passed the main stream by a lofty bridge, we crossed an extremely narrow cleft, from which a large eastern affluent issues. The heavy clouds and drizzling rain, combined with the bare character of the scenery, the frequent apparitions of tall tombstones by the wayside, and the tumbledown, and in many cases ruinous, state of the towers and groups of farmhouses which dotted the slopes, were in dismal contrast to the visions of beauty and unclouded sky which had made the earlier hours of the day so enjoyable. The path first mounted steeply, and then wound at a level round the ravines and promontories into which the hillside was worn by the action of time and weather. Occasionally, we passed a patch of cornland, but the ground was generally stony and uncultivated, and there was no shelter of any kind from the fury of a tremendous storm of wind and rain which attacked us at this moment.

For the last hour of our journey we were exposed to a perfect waterspout; the growing darkness added to the difficulty of guiding our horses against the storm, and we were glad when the stone-houses of Zadelesk, which the Cossack had pointed out to us, some time before, as the best halting-place for the night, were safely reached. The Cossack was now of the greatest use; he took the arrangement of our sleeping-quarters into his own hands, and we were soon invited to remove from the comfortless hovel, in which we had taken refuge, to a snug little room, where mattrasses were found for us, and we passed a very comfortable night. The day's ride had been long; with only short halts, we had been on the road for fully eleven hours.

August 20th.—To our surprise, after the storm of the previous night, the morning was calm and clear. The position of Zadelesk is very curious; perched on a high brow at least 1,000 feet above the Uruch, it is separated from the villages on the opposite side, which are comparatively close as the crow flies, by a tedious climb of several hours. The trough through which the river flows is too narrow to be inhabited, and the hamlets are perched, like swallows' nests, halfway up the steep mountain-walls.

Having paid for our accommodation, we started for the last day's journey over the Caucasian by-roads. The whole valley of the Uruch, from Moska downwards, is so narrow as to merit the title of a gorge; but the principal defile, where the river and the mountains meet for a final struggle, commences a little more than a mile below Zadelesk. A meadow shaded by some fine walnut-trees, and enlivened, when we passed, by a large herd of cattle, affords a pleasant resting-place for those who have won a way from the plains through the long gorge, and thence a glimpse may be caught of the shining tablelands and snowy summits at the head of the Karagam glacier. Immediately behind the meadow towered a splintered comb of rock, with a deep fissure cut down into it, the subject of a legend, in which the Devil, as usual, plays the chief part. The path winds round the side of this spur, and though kept in careful repair, and nowhere difficult for horses, the precipices below us, on our left, were so startling that we preferred walking, to running any risk from a chance slip of one of the animals.

Fairly within the jaws of the mountain, we sometimes had to descend into a lateral cleft, at others to climb over rocky teeth, until we came to a spot where, the roadmakers having apparently given up as hopeless the task of penetrating deeper into the gorge, the path began to zigzag steadily upwards. The Uruch could be seen only from time to time, fretting and foaming between the narrow walls of its prison. On its opposite side the range rose in stupendous limestone cliffs fringed with firs, mingled with a variety of deciduous trees. As we mounted higher, more summits of the snowy chain behind us came into sight, but the converging cliffs of the defile still cut off all view towards the north. A long and hot ascent brought us to a brow crowned by one of the usual tall Mahommedan tombstones, capped by a stone-wrought turban.

Although expecting to see from this point something of our onward course, we were quite unprepared for the unique beauty of the landscape which was suddenly spread before our eyes. Standing, as it were, in the gate of the mountains, at a height of probably not less than 3,000 feet above the level of the Uruch, we looked out over the low country which lies to the north. A broad hilly district, clothed in the densest primeval forest, here separates the mountains and the steppe. In the clearings, few and far between, a thin wreath of smoke revealed the existence of human dwellings; but the country seemed very little removed from a state of nature, to the continuation of which the insecurity and lawlessness, consequent on its position as a border-ground between the Russian posts on the steppe and the inhabitants of the mountain recesses has chiefly contributed. The immense extent of undulating woodland stretching to the horizon, and the rivers—which, unbridged and unconfined, converting their immediate banks into swampy jungles, wander like bright flashes of light across the green landscape—convey to the traveller's mind the impression of a rich virgin country, such as he would rather expect to meet with in the New World than in the Caucasus, the supposed cradle of his race. We remained for some minutes riveted in admiration of the scene, and unable to think of anything within our experience of more civilised countries with which to compare its rich yet melancholy effect. In this extraordinary spot we bade farewell to the mountain fastnesses of the Central Caucasus, and it would have been difficult to quit that no longer mysterious region by an exit more calculated to leave on our minds imperishable recollections of its sublimity.

Our Cossack told us that in the old days a native guard was stationed on the knoll where we were seated, to protect passers-by from the assaults of robbers, who were wont at times to leap out from behind the rocks on their victims, and hurl them over the adjoining precipices. The frequent tombstones we had passed are, in many instances, monuments to these unlucky wayfarers. The path now turned sharply to the east, and crossed a brow where the limestone rock cropped out on the surface, and was strangely split into clefts and crannies, which caused some embarrassment to the horses. Having now fairly emerged from the gorge, we for a time turned our backs on the Uruch, and rode across a steep hillside, until we reached the top of the low ridge forming the watershed between that river and a number of streams which flow down through densely-wooded glens to join the Terek further east. The track somewhat suddenly entered a thick forest of entirely sub-alpine character; the beech was the principal tree, but there was a great variety of foliage, and the usual dense underwood of rhododendron and azalea, some of the plants of the latter being on a level with our heads as we rode between them.

We could not, however, give our undivided attention to the sylvan attractions of the scenery, as our progress was rendered more or less difficult by the swampy nature of the ground, aggravated by the late rains, and by the low branches which, had we not been ruthlessly cropped by a Pätigorsk barber, would have threatened us with the fate of Absalom. The weather most unluckily again turned bad; rain threatened, but held off till after luncheon, which we enjoyed on a freshly-mown meadow. The path continued to follow the watershed, and, after riding along the bare crest, we turned sharply down a spur separating two deep but narrow glens, and again entered the forest.

At the same time the rain began to fall in torrents. How we all safely accomplished the next two hours' ride is still a mystery to me. The rain was blinding, the track was so deep in mire as to be in places almost impassable, the trunks and branches put us in constant fear of concussion of the brain; yet, holding our horses in hand as best we could, we pushed on at a brisk pace headed by the Cossack, who occasionally looked round with an air of surprise at seeing his convoy still keeping up. At last we emerged above the village of Tuganova, built at the foot of the hills on the left bank of a stream which, issuing from a wooded valley, here finds its way to the plain.

A well-built whitewashed house, with verandahs, was conspicuous at the entrance of the village; it is the residence of a native chief, who has been gratified with a military rank by the Russians, and is now known by the imposing title of General Tuganova. He is a Mahommedan, and therefore the owner of several wives, some of whom we admired in passing. It was suggested by the Cossack that we might claim the General's hospitality, but as he also warned us that, should the rain continue, the stream would probably be impassable next morning, we determined to prolong our day's journey to Novo-Christiansky, said to be ten miles distant. The questionable stream, a very rapid one, was barely fordable now; but we managed to get through it safely, though the Cossack's horse went into a hole, and wetted him up to the middle. The wood gradually thinned, until it ceased altogether, and we found ourselves on a dismal steppe, only relieved by occasional mounds resembling artificial tumuli. The ride, on a pouring afternoon, was gloomy enough, and twilight was deepening apace, when we caught sight of the distant church of Novo-Christiansky, a name derived from the late simultaneous conversion of the whole village to the Christian creed—a change, I believe, not unconnected with a remission of taxes. The low houses did not come in sight for some time, and it was quite dark when we rode in through the lanes of mud which serve as streets.

Our Cossack was himself a native of this village, and directed us to a detached building, where we sat in an open verandah, in damp and dismal suspense, while he searched for the key. When this was found we gained admission to a village-shop, behind which were two rooms barely, though for the country pretentiously, furnished. Having been on the road all day, we were now wet, tired, and hungry, and tried to get a fire, a mattrass of some sort to lie on, and something to eat. The two first were not forthcoming at all, and the last did not appear for nearly three hours, so that I had gone fast asleep on the boards long before food arrived, and did not care to rouse myself to partake of it. The wetness of the night was some excuse for the delay in supplying our wants, and the absence of anything to lie on was probably explained by the change in the faith of the villagers; carpets and cushions being apparently creature-comforts so associated with Mussulman faith and rule, and opposed to all Russian ideas, that the first act of a civilised and converted population is to get rid of all such property, or, at any rate, to abstain from producing it for the benefit of strangers.

August 21st.—In the morning, the rain was still falling as heavily as ever, and our horsemen showed a strong disinclination to face the weather, and the perils by water, which they believed must be encountered before reaching Ardonsk. We, knowing that Vladikafkaz was distant only a short day's journey, were anxious to gain it, and felt no inducement to loiter any longer than was necessary amongst our sluggish entertainers the 'New Christians.'

Little can be written about the miserable fifteen miles' ride we had now to get over. The rain fell in torrents the whole time; we had to ford one broad and swollen, but fortunately shallow, stream, and to wade through the heavy mire of the everywhere flat and treeless steppe. At last the grove which shades Ardonsk appeared in the distance, and, crossing a second stream by a bridge, we entered the place, which consists of a straggling village of farmhouses and cottages, adjoining a military station laid out by mathematical rules, and still surrounded by the moat and ditch which were once requisite for its protection. The impetuous Paul, rashly emulating the feat of Remus, attempted to force his horse over the ruined rampart; but the animal falling backwards on the slippery slope, he narrowly escaped an awkward tumble as a punishment for his temerity.

Here we rejoined the highroad once more, and soon discovered a cottage with the double-headed eagle over the door, pointing it out as the post-station. Being wet through and plastered with mud, we made our condition an excuse for remaining at the station, while we despatched the Cossack to report our arrival, and ask the commandant if any carriage with a hood or spring was to be obtained in the place. The answer was, that no such luxury was obtainable at any price, so we reconciled ourselves to the thought of driving the twenty-five miles that still separated us from Vladikafkaz in the common open carts. We found good food and wine at the posthouse, the master of which was a civil Russian. The prospect of a long drive in the rain made us look with envy on the great sheepskin 'boureus' worn by the people of the country, and we finally concluded a bargain for two. A second diversion was afforded by the passage of a regiment of Cossacks, on the way to join those we had seen guarding the Persian frontier on the Araxes; both the men and their wiry little horses looked admirably adapted for such duty, but ill-armed, and too little drilled to be able to maintain an equal contest with European cavalry.

Before starting for our drive, we had to settle matters with our horsemen and the Cossack; the former were put in high good-humour by the addition of a small gratuity to the sum they had bargained for, and became enthusiastic in their gratitude when we granted their request for one of our photographs, which they had seen at Muchol. Despite the friendliness of our parting, we could not altogether forget, though we might forgive, their behaviour on the road before the Cossack joined us, and we were not disposed to underrate his services in removing difficulties, and aiding us in pushing on, through the weather and mud of the last two days. He, too, returned to his home well contented.

The chapter of accidents, which awaited us before reaching our destination, began with our very start; the plank-bridge, by which the road—the highway of communication between Russia and Trans-Caucasia—crosses a ditch in the street, had been allowed to fall to decay, and become impassable for wheeled vehicles. Driving consequently through the water, our wheels stuck in the mud, our shaft-horse refused to draw, and, despite the blows and oaths of the driver, who, up to his waist in water, harangued the refractory animals by turns, we seemed likely to remain a fixture. By the aid of some bystanders, who vigorously thrashed the shaft-horse, the principal offender, over the head, we were at last released, and then, despite the heaviness of the track across the steppe, which is nowhere metalled, made a good pace to Archonsk, the next station, crossing on the way several large streams, by means of long wooden bridges. In fine weather the views of the snowy chain and Kazbek may make this drive interesting, but when these are hid, nothing can be duller, or more monotonous. Archonsk is a decayed military post, now inhabited chiefly by a tribe of lean and hungry pigs.

We had travelled nearly five miles beyond it, when the wheel of one of our 'telegas' flew off, and shivered into atoms; the 'rules of the post' prevented our extricating ourselves from the dilemma, by jumping into a return-cart, which actually passed us, on its way to Vladikafkaz; and we were compelled, after sending the driver of the broken vehicle back on one of his horses to fetch another, to take Paul into our cart, and leave François in the middle of the steppe, with a revolver and the luggage, to await the arrival of a fresh 'telega.' The rain had now ceased, and pursuing a gentle but steady ascent, at a slow pace, we gradually left the verst-posts, and a number of exceedingly tipsy Russian soldiers, behind, and at nightfall entered Vladikafkaz. Long rows of white-washed buildings, used apparently as barracks and government stores, lined either side of the road. We thought the town would never come to an end, when suddenly our horses' hoofs rang on macadam, and in another minute, we were crossing our old friend, the Terek, by a handsome stone-bridge, built by an English engineer.

At a distance of a few hundred yards, on the opposite bank of the river, stands the posthouse, a large and imposing building, which serves both as an hotel, and as a club for the numerous officers stationed at Vladikafkaz, who, at the moment of our arrival, were giving a ball to the ladies of the place. Having secured good rooms, the best dinner they could give us, and a bottle of champagne, to celebrate the conclusion of our rambles in the un-

frequented byways of the Caucasus, we waited, with a
faith not destined to disappointment, the arrival of François and the luggage.

August 22nd to 24th.—Vladikafkaz, which means, in Russian, the Key of the Caucasus—or Terek Kala, the Castle on the Terek, as the Ossetes prefer to call it—is the centre of the military position which Russia has occupied, for many years, at the northern foot of the Caucasian chain. Equidistant between the Black and Caspian Seas, its fortress served as a check on any junction and combined action between the tribes of the Kabarda and the still more formidable mountaineers of Daghestan. When it is remembered that Vladikafkaz is also the key of the Dariel Pass—the great road across the chain, known from Roman times, which, rather to the discredit of modern enterprise, still remains the only one between Anapa and Derbend practicable for wheeled carriages—its strategic importance is at once explained, and the visitor is at no loss to understand the reason of the great piles of barracks which form the principal feature of the place.

The town is prettily situated, on level ground, on both banks of the Terek—open on the north, but sheltered on the south by wooded hills, behind which rise the steeper slopes and higher summits of the Caucasian chain. The place has a thoroughly Russian air. Our hotel stood in an open square: on one side was the bazaar, a row of covered arcades filled with stores, in which Ossete fur-caps were the most tempting wares; on the other stretched a long boulevard, with a shady and graveled walk down the middle. The houses along this are neat buildings; the rest of the town consists of large government stores, offices, and barracks, dropped here and there in the mud, amongst whitewashed cottages, which stand back modestly from the road, as if they felt out of place in such company.

There is, as usual in Russian towns, a shady but untidily-kept public garden on the bank of the Terek, the waters of which only cover a small portion of its wide stony channel. On the slopes of the nearer hills are some villas, in the semi-English style much affected in Russia, and a fortress, important until the last few years as a defence to the town against any sudden inroad of the mountaineers, terror of whom kept the place in an almost perpetual state of siege.

The crowd collected by the performance of a rope-dancer at one end of the boulevard gave us a good opportunity of observing the characteristic costumes of the country. Russian ladies in Parisian toilettes, smoking cigarettes, were mingled with tall regular-featured Ossetes, and comparatively puny and sallow-faced officers, wearing the unbecoming baker's cap so common in Russia. On the outskirts of the crowd hung Persian labourers, in close-fitting skullcaps and ragged dress, come thus far north in search of a livelihood by working on the roads. Sentinels, standing at odd corners, guarded nothing with a careless air, and a military band, lent life to the performance, which was not in itself of a very exciting character.

Our first visit was to the post-office, to obtain our letters. The pleasure of being informed, by the head official, that three letters for me had been received, was soon damped by the discovery that an over-zealous clerk, having seen our names in the Pátigorsk Visitors' List, had sent them on there. Of course we telegraphed to Pátigorsk, but nothing more was ever heard or seen of the mis-sent letters. To our dismay, the first news that met us, when we enquired about a carriage for Tiflis, was that the Dariel road was broken, and that all communication had been interrupted for several days. We began to think we should never escape from floods. The damage had been done in the second stage out of Vladikafkaz, and as a company of soldiers were said to be at work in making the road passable for carriages, we had nothing to do but wait till news came of the completion of their task, before which time the postmaster refused to send out any vehicle.

At first it seemed as if we should be obliged to renew our acquaintance with 'telegas,' but our perseverance was rewarded by the completion of an arrangement for the hire of one of the small 'diligences,' which are a peculiarity of the Russian postal system. These solidly-built and heavy conveyances, which require four or six horses to draw them, according to the nature of the road, contain, notwithstanding their size, only five passengers, besides the driver and the conductor who are attached to each. The inside holds two, or three on a pinch, and there are seats for three more, on what, though slightly different, may be called the banquette. They are fitted up for long journeys, and it is possible to sleep very comfortably inside, as a board lets down from the front, and enables the traveller to lie at length during the night. The presence of the conductor—who, armed with a horn, blows warnings to all lesser traps to get out of the way, wakes up sleepy postmasters, and takes on himself all responsibility in procuring horses—is also a great convenience, especially to the foreigner not speaking Russian. On Sunday evening we saw, to our delight, a carriage drive in from Tiflis, and the road being thus conclusively proved to be open, the postmaster promised us horses at an early hour on the following morning.

August 24th.—The weather, which had been cloudy during the two days of our stay at Vladikafkaz, cleared during the night, and on looking out of window, at sunrise, we were greeted by the double peak of Kazbek, rising high above the ridges on the right of the gap of the Dariel Pass. The view of the snowy chain was far finer than we were prepared for. The mountains west of Kazbek, and the offshoots of the Adai-Khokh group, were well seen, and far away on the horizon, Dychtau once more showed his snowy head. The precipices on the eastern face of this tremendous peak, even from this distance, excited mingled feelings of admiration and respect. Our conductor amused us by piling the roof of the 'diligence' with water-melons, a common article of food in Cis-Caucasia, which he meant to dispose of at a profit to the masters of the stations high in the mountains, where the fruit was not so easily obtainable.

There are three stages between Vladikafkaz and Kazbek, and the distance is about twenty-eight miles. During the first stage the road was well-made, and almost level; after crossing a flat meadow, where we passed a portion of the garrison engaged in drill and rifle-practice, it runs between low hills along the banks of the Terek. The vapours, which almost daily cover the northern plains, had already risen, and it was curious to see them breaking like waves against the steeper slopes before us, while a glimpse up the gap of the Dariel showed that further in the mountains the sky was of an unclouded blue. Shortly before reaching Balta, we passed two carriages, filled by the family of a Russian gentleman, travelling with so numerous a party that they required twelve horses to draw them. The spot where the Terek had swept away the road was a mile or two beyond the station. The accident had arisen from the half-hearted and timid way in which the work had been originally carried out. On the flat, or where the soil is friable, no finer highway could be desired; but directly a hard mass of rock which requires blasting is encountered, the engineers shrink from the bold course that would be taken by the peasants of any Swiss commune, and either climb over the top, or creep round the foot of the obstacle. The latter course had in the present instance been adopted, and it was but the natural result that the Terek, swollen by heavy rains, undermined and swept away the ill-constructed embankment on which the road was carried. The soldiers had done their best, under the circumstances, by throwing down stones and piles of fagots; but the place was only just passable, and our horses had to be taken out, and the heavy carriage dragged across by the men, who left their work for the moment to aid us.

At Lars—the second station—the scenery, hitherto rather tame, suddenly assumes a sterner aspect, and the neat whitewashed buildings of the posthouse look out of keeping with the grim crags that tower above them. There is an ascent of nearly 3,000 feet from this place to Kazbek. We required six horses, the Russian family twelve; but as we had arrived a few minutes first, and the stables did not contain the number of animals required for both parties, our fellow-travellers were compelled to wait. The old gentleman was naturally angry, and produced a 'crown-podorojno'; but, finding that our conductor—by whose readiness we had got the start of him—was possessed of a similar document, he submitted, with much gainsaying, to fate and the law of the post, that when the 'podorojnos' are of a similar class, first come shall be first served. His family, more philosophically disposed, seemed to care little for the delay, the governess and her charges setting out on an expedition up the nearest hillside, with a spirit which excited our hearty admiration.

The famous Dariel gorge differs from most of those traversed by Alpine carriage-roads in one essential feature. It is not so narrow but that the road finds place alongside the river, without making any considerable ascent above it. The traveller is, therefore, exempt from the terrors of profound abysses and yawning depths, which suggest themselves so often to the French tourist in the Alps. That curiously-constituted individual will, however, when he comes to describe this defile, probably apply to it his favourite epithet of 'horribly beautiful,' and, if a classical scholar, will proceed to quote Virgil's 'Sævis cautibus horrens Caucasus,' a passage the Roman poet may well have founded on the report of some friend who had wandered as far as the even then famous ' Portæ Caucasiæ.'

The unimpressionable Anglo-Saxon, now that the once real danger of being picked off by a mountaineer in ambush behind some neighbouring crag no longer exists, will feel no other emotion than one of vague delight in gazing up to the gigantic cliffs amongst which he finds himself. Their bold and broken forms must arrest the attention of even the most indifferent observer of nature. The mere fact of the existence of a carriage-road is some detraction from the impressiveness of a mountain-gorge, and, partly perhaps for this reason, we felt indisposed to rank the Dariel beside the ravines of the Tcherek and the Uruch we had lately traversed, yet we agreed unanimously that it had nothing to fear from a comparison with the finest defiles of the Alps. The road deserts the line of the old horse-track—a mere shelf cut in the rock on the left bank of the river—and, crossing to the opposite side, winds round the huge bastions of basalt-crag, which rise tier upon tier to a height of at least 5,000 feet above the level of the Terek.

After passing the narrowest part of the ravine, the fortress of Dariel comes into view—a low brick building loopholed for musketry, and commanding, by means of two projecting towers, the narrow pass. A flat-topped table-

shaped mass of rock on the opposite side of the valley was,
we were told, formerly occupied by a more ancient fortification, now entirely destroyed. The ascent here becomes
very rapid, and the Terek falls in a succession of cascades.
Beyond Lars, the mountain-walls on either hand are unbroken by any deep cleft, until, through a sudden opening
on the right, the upper portion of the glacier of Devdorak

Fort of Dariel.

and the summit of Kazbek come into view, and, to an eye
unaccustomed to scenery on such a scale, seem close at
hand. The tower of rock on the ridge between the two
glaciers is easily recognisable. Close to the spot at which,
two months before, we had crossed the Terek on to the
highroad, we met one of our old porters, who hailed us
with enthusiasm. At Kazbek posthouse the German-

speaking woman who attends to the wants of passers-by, entertained us with an account of some people who, since our previous visit, had come from Tiflis to ascend the mountain, but had been content with looking at it, and passing on. At Kobi the postmaster was, as usual, tipsy. Night drew on, and rain began to fall as we drove up the pass, and, owing to the absence of moon, we were obliged to sleep at the station near the summit, which is not so well-fitted-up as most of those on this road.

August 25th.—We started at 5 A.M., and after twelve hours' rapid progress, without any delays from want of horses, reached Mscheti, the station nearest Tiflis, where, as my readers may recollect, the Dariel and Kutais roads unite. Despite the double drain inevitably thrown on the resources of the establishment, the supply of horses is only the same as at other stations on the road, and travellers are constantly obliged to submit to the inconvenience and annoyance of stopping at an early hour, when only twelve miles distant from their destination. On the present occasion, after failing in an endeavour to hire peasants' horses, we were forced to make up our minds to the impossibility of reaching Tiflis that evening, and to take up our quarters on the floor of a prettily-decorated room intended for the use of the Grand Duke, but in which travellers willing to pay for the accommodation are allowed to spend the night.

August 26th.—At an early hour in the morning we drove into Tiflis, and aroused the people of the ' Hôtel d'Europe,' who had almost given up expecting our return. During the three following days we were fully occupied, first in settling our plans, and afterwards in making the arrangements necessary to their execution. Our original scheme had included a visit to Daghestan and the Caspian, but the time taken by our explorations in the central chain prevented us from fully carrying it out. We can therefore say nothing of Kakhetia, a broad valley teeming with corn and wine, and overlooked by the snowy mountain-wall of the Eastern Caucasus, which was described to us by some of the residents at Tiflis, in the most glowing terms, as little short of an earthly paradise. We had also to give up all thoughts of attacking Busardjasi (a peak of 14,700 feet, surrounded by others exceeding 13,000 feet, all of which are still unclimbed), and of exploring the district of which it is the centre, which we had been assured by Herr Abich we should find well worthy of a visit.

The question now before us was by what route we should return to the coast of the Black Sea, and our enquiries resulted in the determination to send some of our luggage with François direct to Kutais, while we, leaving the highroad at Suram, followed the valley of the Kur to Borjom and Achaltzich, and thence rode across the hills to Kutais. To effect the arrangements necessary for the execution of this plan we had to pay many visits to the various official bureaux, where the employés seemed to have acquired a skill in the art of 'how not to do it,' which could scarcely be surpassed even at home. At last the various 'podorojnos,' police orders, and passports were all procured, and we had only to consign our heavy portmanteaux, and a box of purchases, to the care of the agent of the Black Sea Steam Navigation Company, through whom we subsequently received them in England.

In the intervals when we were not engaged in the preparations for our homeward journey, we found time to call on Herr Radde, one of the few residents who happened to be in Tiflis at this time of year. We also visited the studio of a Russian painter, who has boldly taken for his subjects the scenery of the Dariel and the highest mountains of the Caucasus, and were glad to carry away with us some of his pictures, as reminiscences of a country with which European artists are as yet unacquainted. In the evenings we went to the gardens in the German quarter, known respectively as 'Mon Plaisir' and 'Sans Souci,' where a crowd of townsfolk, sitting under cover, sip their tea, and listen to the strains of a good band; or, if so disposed, wander in couples down long alleys arched over with trellised vines, from which the grapes hanging in ripe clusters seemed, at the present season, almost ready to drop into one's mouth. Some of the accounts we heard of the effect produced by the news of our ascent of Kazbek were rather amusing. It had apparently caused much discussion at Tiflis among the citizens, who had all their lives asserted the impossibility of reaching the summit. We were told that, on the first intelligence being received, a person high in authority had remarked, that it was strange that a mountain which had been declared for sixty years inaccessible by Russian officers, should be ascended by Englishmen in a few days. The answer of the insulted officers was prompt and ingenious: 'We could have said we had been to the top as easily as the Englishmen!' I do not think there was a single Russian in Tiflis, unconnected with the Government, who believed in the truth of our story.

We amused ourselves, during our stay at Tiflis, by contrasting our present ideas of the Central Caucasus with those with which we had left the same place two months previously, and in endeavouring to ascertain in what particulars the impressions acquired by reading or hearsay had been reversed or modified by actual experience. We found the process a profitable one, and I think a summary of the results obtained will not be without interest to the general reader.

The published accounts of the Caucasus, that we had met

with before leaving England, had been more or less vague and unsatisfactory. Their authors had, as a rule, kept at a respectful distance from the giants of the chain, and, indulging chiefly in ethnological researches, confined themselves, when they approached mountain scenery, to generalities, useless to the mountaineer, anxious to form some idea of the character of the range, the height of its peaks, and the relations of its groups. Those who did give information on the subject contradicted one another in the most emphatic manner, and only increased the perplexity of the reader. We learnt from one writer: 'The mountains of the Caucasus are not peaked, as in the Alps, but are either flat or cup-shaped; the existence of glaciers is uncertain.'* We read in another: 'Neither the Swiss Alps, the Taurus, Atlas, Balkan, Apennine, or any of the well-known mountains of Europe, have such furrowed and broken, rocky and snowy precipices, or such bold peaks, as the giants of the main chain of the Caucasus. The Orientals have rightly named these mountains the "thousand-pointed."'†

The first-quoted opinion seemed the most popular, and many of our friends in England smiled at our idea of setting out to climb in a region where, as they believed, there were no valleys or steep-sided summits, and where nothing was to be seen except two large volcanoes, rising from a lofty plateau, and culminating in snowslopes, the ascent of which would be equally laborious and uninteresting.

These accounts so far imposed upon us that, when we reached Constantinople, the chief impression in our minds was, that the Central Caucasus consisted of a long watershed, devoid of prominent peaks, and dominated at either end by a huge dome—the eastern known as Kazbek, the

* Keith Johnston's 'Dictionary of Geography.'
† Travels in Georgia, Persia, and Kurdistan (Wagner, 1856).

western as Elbruz. In fact, we were led to believe that the architecture of the mountain region bore some resemblance to that of the Exhibition Building of 1862, and that for beauty of outline it could no more be compared to the Alps than could the Brompton shed to a Gothic cathedral.

The first ray of light that dawned upon our minds was the information we gained from Mr. Gifford Palgrave, who had been twice to the base of Elbruz, and discoursed with enthusiasm of the granite peaks and foaming torrents he had seen on the way. When we reached Batoum, and for the first time saw the mountains with our own eyes, our rising hopes were converted into certainty. The serrated array of rocky teeth and icy cones which stretched along the northern horizon, convinced us, at once, that climbing sufficient to satisfy the greediest mountaineering appetite would be found there, and the way in which ridge behind ridge rose up to the snowy chain disposed for ever of the plateau theory, and satisfied us that many landscapes of exquisite beauty must be hid within the folds of the mountains.

Our journey from Kutais to Tiflis did not add much to this knowledge, nor did we gain any practical information from our intercourse with Russian officials, whom we found, with the exception of General Chodzko and a few others, entirely unacquainted with the nature of the mountain districts west of Kazbek. The German savants resident at Tiflis, and in the employ of the Government, were far better informed. Herr Radde had visited the upper valleys of Mingrelia, and had himself attempted to ascend Elbruz. Herr Abich had examined both summits with the accurate eye of a man of science, and gave us hints which we afterwards turned to good account. The purchase of the Five Verst Map enlightened us on some matters of detail, indicating the existence of large icefields, and confirming our belief in the practicability of our proposed route, by marking numerous passes across the chain, and showing groups of villages near the heads of most of the valleys. The map also first revealed to us two mountains, Koschtantau and Dychtau, respectively 17,000 and 16,000 feet in height, yet hitherto unknown to English geographers.

The real character of the tribes, and the extent to which Russian rule was a reality in the more remote districts, was another topic which interested us nearly as much as the nature of the country. On this subject we had read contradictory reports. The majority of travellers described the natives of the mountains as robbers, in whose word no trust could be placed—barbarians by nature, and incapable of civilisation. On the other hand, they were painted, by a few enthusiasts, as noble patriots, whose only crime was to have been defeated in an unequal struggle against an invading despotism. We were unanimously assured at Kutais and Tiflis that the subjugation of the inhabitants was complete, and that there was no longer any risk to life or property in travelling amongst them. Each informant had his own view as to their character, but, satisfied with the fact that our journey would not be rendered impracticable, we were content to wait and judge for ourselves.

Such was our knowledge of the Caucasus and its inhabitants on leaving Tiflis for Kazbek posthouse. I shall now proceed to the results of our day-by-day experiences during the two months we spent in the mountains. A definite idea of the scenery of the Caucasian chain will best be formed by comparing and contrasting it to that of the Alps.

The first feature that strikes the traveller is the singleness of the Caucasian compared to the Alpine chain. I do not mean that it is one long snowy wall, and nothing more. It is single contrasted with the Alps, in the same way that the Pennines are a single chain, although they possess spurs like the Weisshorn, and minor ranges like those that form the southern boundary of the Val Pelline. This characteristic is proved from the fact that, from elevated points north or south of it, the same summits are generally visible, whereas, as is well known, the observer at Milan or Lucerne, Salzburg or Venice, sees from each an entirely different range of snowy summits. From this cause the panoramas seen on the highest peaks of the Caucasus differ from those of the Alps, in the fact that the portion of the horizon occupied by mountains is far less in the former than in the latter. Whether this is to be considered a recommendation, or a fault, must depend on individual taste; but no one can deny that if it had been desired to enhance by contrast the stern beauty and bold outlines of the central chain of the Caucasus, no better means of doing so could have been found than by putting beside them the boundless plains of the steppe, or the wavelike ridges of the Mingrelian hills.

Let us now descend from the mountain-tops, whence we have naturally begun our survey, and take a closer view of the individual features of the country. As a whole, Caucasian must, I think, rank above Alpine scenery. There is nothing in Switzerland or Tyrol that can compare with the magnificent grouping of the Suanetian ranges, or with the gorges cut by the northern rivers through the limestone ridge which bars their way down to the steppe. In the Caucasus the slopes are steeper, and the usual character of the peaks is that they shoot up from the valleys at their base, in unbroken walls of rock and ice, to which the cliffs of the Wetterhorn afford the nearest parallel.

G G

Enormous cornices of ice are frequent, and sometimes crown the highest peaks, presenting an insuperable obstacle to the climber. The mountain-sides, owing to their precipitous character, afford precarious resting-places to the winter-snow, and avalanches, which choke the upper glens to an extent rarely seen in Switzerland, are consequently of frequent occurrence.

Another peculiarity of the Caucasus is the constant appearance of red snow, which in the Alps is often heard of, but seldom seen. Here it is met with every day, and the effect produced is as if the whole surface of the slope had been sprinkled with brickdust. We did not sufficiently explore the glaciers to be able to form a conclusive judgment as to their extent; but there can be little or no doubt that the number of square miles covered by snow and ice is less than in the Alps, though there are many glaciers worthy of comparison with any Swiss rivals. Owing to the steepness of the chain, the icefalls are loftier, and in every respect finer, than those of the Alps, and they are rendered more attractive to the eye by the general purity of their surface.

So little having been known up to the last few years of the existing glaciers of the Caucasus, it is not surprising that we have as yet scanty information as to the traces of the glacial epoch in this region. So recently as 1858, Herr Abich declared that the Caucasian chain showed no marks of its influence; since that date, however, the learned traveller has seen reason to change his opinion, and has himself borne witness to the existence of traces of vast glaciers in the upper valleys of the Ardon and the Ingur. Mons. E. Favre of Geneva, who was in the country at the same time as ourselves, and to whose kindness I am indebted for the following details, recognised marks of glacial action on an extended scale on the Krestowaja Gora and in the Dariel gorge. In the neighbourhood of

Vladikafkaz, and on the steppe to the north of it, he found numerous erratic blocks, generally of granite, and from fourteen to sixteen feet in thickness; and in the valley of the Baksan, fifteen miles from the present glacier, and two miles above the village of Urnspieh, there is a moraine, 200 feet in height, principally composed of granite blocks. Sufficient data have already been collected to justify the assertion that the present glaciers of the Caucasus, like those of the Alps, are only the shadows of their former selves.

Fine as Alpine forests often are, they can bear no comparison with those of the Caucasus. Lest it should be thought I have overstated the effect likely to be produced by the woodland scenery of Mingrelia on an European mind, I shall take the liberty of quoting the words in which Herr Wagner sums up his eloquent description of a ride near Kutais: 'Every spot that is not occupied by perennial plants presents one tangled growth of grasses, flowers, annuals, and every variety of creeper. Higher up, among the trees, the eye is soothed by the numerous shades of green, from the sombre verdure of the fir, tamarisk, and cypress, to the lustrous foliage of the laurel, and to the silver-green of the Colchian poplar, whilst the purple clusters of the grapes peep out beneath every branch. "Why, this is like Paradise," exclaimed my companions, in one breath, at the sight of such glorious profusion.' In richness of flora the Alps must also yield to their rivals: the azalea and rhododendron make the 'alpenrosen' seem humble, and even the gentian looks bluer when brought into immediate contrast with beds of snowdrops, while there is nothing nearer home to compare with the gorgeous magnificence of the Caucasian tiger-lilies and hollyhocks.

Hitherto the comparison has been in favour of the

'new love,' which, without wishing to persuade them to be 'off with the old,' we desire to introduce to the notice of lovers of Alpine nature. I must now call attention to the deficiencies of the Caucasus, and the points in which it is manifestly inferior to its better-known rival. A total absence of lakes, on both sides of the chain, is the most marked failing. Not only are there no great subalpine sheets of water, like Como or Geneva, but mountain-tarns —such as the Dauben See on the Gemmi, or the Klonthal See near Glarus—are equally wanting. There is no firstclass waterfall in any of the valleys we visited, nor did we hear of any elsewhere. Certain districts, notably the headwaters of the Terek, are duller than anything in Switzerland, and their treeless monotonous glens are defaced, rather than enlivened, by the dingy and ruinous character of the native dwellings. Add to this list of defects that, on the north side, the mountains sink abruptly into a bare and featureless steppe, and that the only halting-places within reach—Pätigorsk and Kislovodsk—cannot vie in attractions with the numerous tourist-haunts on the north side of the Alps, and we shall, I think, have fairly gone through the principal charges to which Caucasian scenery is liable.

Readers seeking geological information in old scientific works on the Caucasus must beware of the hasty generalisations in which they indulge, on the strength of an acquaintance with perhaps only one valley of the chain. It is unfortunate that no member of our party was skilled enough to make his observations of value; but we have the authority of Herr Radde, as well as the evidence of our own eyes, for stating that the central chain is chiefly composed of granite, and that the rocks of both Kazbek and Elbruz are igneous. I do not think that anyone who has made close acquaintance with Elbruz will doubt its having once been an active volcano. The limestone ridge on the north of the watershed is more abrupt, and the gorges cut through it have bolder features than those of the secondary ridges of Mingrelia, where the rock is more friable, and steep slopes take the place of cliffs.*

I have hitherto spoken only of that portion of the country with which we became personally familiar, the 120 miles between Kazbek and Elbruz, and it must therefore be borne in mind that scant justice has been done to a chain the entire length of which, from Anapa to Baku, is 700 miles. Those who may be disposed to carry on the work of exploration have, consequently, a large field open to them. West of Elbruz there is said to be much noble scenery; the glaciers are few and small, but the chain bristles with sharp peaks, between 10,000 and 12,000 feet in height. The depopulation of this district after the late Abkhasian revolt will prove a serious difficulty, to be taken into account by future travellers. Eastwards, between the Dariel Pass and the Caspian, stretch the highlands of Daghestan, a region of flat pasturages, cut off from one another by profound gorges, and dominated by at least three snowy groups, rising to a height of over 13,000 feet.

Having said thus much of the natural features of the Caucasus, we may now review its inhabitants. The diverse character of the mountain-tribes renders any general description of them a work of extreme difficulty.† Even in the small portion of the chain we visited, leaving

* According to Herr Abich, the snow-limit in Suanetia is 9,500 feet. The same author fixes the limit of the forests at 7,300 feet. Herr Radde, after numerous observations, estimates it at 7,600 feet. The average height of the base of the great glaciers as yet measured is 7,200 feet.

† Readers who desire further details as to the tribes of the Caucasus, will find them in Wagner's 'Travels in Persia, Georgia, and Kurdistan,' and Haxthausen's 'Transcaucasia,' works translated from the German; or (in French) in Dubois de Montpereux' 'Voyage autour du Caucase,' and a more recent work, 'Lettres sur le Caucase.'

out Abkhasia, all but a corner of the Tcherkess country
(the true Circassia), and Daghestan, the scene of Schamyl's
final resistance and capture, we encountered three entirely
distinct races, speaking widely different languages. These
were the Georgian tribes of the southern valleys, the
Tartars of the north, and the mysterious Ossetes, who
have long been a puzzle to ethnologists. The language of
the latter, according to Sir Henry Rawlinson, is the most
nearly allied to Sanscrit spoken west of the Indus. The
other mountaineers use dialects of Tartar and Georgian so
diverse that the people of one valley often have difficulty
in understanding those of the next, although nominally
speaking the same language.*

The religions of the Caucasus are as various as its
languages. As a rule, whatever religion exists on the south
side of the chain is called Christian, and on the north
Mahommedan. The Ossetes, as usual, must be excepted;
they were converted to Christianity in the days of Queen
Thamara, but afterwards relapsed into their former pagan-
ism, which is at the present day again overlaid by a slight
varnish of nominal Christianity. This re-conversion, if it
deserves the name, took place about the time of Herr
Wagner's visit to the country (1843–4), and he gives an
amusing account of the means employed by the Russian
missionaries to effect their end. 'The Russians' (says this
writer) 'have made many efforts to win back the Ossetes

* The Caucasus has in all ages been famed for its variety of languages. Pliny
tells us that in Colchis there were more than three hundred tribes speaking
different dialects, and that the Romans, in order to carry on any intercourse
with the natives, had to employ a hundred and thirty interpreters. This is
probably an exaggeration, but there seems no reason to doubt Strabo, who
informs us that in his day no less than seventy dialects were spoken in the
country, which even now is called 'the Mountain of Languages.' We find
archaic forms of various Georgian, Mongolian, Persian, Semitic, and Tatarian
languages, as well as anomalous forms of speech, which bear no affinity to any
known tongue of Europe or Asia.—See Max Müller's 'Lectures on Language'
and Rev. J. Taylor's 'Words and Places.'

to Christianity. This was easily accomplished with a people indifferent about religious matters, especially as a linen shirt and a silver cross were given to every Ossete who underwent baptism. The pious zeal of the new converts was greatly excited by these means, and there was no end to the number of neophytes who aspired to the rite of baptism, till at length it came to pass that one immersion was not reckoned sufficient, and that many Ossetes, in order to become genuine Christians, and at the same time the owners of a respectable amount of linen, received the sacrament five or six times following.' He adds: 'If the Russian Government had permitted other Christian confessions to hold intercourse with the mountaineers of the Caucasus, their Christianity might possibly have been something better than "sounding brass and a tinkling cymbal."'

One of the peculiarities of the country is that the superiority of the Christian over the Mahommedan population, commonly seen in Syria, is entirely reversed. In the Caucasus the traveller will be compelled to contrast the truthfulness, industry, and courteous hospitality of the Mahommedans north of the chain with the lying, indolence, and churlishness of the Christians on the south. The Georgian races who inhabit the upper valleys of Mingrelia are, as a rule, too lazy to take advantage of the rich natural gifts of the country they inhabit; they are greedy of ill-gotten gain, and careless of life in its pursuit. This conclusion as to their character is the result of our own experience, but it is confirmed by that of other travellers, even from remote times. Thus Chardin, writing nearly 200 years ago, says: 'The women of Mingrelia are extremely civil, but otherwise the wickedest in the world,—haughty, furious, perfidious, deceitful, cruel, and impudent—so that there is no sort of wickedness they will not put in execution. The men are endowed with all these mischievous qualities, with some addition. There is no wickedness to which their inclinations will not naturally carry them, —but all are addicted to thieving. That they make their study—that they make their whole employment, their pastime, and their glory. Assassination, murder, and lying are among them esteemed to be noble and brave actions, and for all other vices, they are virtues in Mingrelia.' *

Haxthausen, whose work was published in 1854, writes: 'The Russian officers, civil and military, all agreed in describing the people of this country, especially the Imeritians, as thoroughly depraved, immoral, thievish, mendacious, and quarrelsome.'

Malte-Brun, in his 'Geographical Encyclopædia,' which contains much correct information on the Caucasus, thus describes the Imeritians: 'The indolence of the inhabitants allows the rich gifts of nature and the climate to perish in a most useless manner.' He says of the Mingrelians: 'They live surrounded by women, who lead a life of debauchery, often eat with their fingers, and bring up their children to lying, pillage, and marauding.'

The Tartars of the Kabarda are in most qualities the reverse of their southern neighbours. Rich in flocks and herds, and cultivating cornland sufficient to supply them with daily bread, they pass a peaceable and patriarchal existence, and are ever ready to extend towards travellers that hospitality which they regard in the light of a religious duty. One unamiable trait they share with all the Caucasian races we came in contact with—a desire to drive a hard bargain in matters of business.

The distinction here drawn between the character of the so-called Christian and the Mahommedan tribes is so marked, that no honest traveller can pass it over in silence. The explanation of the fact must be sought in the degraded

* I am indebted for this quotation to Mr. Usher's 'London to Persepolis.'

character of the native Georgian and Armenian Churches, and in the evil wrought by Russian proselytism, which endeavours to effect wholesale conversions by holding out baits of worldly advantage. Converts thus made abandon their old religion without gaining anything in its place. For the present, the heads of the Greek Church may more profitably employ its energies in sending missionaries to lead their ruffianly co-religionists in the southern valleys to a better mode of life. When Mahommedans have no longer to maintain a guard, to protect their flocks from Christian thieves, there will be more hope of their adopting the religion recommended to their notice.

I now turn to the relations existing between the Russian Government and the mountain-tribes. Our observations fully confirmed all we had been previously told respecting them. Since the conclusion of the Crimean War, the whole country has been fairly conquered, and the inhabitants have learnt from experience that any rising will be promptly put down and summarily revenged. The European traveller need no longer fear open robbery, except in Suanetia, and in this district it is owing rather to want of will than of power that the Russians leave the village communities to their native misrule. Selfishly speaking, the policy of the Government in abstaining from garrisoning the upper valleys is prudent; it would gain nothing, and spend much, by acting otherwise. It matters little to Kutais if the inhabitants of two Suanetian villages like to cut one another's throats, and the maintenance of an armed force in the district would, except from a philanthropic point of view, be unprofitable. That such conduct, however, shows a neglect of the first duty of a Government, even Russians must admit, if they believe, with us, that the extension of any organised system of justice, however imperfect, is a blessing to regions formerly a prey to the misrule and exactions of petty princes, and still the scene of constantly-recurring robberies and murders. The evidence of our Mingrelian servant, whose prejudices were certainly not Russian, was conclusive on this point. The picture he drew of his native district, Sugdidi, on the Lower Ingur, during the Crimean War, when Russian rule was relaxed, was indeed deplorable. Robberies, as often as not accompanied by murder, were of daily occurrence; the culprit in most cases escaped, shielded by the influence of the petty chieftain whose vassal or serf he was. A man with a reputation as a successful murderer was too useful ever to feel the lack of princely favour. The peasant-farmer, with the knowledge that another would reap the fruits of his toil, and that a large portion of his crops would go to swell the contents of the nearest chieftain's barn, had no inducement to agricultural improvement.

This state of things is now at an end. Offences against life and property are promptly punished, and though small disputes still come before the native princes, an appeal is possible to Russian officials. The ukase for the emancipation of the serfs, the operation of which was specially delayed in Mingrelia, is just taking effect. In return for these advantages, the inhabitants pay a house-tax varying, in the mountain districts, from five to ten roubles (fifteen to thirty shillings) per annum. This does not seem regarded as a grievance, but we heard complaints of the increased price of imported goods, owing to the high tariff maintained by the protective policy of the Moscow merchants. Georgia enjoys a special immunity from the conscription, founded on the terms on which it was handed over to the Czar by the last of its native princes, and the Caucasians are, as a rule, exempted from compulsory service.

The dark picture given above of the past condition of the country is, of course, far from being universally applicable. It is drawn from one district, towards the Black Sea coast; but the whole mountain region has always been more or less given over to lawlessness, and the dwellers in the plain had probably good ground for attributing to their neighbours of the mountains a belief in the following legend: 'This wild race pretend, that after God created the world, an edict was published, by which all people were summoned to take possession of their several portions. All mankind had an appointed share, except the inhabitants of the Caucasus, who had been forgotten. Upon putting in their claim, which the Deity acknowledged to be just, they received permission to live at the expense of their neighbours, and assuredly they reap ample profits from the presumption of such license.' Like our own Highlanders of former days, the mountaineers of the Caucasus look on the wealth of the lowland population as their lawful perquisite, and their final subjugation will be a necessary consequence of the progress of civilisation.

Though the politician may, with reason, regret that the Russian armies, no longer conscious of formidable foes in their rear, can now, from the highlands of Armenia, look down over the Valley of the Euphrates, the traveller in the Caucasus, out of temper, as he will often be, with the corruption and stupidity of all but the highest class of officials, must not forget that, but for their presence, he would be unable to penetrate at all into the interior of the country.

Little or nothing has been said in the course of this narrative about the wild animals to be found in the Caucasus. We met with chamois occasionally, but never in any great number; we twice saw dead bouquetin, and we noticed the track of bears in the forest; we also found

some cubs in captivity in Suanetian villages, but this was
the sum of the quadrupedal life which came under our
eyes. A few eagles and a great number of cuckoos, vocal
despite the lateness of the season, were the most remarkable members of the feathered tribe which attracted our
observation. Nevertheless, those who may visit the Caucasus for the sake of sport will probably find it. A
sportsman, ambitious of a bullfight, will meet with a
worthy foe in the gigantic 'auroch,' which still haunts
the valleys west of Elbruz, and of which a stuffed specimen
may be seen in the museum at Tiflis.

That bears abound is proved by the complaints we heard
of their ravages, and by the frequency of their tracks;
chamois and bouquetin must be sufficiently numerous, if
the account given to Herr Radde, by the natives of the
upper Zenes-Squali, of a winter-drive on snowshoes, when
thirty-three were slain, may be believed. In the swamps
bordering the rivers that join the Terek, as well as in the
jungle of the Kur, wild boars make their lairs, and the
traveller obliged to spend the night at Mscheti is frequently disturbed by the cries of wolves and jackals.
Pheasants still exist on the banks of their native river,
the Phasis, and are said to be so plentiful on the north of
the chain, as to be sold by the Russian soldiers to their
officers for fourteen copecks (or about fivepence) a brace.
Radde gives an account of the snaring of ptarmigan in
Suanetia, where they are found upon the mountain-slopes
even in winter. In summer they raise their broods
within the forest boundary.

Much as we wish to persuade travellers, and more
especially mountaineers, to abandon for a season their old
Swiss loves, and to start in quest of the fresher charms of
the hitherto-neglected maiden peaks of the Caucasus, we
must, in fairness, point out the principal difficulties that

will be met with in such an enterprise, and how best they may be encountered. For some years to come travellers in the Caucasus will find letters of introduction to the government officials a useful, if not essential, part of their outfit. Without them they may be objects of suspicion, and their purpose in desiring to penetrate the fastnesses of the mountain-tribes may be misunderstood. Some people say that a uniform is absolutely necessary in Russia; though this is not the case, those who hold any position entitling them to wear an official dress will do well to take it. In the East even more than in the West, and above all among Russian officials, 'fine feathers' are considered to make 'fine birds,' and even an old Volunteer tunic would protect its wearer from much rudeness from postmasters and sub-officials.

The selection of a starting-point will depend upon the tour proposed by the traveller. Tiflis is far from the mountain-chain, but the fact that there only can the necessary maps be bought, will induce even those to whom a city combining, in such a striking manner, the discordant elements of European and Asiatic life is not a sufficient attraction, to visit the Trans-Caucasian capital. Vladikafkaz is the best base for the exploration of the northern valleys. It is the residence of the commandant of the district, and the Cis-Caucasian officials appreciate better the aims of an explorer, and are more practical in the aid they afford him, than those at Tiflis. Kutais is admirably situated as a starting-point for the Mingrelian valleys, and next year will probably be brought nearer to the Black Sea coast, by the partial opening of the Poti-Tiflis Railroad; while Pätigorsk is a convenient Capua, within two days' ride of the base of Elbruz.

The first necessity for a journey in the mountains is a servant ready to rough it, and sufficiently conversant with the native dialects to act as interpreter. A knowledge of cooking, such as our man possessed, is a great additional recommendation. A light tent and a cooking apparatus are essentials, as well as the usual requisites for travel in uncivilised countries, which I need not catalogue here. The difficulties of mountaineering inherent to such a country as the Caucasus are obvious. The peaks are, generally speaking, extremely formidable; the natives, except at Uruspieh, are useless above the snow-level, and it is often impossible to leave luggage at the mercy of villagers while making an ascent. The climate is changeable, and the rainfall, owing to the position of the chain between two seas, is frequently excessive. Thus the impediments to a mountain tour are very serious, though not, in my opinion, sufficient to counterbalance the advantage and pleasure to be derived from a journey in a country surpassing, both in freshness, grandeur of natural scenery, and ethnological interest, any other so accessible to English travellers.

The expedition will, of course, differ much from a run to the Oberland or Zermatt, and it should be undertaken only by men prepared to face daily-recurring difficulties with good temper and perseverance. A party of five or six, accompanied by not less than two firstrate guides, of which two of the members have botanical or artistic tastes, and would be content to remain below, or to cross a lower ridge with the luggage, while their friends attacked 'peaks, passes, and glaciers,' would have the best chance of success; if favoured with fine weather, and with the help of a Cossack in all dealings with the villagers, they might effect a great deal.

Before bidding farewell to the Caucasus, I must remove any impression the previous pages may have given that either Kazbek or Elbruz are in themselves difficult mountains. First ascents are proverbially the hardest. On

Kazbek we had to contend against severe wind and cold, and total ignorance of the mountain, which made us go up the wrong way. On Elbruz we encountered a tempest, which, but for the entire absence of other difficulties, would have rendered the ascent impossible. On both occasions we might have imagined that we were wrestling against 'principalities and powers.' The icy wastes of the Caucasus have been peopled throughout all ages with invisible occupants. In this region dwelt Gog and Magog; here Oriental fancy has placed the abode of the Deevs, a race of pre-Adamite monarchs, and the retreat of the Peri and the Genii. It was to this snowy prison that Solomon consigned the rebel Afrites, and it would have been strange indeed had not the Djin-Padishah, or Ruler of the Spirits, who dwells on Elbruz, summoned 'the Prince of the Power of the Air' to his aid, to resist the strange company who, armed with rope and ice-axe, ventured to intrude on his dominions.

According to their best biographers, giants and gnomes seldom fight a second time. After their power has been once successfully defied, they either tamely expire, or retreat to some more remote fortress. The Djin-Padishah has, for the present, probably taken up his abode on Mount Everest, whence, let us hope, he may soon be dislodged, and dismissed to the North Pole, or some equally remote and apparently unattainable spot. But, abandoning allegory, I think we may fairly assume that, short of actual wet weather, in which no one would attempt a first-class peak, we encountered, in our own attacks on both Elbruz and Kazbek, every obstacle that either mountain possesses or can summon to its aid. Any mountaineers whom this account of our journey may set 'thinking on the frosty Caucasus,' may rest assured that in fine weather they cannot fail to reach the summits of both. Few of our followers are likely to rest content with this measure of success, but, however formidable may be the difficulties to be overcome in climbing the other great Caucasian peaks, it is something to know that the two most famous mountains of the chain are within the reach of all those who possess the physical endurance necessary for an ascent of Mont Blanc.

CHAPTER XIV.

TRANSCAUCASIA AND THE CRIMEA: HOME THROUGH RUSSIA.

Borjom—Bad Road—Beautiful Scenery—Achaltzich—Across the Hills—Abastuman—A Narrow Valley—The Borst Forest—Panorama of the Caucasus—Last Appearance of Kazbek and Elbruz—A Forest Ride—Bagdad—Mingrelian Hospitality—A French Baron's Farm—The Rion Basin—Kutais—The Postmaster—Poti—A Dismal Swamp—Soukhoum-Kalé—Sevastopol—The Battlefields—The Crimean Corniche—Bakhchi-Sarai—Odessa—A Run Across Russia—A Jew's Cart—The Dnieper Steamboat—Kieff—Picturesque Pilgrims—The Lavra—Sainted Mummies—A Long Drive—Vitebsk—St. Petersburg—Conclusion.

August 29—30th.—We were able to retain the same 'diligence' and conductor we had brought from Vladikafkaz for our drive to Achaltzich. Having already traversed the road between Tiflis and Suram, we determined to start in the afternoon of Saturday, and travel through the night, by which means we hoped to arrive at Borjom at midday on Sunday. At Macheti we thought ourselves lucky to escape with only a couple of hours' detention; after this all went well, and we arrived at Suram in time for breakfast. Here we parted from François, who was despatched, with some of the luggage, to make his way in 'telogas' to Kutais, by the aid of a 'podorojno,' and the knowledge of about a dozen words of Russian.

From Suram to Borjom is a distance of eighteen miles, divided into two stages. Most of the valleys, by which the streams rising in the tablelands of Armenia force their way northwards, are deep, narrow, tortuous, and well-wooded,

and that of the Kur forms no exception to the general rule.
The road, or rather track (for made road there is none),
is wonderfully bad, when one remembers that it is the only
communication between the capital and summer residence
of the Governor of the Caucasus. It climbs up and down
steep hills in the most reckless manner, and if there is one
thing more than another its constructors appear to have
aimed at, it is the production of steep pitches with sharp

Grand-Ducal Villa at Ikerjom.

corners at their feet. After waiting two hours for horses
at the roadside shed which serves as the halfway station,
we were given over to the mercies of a bumpkin, who
had apparently never driven anything but a cart, and who
was with the greatest difficulty persuaded to put on the
drag in going downhill, even after one of his horses had
fallen.

The scenery is exceedingly pretty; the Kur flows in a clear rapid stream along the bottom of the glen, the sides of which, broken here and there by masses of crag, are clothed in thick pinewoods. Beside the road grow copses of the wild rose, which is indigenous to this country. Borjom is situated at the point where the Kur receives a small tributary flowing from the eastern hills. The grand-ducal villa, a modern construction in the châlet style, is on the left bank of the river; the hills rise immediately behind it, in steep slopes. Borjom itself is on the further side of the Kur, which is here crossed by a substantial bridge. The village is of quite modern date, and consists of a few shops and a number of low cottage residences with large verandahs, in which their occupants seem to pass the greater portion of their existence. The mineral spring, the waters of which are of a ferruginous character, bursts out of the ground in a lateral glen watered by a small stream, and overhung by lofty cliffs. A bath-house has been built, and the grounds in the neighbourhood laid out in lawns and garden-walks.

Borjom, owing to its being the chosen retreat of the Grand Duke, has become the most aristocratic of the Caucasian bathing-places, and a recent Russian writer goes so far as to call it the Baden-Baden of the East, and to reproach it with excessive luxury and extravagance, a charge which certainly did not suggest itself to our minds. We were driven to the front-door of a large building at the mouth of the ravine in which the source is situated, and had already alighted, when a domestic stepped forward and informed us that that part of the building was reserved for 'les hauts employés,' information which did not impress us with so deep a sense as he seemed to expect of the impropriety of which we had been guilty. Having been driven round to a side-door, we were allowed to enter,

and shown a large room, ill-provided, as Russian rooms
generally are, with sleeping accommodation, but otherwise
comfortable. We spent the evening in sauntering about
the place, which is so overshadowed by trees and rocks
that scarcely any sunshine can reach it—a circumstance
which, in the hot climate of Georgia, has contributed
greatly to its reputation as a pleasant summer retreat for
those who do not require any very violent course of
mineral waters.

At Pätigorsk at least two-thirds of the society are real
invalids; here we saw scarcely any, and the Russian young
ladies who raced about the gardens, and chatted together
in excellent English, afforded a more pleasing spectacle
than the sickly officers and decrepit old men of the Cis-
Caucasian Spa. An excellent military band, by far the
best we heard in the Caucasus, played in the gardens
about sunset, and we had the satisfaction of hearing the
'Mabel Waltzes' (the popularity of which seems unbounded
in Russia) and the overture to 'The Huguenots' performed
in a masterly style. A question seemed likely to arise as to
the possibility of procuring horses in the morning to go
on to Achaltzich, but, owing to an officer for whom some
were ordered being unable to start, we succeeded in getting
the requisite number.

August 31st.—The first stage is a long one of twenty-six
versts (seventeen miles) and the road is extraordinarily hilly.
The morning was lovely, and we fully enjoyed the pretty
scenery of the winding valley, where the Kur flowed between hillsides clothed with thick oak and fir-forests, from
amongst which rise the ruins of old castles, commanding
in former days this entrance to Georgia. We were reminded at every turn of the Jura, to which, in its relation
to the higher neighbouring chain, the Georgian hill-country
may, *mutatis mutandis*, be very fairly compared. The

situation of Atskur, the halfway station between Borjom and Achaltzich, is extremely picturesque. The houses are grouped at the base of an abrupt crag, crowned by the extensive remains of an old Georgian fortress, which commanded alike the entrance to the defile and the bridge over the Kur. The type of the buildings, and of the men we met on the road, had already changed; the 'baschlik' and cartridge-breasted coat had disappeared, and we saw in their place the turbans and dress of a Turkish race.

Atskur marks the limit between the wooded and bare country; beyond it the landscape became more open, and we found ourselves again in the region of rolling hills which extends as far as Erivan and Erzeroum. The road was heavy, and our cattle were weak poor-spirited brutes, though far superior to the driver, who was the most incompetent man for his post we had yet come across. We had constantly to get out and walk, and even thus the heavy carriage had several narrow escapes of rolling back again, horses and all, when halfway up one of the steep hills which occurred every quarter of a mile. The Kur forces its way through a narrow cleft in a low range, and the road is carried over the brow. Above this the valley opens out into a broad cultivated basin, and there is nothing to attract attention until the green roofs and white walls of the Russian quarter of Achaltzich come into view. Our carriage might never have reached it, had not the conductor forced the miserable postilion to dismount, and himself urged on the horses.

Achaltzich, though not an imposing, is an interesting place; it is situated on the banks of a tributary of the Kur, which divides it into two quarters, the Turkish and Russian —the former exactly similar to every other military colony in the Caucasus; the latter a mass of grey flat-roofed houses, rising tier above tier against a steep hillside, the top of which, a bold bluff commanding the valley, is crowned by a fortress. It was to gain possession of this position that the Turks, under the leadership of a Pasha more brave than prudent, gave battle to the Russians in the winter of 1853, and suffered a signal defeat, only prevented from becoming a disastrous rout by the bravery and promptness of some English officers with the Turkish army, who, by a judicious use of two field-guns, put a stop to the Russian pursuit.

Immediately south of the town (3,376 feet above the sea), the bare slopes rise to a rounded summit, 8,402 feet in height. We found sleeping-quarters in a restaurant, chiefly frequented by Russian officers. Being a saint's day, the bazaar was shut up, and, cut off from this source of amusement, we took refuge in the never-failing public garden and band. Although one of the tracks leading to Abastuman, a bathing-place in the mountains which divide this district from the basin of the Rion, is called a post-road, we heard such bad accounts of its condition that we preferred to dismiss our 'diligence' and procure horses to take us all the way to Kutais—an additional inducement to this course being that we were more sure of obtaining the requisite number of animals here than at Abastuman.

September 1st.—It is at Achaltzich that the greater part of the silver filigree-ware sold in the shops of Tiflis is fabricated, and one of the workmen brought some of his goods to show us in the morning. They were exceedingly pretty, and nearly a third less in price than at Tiflis, but there was not so large a supply to select from. The man with whom we had, without any difficulty, made a bargain for horses, appeared in due time, and we set out, on some ungainly but enduring animals, for our ride across the hills to Kutais. The first day's journey was to be a very short one. The direct horse-road to Abastuman crosses the spur on which the fortress is built, turns up into the hills, and after a long ascent—which in parts reminded us of Syria, except that the features of the surrounding landscape were on a larger scale—reaches the top of a wide down, covered with cornfields and dotted with villages, the inhabitants of which, picturesque Turks dressed in the brightest colours, were enjoying a midday rest from field labour, clustered in groups, any one of which would have made the fortune of the artist who faithfully reproduced it.

From the heights we had gained, we looked down on a wide basin of cornland, with numerous villages, which, by a judicious arrangement of the frontier-line, have been just included within the Russian Empire. The rounded hills rising beyond it are in Turkish territory, and a long day's ride over them would bring the traveller to the gate of Kars. The view at the present season was very striking; the valleys and cultivated slopes stood thick with corn, and shone golden in the cloudless sunshine, and the far-spreading downs above them seemed to bask in the universal blaze of light and heat. We crossed several deep hollows, and passed a large village surrounded by fruit-trees, before descending finally to the banks of a little stream, just where it issued from the wooded limestone chain. The road henceforth follows the water through a narrow winding glen, clothed in firs, where an old castle, perched like an eagle on a lofty crag, looks down on the passers-by. The scenery is no more than pretty, and its features seemed puny and tame compared to the wide landscape we had just left.

The bathing village of Abastuman is entered almost before it is seen; it consists of a row of houses along the roadside, one of which—very unlike most Russian buildings, and reminding us of a small Swiss inn—serves as an hotel. We had plenty of time during the afternoon to stroll about the neighbourhood, and contrast the attractions of Abastuman with those of the other Caucasian watering-places. Situated in a basin surrounded on all sides by wooded hills, the place is necessarily without any distant view, and depends on its home-scenery. The stream is crossed by rustic bridges, and the small arbours, which crown the rocky points projecting everywhere from the hillsides, give a cockney air to the place which is the last thing one would expect in so retired a corner of Asia. Owing to its elevation (4,178 feet above the sea), Abastuman enjoys a comparatively cool climate during the summer months, when it is much frequented by the Russian residents both of Kutais and Achaltzich, the latter town being fearfully hot. It cannot, however, claim to rival Borjom in the rank and fashion of its visitors, and in one essential attraction it is far inferior. The band was small and indifferent, but the audience seemed little critical, and crushed any sense of its deficiencies by dancing vigorously to a sorry performance of the 'Guards' Waltz.'

September 3rd.—The morning was again lovely, a matter of some importance, as we were about to cross the mountain ridge dividing us from Mingrelia. We set out at 5 A.M., and rode for several miles through a narrow valley, where the road constantly crosses and recrosses the stream. There is plenty of wood, but nothing of a peculiarly large or striking character. Where the two rivulets forming the sources of the Abastamanska unite their waters, our way struck up the hillside, and we gained a considerable height, by a long series of zigzags. A carriage-road over this pass has been traced, and partially cut, from Achaltzich to the summit, and a little way down the other side. A few bridges have also been built, but some of them have already been carried away by the floods. The track, left half finished like most Russian engineering works, is already falling into disrepair. At its present rate of progress, years must pass before the arduous work in the long Mingrelian valley of the Chani-Squali is brought to a termination. Having reached the top of the spur, the road kept along a tolerably broad ridge between the two glens, affording views, now over one, now over the other. The eastern basin was the most extensive, and we continually remarked the admirable grouping and forms of the ridges that surrounded it. The road passes through an extensive tract of forest desolated by fire; there are few gloomier sights than a burnt forest, and beyond the crop of weeds which covered the ground, nature had done nothing to repair the desolation. The tall charred trunks stood up, brown and leafless, and no younger trees had as yet sprung up amongst them.

We were glad to reach the point where the ridge merges in the watershed of the mountains, a few hundred feet below the broad gap which forms the pass. The forest ceases at about the same level, and the final ascent is by long zigzags over a grassy slope covered with rank herbage. Passing a solitary house occupied by several Cossacks, and the ruins of a large encampment used by the detachment formerly engaged in cutting the road, we pushed eagerly up the crest, anxious to resolve the question which, since leaving Tiflis, had been a source of alternate hope and fear—whether we should gain from this point a clear view of the Caucasian chain. The sky overhead was of unclouded blue, but, knowing how soon the vapours drawn up by the morning sun from the Mingrelian marshes condense into clouds, we feared that the mountains might already be partially obscured. Our delight therefore was unbounded when, as we crested the ridge, not only the vast basin of the Phasis lay spread beneath, but beyond, and standing out sharp and clear against the northern horizon, the icy wall of the Central Caucasus met our eyes, distant at its nearest point eighty-five miles. By following the ridge a few hundred yards to the west, we gained a brow, whence Kazbek, previously hidden by a neighbouring eminence, was added to the view, and then sat down to examine more fully the details of the vast panorama.

The foreground was of exquisite beauty; forested ranges fell gradually from our feet to the Mingrelian plain, which was flooded by a transparent purple haze. On the further side an army of green hills clustered round the knees of the snowy giants of the central chain, which were ranged in line along the horizon. Directly opposite our viewpoint was the great wall of rock and ice which towers over the sources of the Ingur, terminated on the west by the graceful snow-cone of Tau Tötönal. Equidistant from this mass rose, on either hand, the clustered peaks above Gurschavi, and the solitary Uschba. The latter mountain looked taller and more terrible than it does even when seen close at hand, where its gigantic proportions, and the comparative insignificance of its neighbours, are not so fully revealed. Elbruz, huge and rounded, asserted as usual its supremacy, at least in height, over all the other summits. Further west there was only one peak, a remarkable obelisk of rock, which attracted our attention. In the far east, the snowy sides of Kazbek, bathed in a flood of morning sunshine, gleamed on us for the last time.

We had before us a panorama, extending over 150 miles, of the Central Caucasus, Kazbek and Elbruz being each 105 miles distant in a direct line. These figures give but a weak idea of the extent, and tell nothing of the splendour, of the glorious vision. To describe it in all its beauty would be impossible, and, even were it otherwise, fear of disappointing the next traveller, who, having read these pages, may follow the same path, would make me hesitate in the attempt. In the first place, few can hope to be favoured with such a day; in the second, the scene will not produce in every traveller the same feelings that it did in us, to whom every summit and glacier in the long snowy ridge were full of memories. The ordinary observer, unused to the scale of great mountain scenery, will see nothing but curious crags in the huge cliffs of Úschba, and will pass over without a second glance the strips of glistening silver which seam the mountain-wall: we, who recognised in one of them the frozen cascade above Adisch, in another the glacier above Glola, from the icefall of which we had retreated in despair, lingered to take a long and affectionate farewell of old friends seen beyond hope once more. Among the numerous Russian officers who had passed this way with whom we conversed, we did not find one who seemed aware that the great mountains were visible from the pass, and travellers like them, incapable of interpreting rightly the images presented to their eyes, will perhaps accuse us of making an absurd fuss over a distant horizon of snow and a few jagged rocks, which seem to them rather to spoil the sweep of the skyline.

We paid comparatively little attention, at the time, to the view looking back into the Turkish territory, but it must not be left wholly unnoticed. There, beyond the wooded and broken spurs of the chain on which we stood, the highlands of the province of Kars, a succession of rounded hills, with a few patches of snow still lingering on their summits, stretched away to the horizon.

A long ride still lay before us—our horsemen urged us

onwards, and we were forced reluctantly to turn away and commence the descent. For some distance the new road is cut in steep and long zigzags; it plunges almost immediately into a grove of noble pines and firs, with long mossy streamers hanging from their branches. The track, reduced to a rough bridle-path, crosses a rivulet, and makes a second and deeper plunge to the bed of the Chani-Squali, the course of which it follows henceforth to the Mingrelian lowlands. We imagined that we had already exhausted the charms of sylvan scenery, but the forest in which we now found ourselves surpassed in the richness and variety of its foliage any we had yet seen. To enumerate the trees would be not only to exhaust, but to make several additions to, the list of those found in Central Europe; the beech, the elm, and the alder, which here grows to an enormous size, were the most conspicuous. Long wreaths of ivy hung from their branches, and twisted round their stems, and the ground was covered with a dense undergrowth of box, holly, laurel, azalea, and rhododendron bushes. Long grasses and ferns, some rising to the height of a man, filled the glades; others, small and delicate, grew in the crannies of the mossy cliffs. The stream foamed at the bottom of the deep glen in a succession of falls and rapids; the path, following and frequently crossing it, grew worse and worse, and our horses found difficulty in picking their way along it. A causeway of logs had in many places been laid upon the swampy ground, and the track, poached into holes between the timbers by the feet of passing animals, was converted into a succession of ridges and furrows similar to an American 'corderoy.'

We wandered on for some hours through the glades and thickets, halting at times to admire some exquisite vista, in which the snowy peak of Tau Tötönal, framed between the green hillsides, seemed to float in blue haze rather than to belong to earth. The narrow trench gradually expanded, leaving space for occasional patches of cultivated ground. Cornfields were in time succeeded by plantations of the tobacco-plant, the bright-green leaves of which are in their natural state always a pleasing sight. At the corners of many of the enclosures, which are generally surrounded by rough fences, we noticed raised wooden platforms; these are said to be look-out posts, where a watchman keeps guard against the depredations of the bears which abound in the forest. Below the junction of the two glens the valley widens, and is dotted with numerous clusters of cottages. Fruit-trees now become plentiful; the plum, the pear, and the medlar grow wild, and the vine trails its long branches over the forest trees.

The new road was in course of construction, and we found parts in a sufficiently forward state to enable us to ride along it. The valley, having trended north-west for some distance, turned due north, and a village stood on the opposite bank of the stream. From this point, the hills sank rapidly, and our horsemen pointed out the position of Bagdad in the distance. It was dark, and the woodcutters' fires blazed out cheerily, high upon the hillsides, before we reached our resting-place. Bagdad is situated close to the point at which several valleys open on to the plain. The village consists of one street, with houses on either side; there were plenty of people about, but they one and all refused us shelter for the night; we were getting angry and perplexed at this final specimen of Mingrelian manners and hospitality, when one of the peasants suggested that a French baron lived half a mile off, and that we might find lodgings with him. We guessed rightly that there could not be two French barons near Kutais, and that this one must be the Baron de Longueil, whom we had met six months before on board the Rion steamer. Under the circumstances, the best course was evidently to accept the proposal, and claim the hospitality of the only civilised being within reach.

Though the Baron was away from home, 'Madame' received us with the greatest kindness, and gave us both food and beds, luxuries which, ten minutes before, we had seemed little likely to obtain on this side of Kutais. From the account we heard of it, farming in Mingrelia does not seem to be so wholly delightful an occupation as the natural fertility of the soil might lead one to suppose. 'Madame' complained bitterly of the laziness and dishonesty of the native servants, and of the excessive difficulty of transport, Kutais, though only about thirty miles distant, being often rendered inaccessible for weeks in winter by swollen streams and the horrible state of the roads.

September 4th.—When daylight came, we saw that the Baron had built himself a pretty little villa, ornamented with a verandah overgrown with creepers, and some attempt at a garden. Bidding a grateful adieu to our kind hostess, we remounted our horses, and started to ride across the flat country that separated us from Kutais. The sky was overclouded, and we could not but congratulate ourselves on our good-luck in having had so perfect a view the previous morning. The road leads at first across glades of turf, and between copses of fruit-trees overhung and knitted together by wild vines, and passes through several villages. Mingrelian hamlets are all exactly alike, and it would be impossible to improve on Mr. Palgrave's description of one: 'The houses are neither ranged in streets, nor grouped in blocks, but scattered as at random, each in a separate enclosure. The houses themselves are one-storied, and of wood, sometimes mere huts of wattle and of clay; the enclosures are of cut stakes, planted and interwoven lattice-wise. Old forest trees, fresh underwood, bramble, and grass grow everywhere, regardless of the houses, which are in a manner lost among them; one is at times right in the middle of a village before one has even an idea of having approached it.' On the country roads in the neighbourhood of Kutais, rude vehicles, half-sledge, half-cart, may frequently be met, drawn by two oxen,

Mingrelian Wine Jar.

and laden with one of the huge earthenware jars used for storing wine, which will scarcely fail to recall to the traveller's mind the story of the Forty Thieves in the 'Arabian Nights.'

For nearly an hour we traversed a thick beechwood, and emerged from it at the point where it is necessary to ford not only the stream of the Chani-Squali, but the larger and far more formidable Quirila. Three months earlier we should have found a new bridge just completed, but after a career of usefulness, short even for this country, the unhappy structure had been left standing high-and-dry, while the river flowed in a new channel fifty yards further south. To judge them by their works, Russian engineers seem incapable of anything thorough, and the amount of money wasted during the last few years in the Caucasus would be difficult to calculate. Even the late Czar Nicholas is said to have sighed over the constant call for fresh grants for the construction of the Dariel road, which is not finished yet. With more skill and less jobbery a Swiss canton would have made it in a quarter of the time, and at half the expense it has cost the Imperial Government.

The Quirila was fortunately not in flood, and we waded without difficulty through its broad clear-flowing current. The hard road into Kutais is half completed, and follows the left bank of the Rion at no great distance from the river. The journey divides itself naturally into two stages; the first across the marshy lowlands, from which a steep bank leads to an upper level covered with dense oak copses, through which the track runs straight as an arrow for several miles, until, bending to the left, it descends into Kutais. The first view of the town from this side, with its large white-walled green-roofed buildings and domed churches, set in a framework of hills, and watered by the rapid stream of the Rion, is exceedingly pretty. We trotted through the streets, and dismounted at the door of the 'Hôtel de France,' where we found François, who had arrived safely with the luggage two days previously.

There was some difficulty in finding rooms, for the hotel was crowded; the Grand Duke Michael, after escorting the Grand Duchess to the seaside, had returned to Kutais, and set out from thence early on the previous day to ride through the Radscha and over the Mamisson Pass to

Vladikafkaz. Many of the officers, who had come from country posts to meet him, were still in the place, and there were, besides, an unusual number of travellers awaiting the departure of the next steamer from Poti. Amongst them were two gentlemen, who had, like ourselves, been engaged in exploring the Caucasus—Mons. Favre, the son of the well-known Genevese geologist, and Mons. Desrolles, an entomologist, whom the natives had facetiously nicknamed the 'Father of Flies.' The extent of their excursions in the mountains had been to cross the Mamisson Pass and ride up to Uruspieh. We also met an English gentleman and his wife, who had made their way across the Caucasian isthmus from Petrovsk, on the Caspian. The journey from that place to Vladikafkaz, occupying three days, had to be performed in 'telegas,' and is one which few ladies would care to undertake.

September 5th.—We discovered during the day two new attractions in Kutais—its photograph shop, and its jet. We purchased a considerable number of 'cartes-de-visite' of the peasantry of the surrounding districts, executed with an eye to the picturesque in the grouping and accessories, which did great credit to the enterprising artist. The jet, which is somewhat softer in substance and more brittle, but otherwise similar to that sold in England, is hawked about the streets in long chaplets, and may be bought for very low prices; we were assured that all the beads are hand-cut. Its native name is 'gicher,' which, in Armenian phrase, also means ' night.'

Before leaving Kutais, we had the pleasure of making the acquaintance of Mons. Khalissian, an Armenian gentleman who has spent much time in exploring the neighbourhood of Kazbek, and lived for some weeks encamped at the foot of the Devdorak glacier, engaged in scientific researches, of which he means to publish the results, accompanied by a map of the glaciers on the eastern flank of the mountain. Much of the information we derived from him has been embodied in previous pages.

We had intended to remain two days, and spend the second in an excursion to the celebrated monastery of Ghelathi, founded at the end of the eleventh century by a Mingrelian sovereign, which is celebrated alike for the interest of its architecture, and the venerable images and ecclesiastical wealth it contains. We heard, however, such disquieting reports of the irregularity of the Sunday steamer on the Rion, that it seemed more prudent to follow the example of most of our fellow-travellers, and start twenty-four hours sooner than we had proposed. To this end an arrangement was concluded with the postmaster, the same man who had given us trouble on our former visit, by which he undertook to provide a 'tarantasse' and horses at midnight on Friday—the usual time for starting, as the accommodation at Orpiri is bad, and the Rion steamer leaves early in the morning. The money for the horses was paid, and we believed the affair settled; but at the appointed hour no horses came, and on sending to the post, we were told that the official had gone off to Orpiri, leaving no instructions, and that if we wanted horses we must pay over again, and make a present besides for the favour of having them. The postmaster who had thus sought to cheat us was described by a gentleman of Kutais as a ' brigand du premier ordre,' and, unwilling to become his victims, we visited two higher officials, from one of whom we received an order that horses should be given us directly, and no further payment asked. Let all travellers beware of that pair of harpies, the mistress of the 'Hôtel de France,' and her friend the postmaster at Kutais!

We drove to Orpiri in pouring rain. During our voyage down the river, the weather, though cloudy, was fine—a fortunate circumstance, as, owing to the shallowness of the water at this season, the ordinary boats cannot get up to Orpiri, and the passengers and cargo are transferred halfway from one boat to another. At Poti we went to the 'Hôtel Jacquot,' which is clean and comfortable. No reader of 'Martin Chuzzlewit' could fail to be struck with the resemblance to Eden of this miserable spot. Its situation, in a swamp rather below the waters of the Rion, which are only prevented by embankments from sweeping away the place, combines almost every disqualification for a commercial town. The hotels (there are three), the office of the steamboat company, and a few houses of the better sort, are planted at irregular intervals near the quay. Behind them is the main and only street, which consists of a causeway running between two rows of log-shanties, raised on piles above pools of fetid water, and ending abruptly in a dismal swamp. Every road has two large ditches, brimming with stagnant slime on either side, crossed by little bridges, which, as one of our party found by unpleasant experience, are easily missed in the dark. What wonder that fever and ague are written in the faces of the dismal gathering of officers and employés, to whom, in what the residents are pleased to call a public garden, a melancholy band nightly discourses doleful tunes! All the real merriment and music of Poti is confined to the frogs, and they, to judge by the noise they make, lead a merry life of it. All night long their ceaseless chorus resounds through the place, and it is asserted by the inhabitants— though I cannot wholly credit the story—that the sound, when the wind blows that way, is audible even at Constantinople.

So long as Turkey keeps Batoum, Russia is reduced to make the best of Poti, as the port of the Caucasus. Soukhoum-Kalé, which seems at first sight preferable, is little less unhealthy, and is besides rendered difficult of access from the interior by the numerous streams (of which the Ingur is the largest) flowing out of the mountains, none of which the Russians have as yet succeeded in bridging. The future Transcaucasian Railway is to begin at Poti, and endeavours are being made to deepen the bar of the Rion, and convert its second mouth into a harbour. Colonel Schauroff, the officer in charge of the works, is firmly persuaded of the eventual success of the scheme which he has himself originated, and is endeavouring to execute; but, as is often the case, his superiors do not share his convictions, and their half-hearted support and scanty doles are likely to delay indefinitely the completion of the proposed works.

On the morning of the 6th we made the first break in our party. François left us on board the Batoum steamer to find his way home to Chamonix by Constantinople and Marseilles. Later in the day we embarked with Paul on board the small but prettily-fitted boat which was to convey us to Soukhoum-Kalé. The sunset was fine, but clouds hung over the Caucasian chain, and deprived us of our last chance of seeing Elbruz, which in clear weather is plainly distinguishable from shipboard. At daybreak on the 7th we were at anchor in the bay of Soukhoum-Kalé, alongside the larger steamer to which we were to be transferred for the further voyage. The town, a small seaport, is situated at the head of a southward-facing bay; a short distance inland the country rises in graceful wooded hills, but Soukhoum-Kalé stands on the level marshy shore, and is very unhealthy. Its only sights are some wonderful weeping-willows in the main street, dignified as a boulevard, and a botanical garden, or rather plantation of exotic trees. There was much ripe fruit there—grapes, pears, and plums—well guarded by a sturdy youth, who assured us they were all reserved for the consumption of a General, and whom our felonious attacks on his treasures excited to desperation. This spot witnessed some hard fighting in 1864, when the Abkhasians broke out into open revolt, overpowered and murdered every man at some of the Russian outposts, and attacked Soukhoum in force.

Our steamboat, the 'General Kotzebue,' one of the finest in the Black Sea Company's fleet, left on the evening of the 7th. All night and the next day we were running along the Caucasian coast. The shore is lined with grey or white cliffs, behind which the mountains rise in long wooded ridges broken by valleys, through which numerous streams find a way down to the sea. Though on the whole fine, the character of the scenery was not so grand as we had been led to expect. After a short stay at Novorosiski, an unattractive-looking Russian colony in a deep bay, we continued our voyage, the coast of the Caucasus gradually fading in the darkness. During the night we arrived off Kertch, where the boat remained till midday, allowing time to visit the town, rebuilt since the war; the museum, the chief treasures of which have been removed to St. Petersburg, and one of the tumuli in the neighbourhood, containing a curious stone chamber.

The run from Kertch to Sevastopol occupied twenty-six hours, including stoppages at Theodosia and Yalta. Beyond the latter place the coast-scenery is superb, and the magnificence of the weather enabled us fully to enjoy it. We had pleasant and amusing society on board. Besides the travellers whom we met at Kutais, there were two Russian Generals, types respectively of the two extremes met with in the Imperial Service; a Georgian youth, splendidly dressed in his full national costume, and a less showy boy (a son of the Suanetian prince who murdered the Governor of Mingrelia), both of whom were going to complete their education at the University of Odessa.

On the afternoon of September 10th, after running across Balaclava Bay, and rounding Cape St. George, we entered the harbour of Sevastopol, at the mouth of which stands Fort Constantine, looking as strong as ever, though its southern brother is utterly destroyed. The interior of the town presents a scene of destruction for which we were quite unprepared. Not only are the dockyards and government buildings blown to pieces, but the main street is deserted and grass-grown, and the houses that line it, built of white stone, stand roofless and shattered wrecks. Nowhere but at Pompeii have I seen such desolation. The population has fallen from 80,000, before the war, to 8,000; it is now rising again, owing to the recent establishment of the shipbuilding yards of the 'Black Sea Steam Navigation Company' in the Admiralty Creek. Their new machinery-sheds, and the adjacent barracks, are the only signs of life about the place.

We were surprised to find the lines of the Russian defences so perfect; the lower story of the Malakhoff tower still stands, surrounded by the big ditch and high mound; the salient angle of the Redan looks fresh and sharp, and a dismounted cannon lies in one of the embrasures. On the heights outside the town, the trenches are easily traceable, and at a greater distance, where the huts stood, the ground is strewn with fragments of broken bottles and old shoe-leathers. The French dead have been, as far as possible, collected into one cemetery, which is planted with trees, and placed under the charge of a resident guardian; but the bodies of our countrymen lie scattered over the downs, in more than fifty small enclosures, each surrounded by a low wall. At the time of our visit, these graveyards were covered with a dense growth of weeds, many of the tombstones were broken and the inscriptions erased, and we saw everywhere proofs of a carelessness and neglect which are discreditable to the English nation.

The battlefields of Balaclava and Inkerman are marked by simple stone obelisks. It is difficult to recognise the 'Valley of Death' in a slight depression between two grassy knolls; the heights of Inkerman are more like what fancy pictures them, and the ravine up which the Russians came to the assault is striking, apart from its associations. On a sloping hill, above the forts on the northern side of the harbour, is the great Russian cemetery. The simple fact, that from 250 to 300 dead lie under each of the large nameless tombstones that line the central avenue, gives some idea of the numbers buried there. Prince Gortschakoff's monument stands at the top of the enclosure; though he survived the siege for several years, the inscription states that he wished to 'lie with those brave companions in arms, by whose valour the enemy was prevented from penetrating further into fatherland.' On the brow above the cemetery, a handsome memorial chapel has been erected. The building is an attempt, not wholly successful, to unite a monument and a chapel; externally it has the form of an irregular pyramid surmounted by a large cross. The interior, a Greek church of the usual form, is in course of decoration with a series of frescoes by native and Italian artists, which seemed to us of considerable merit.

English travellers are strangely indifferent to the attractions of the Crimea. Setting aside for the moment natural beauties, its historical interest well repays the trouble of a visit. Whatever monuments may be raised elsewhere, Sevastopol itself will for many years to come remain the greatest memorial of the struggle which centred round it. Great battles are fought, and little trace remains; it is in the ruin caused by sieges that war stamps its most lasting mark. In the bullet-riddled walls of the once handsome buildings, in the laboriously-wrought labyrinths of lines and counterlines that encompass them, in the shattered forts and demolished dockyards on the water's edge, and, more than all, in the crowded burial-grounds on the heights, it is easy to read the story of the siege; and, in gazing on them, one is led to appreciate both the importance of the result, and the cost at which it was obtained.

Travelling in the Crimea is rendered more agreeable than in most parts of Russia by the excellence of the roads, combined with the civility and promptness met with at the post-stations. These unusual phenomena are in a great measure due to the fact, that this is the only district of Russia where pleasure-travellers are understood, and somewhat also to the pervading influence of the Woronzoff family, at whose expense the greater part of the coast-road was constructed. It is a very pleasant drive from Sevastopol to Yalta, Simferopol, Bakhchisarai, and back. The entire distance can be got over in three days, but five are the least that should be allowed, as it is desirable to leave time for a visit to the villas on the coast, and Bakhchisarai and its neighbourhood afford employment for a long afternoon.

Balaclava is the first post-station. While changing horses we had time to climb to the old Genoese tower, and look down, on one side on the landlocked creek, on the other on the iron-bound coast on which the ill-fated 'Prince' struck and went to pieces. It is after a long inland climb that the road, in a gap between two wooded hills, reaches the Gate of Baidar, a classical archway built to mark the spot whence the traveller gains his first, or last, view of the Garden of the Crimea. From the water's edge, 1,500 feet below, a long slope of garden, wood, and vineyard, dotted with villas, runs up to the foot of a tall range of grey limestone cliffs. The Russian Corniche, as the post-road from this point to Alushta (a distance of sixty miles) has been aptly called, need not shrink from comparison with its more famous rival. Few will be found to depreciate the beauty of a series of landscapes which unite, in constantly-shifting proportions, the charms of bold rock-scenery and rich vegetation, enhanced, as far as such scenery can be, by human aid, and set in a frame of blue sky and still bluer sea. No one who is fortunate enough to travel here in the vintage-season will despise the merits of the grapes, which are sold at a few copecks the pound.

Yalta is the Mentone of the coast. Here however, as is generally the case in Russia, the English rather than the Italian style has been adopted, and the little town, seen from the sea, with its prim houses and square-towered church, reminded us more of the Isle of Wight than of anything on Mediterranean shores. The three principal estates are all near Yalta. At Alupka, Prince Woronzoff's seat, the house is an odd mixture of the feudal and Saracenic styles, while the grounds are laid out entirely in the English manner. Orianda, the property of the Grand Duke Constantine, occupies the finest position, but is the least interesting; Livadia, the Emperor's villa, is more in the châlet style, and has attached to it a chapel, small, but exquisitely decorated. We remarked that the Czar has some fine watercolour drawings of the Caucasus in his study, the general fittings of which, as of the rest of the house, are very plain, and, with the photographs of the Imperial Family hanging on the walls, give an unexpected but pleasant impression of homeliness, and absence of court restraint.

The road to Simferopol continues along the coast as far as Alushta—then turns inland, and crosses a well-wooded ridge of 3,000 feet, a spur of Tchatyr-Dagh (5,125 feet), the respectable monarch of Crimean heights. Sleeping at Simferopol, an uninteresting town, we drove on next day to Bakhchisarai, crossing halfway the brook Alma, considerably above the battlefield. The town, picturesquely situated in a narrow glen, in the centre of a wide desolate steppe, is entirely Turkish in character, and forms a striking contrast to the Russian style of the rest of the Crimea. The principal attraction to visitors is, however, the residence of the Tartar Khans, used as a hospital during the war, but which has since been tastefully restored, at the expense of the Government. It is a very perfect specimen of an Oriental palace, and the gaily-decorated ceilings and brilliant stained-glass make the deserted rooms look bright and cheerful. A soldier acts as cicerone, but he was so gloriously intoxicated at the time of our visit, that little information could be got from him. In the neighbourhood is a curious monastery, hollowed out of the rock, and a village of Karaite Jews, a sect the origin of which seems doubtful, if not unknown, and who are accordingly supposed to be a remnant of the lost Ten Tribes.

We returned to Sevastopol on the 16th, and on the afternoon of the 17th embarked for Odessa. The steamer, calling at Eupatoria on the way, makes the passage in twenty hours. At Odessa the Eastern element is altogether wanting, and even the Russian is unobtrusive; it is, in fact, a Western city. Its character is no doubt due to its having grown up under the patronage of a French exile, the Duc de Richelieu, and a Russian Anglo-maniac, Prince Woronzoff. It is at present the best-paved and best-lighted town in Russia, and boasts a handsome boulevard and an opera. Its commercial prosperity will no doubt be largely increased by the opening of railway communication with Moscow.

Situated on the brow of a cliff, and at the edge of a sandy plain, Odessa has few attractions for the passing traveller, who, as soon as he has sauntered through the streets, visited one or two churches, the Jewish synagogues, and the boulevard, will be glad to continue his journey. This, now that the railroads are finished, will be found no difficult matter, but for us it was different. The line to Kieff was as yet unopened, and at one time the difficulties of making our way across Russia to St. Petersburg seemed so great that we had almost decided to fly to Istamboul and return home by Athens, when a piece of intelligence, which afterwards proved untrustworthy, made Tucker and myself revert to our original plan. Moore, however, could not spare longer time, and left us to cross Europe by a more direct route, by way of Lemberg, Czernowitz, and Cracow, by which he succeeded in reaching Paris in 6½ days' hard travelling, including a detention of twenty-four hours at Czernowitz. Paul also, who had served us well and faithfully during our Caucasian wanderings, was now dismissed, to return to his Mingrelian home; thus Tucker and I were left to end our journey, as we had begun it, by ourselves.

On the evening of the 21st September, we went by rail to Birzoula, where the new Kieff Line branches off from that to Elizavetgrad; it had been partially opened, and we had hopes of being forwarded along it, which however proved illusory. After a vexatious delay of twenty-four hours, we went on again by the Odessa train, and reached Elizavetgrad on the morning of the 23rd. This part of Russia consists of nothing but bare rolling downs, on which much of the corn shipped annually from Odessa is grown; the scenery is consequently very uninteresting and monotonous. The towns and villages are situated in depressions, watered by small streams. Elizavetgrad, planted in one of these wrinkles of the steppe, is a large but unremarkable place, the headquarters of the cavalry in Southern Russia. We found a man at the railway-station who talked German, and who undertook to procure a carriage to take us to Krementchuk, eighty miles distant, where we hoped to catch the Dnieper steamboat on the following morning. A springless waggon, covered with a tilt, and drawn by four horses, was made ready, and after a tedious drive of twenty hours, we reached Krementchuk at 8 A.M. on the 23rd, an hour after the time fixed for the steamboat's departure. Fortune, however, befriended us for once, and as we drove over the long bridge of boats, we saw the little steamer still lying beside the wharf, and, urging our driver to quicken his pace, we made our way through the loose soft sand which covers the banks, and got on board ten minutes before she started. The steward and waiter spoke German; the cabin, though very small, was clean, and the fare good.

The voyage up the Dnieper from Krementchuk to Kieff occupied two days, for we lay-to at night, owing to the absence of moon, and the difficulty of the navigation. There is little or no scenery on the river, which for many miles runs through a level country between low sandy shores; nearer Kieff, the right bank rises into bold bluffs, crowned here and there by the pagoda-like churches of Russian villages. The river-boats are very picturesque objects, with a tall tapering mast bearing a huge sail, and a long pennant (generally crimson) flying from the top. Although on the evening of the second day we caught sight of the burnished cupola of the Lavra, reflecting the last rays of the setting sun, it was long after dark before we passed under the central span of the great suspension-bridge, and anchored alongside the busy quay of the Podole, or lower town of Kieff. A 'droschky' soon carried us up the steep hill to the upper town, and after some difficulty in making ourselves understood, we procured rooms in the 'Hôtel d'Europe.'

We spent two days in Kieff, which is the most picturesque and one of the most interesting of the great cities of Russia. The town, built on the top of two lofty bluffs, commands a wide view of the plains stretching far away eastwards in the direction of Koursk; as a matter of course the public buildings are large, the streets wide, and the open spaces numerous. The peculiar character of the town is due to the multitude of churches with green, gilt, or silvered cupolas, and the number of trees interspersed amongst the houses. The public gardens are, for Russia, exceptionally pretty and well-kept; in a dell in their centre, sheltered by wooded banks, is a large café and pleasure-ground, where two excellent bands performed in the evening. The cathedral is very old and curious, containing some fine eleventh-century mosaics and the tomb of Yaroslaf; but the great attraction to visitors is the famous fortress-convent, or Lavra, considered the holiest in Russia, to which crowds of pilgrims draw together from the farthest parts of the Empire. We were lucky in visiting it on a saint's day, when every corner was crowded with peasants in the most picturesque costumes—men in heavy jackboots, bright-coloured shirts fastened in by a belt at the waist, and low-crowned hats—and girls with gaudy necklaces and wreaths of paper-flowers round their heads, some of them fresh and pretty-looking, though all more or less of the flat-faced Russian type.

The churches are gorgeous with silver-plated pictures of saints and jewelled relics. At the time of our visit a large and excited crowd were pushing and jostling in the eager struggle to approach and kiss these holy treasures. In the catacombs, a long series of cellar-like vaults hewn in the rock, a multitude of saints and pious virgins, each in a separate niche, lie in open coffins robed in gorgeous silks, the faces veiled, but a shrivelled finger protruding to receive the kiss of the orthodox. The pilgrims, who accompanied us through the vaults, laid small offerings on the bodies of their favourite saints, which were collected by the square-capped monk who brought up the rear of the party. English visitors are few and far between at the Lavra, and we excited a good deal of curiosity, though we met with nothing but civility.

Our first enquiries as to the best mode of continuing our journey were met by the unpleasant news that, only ten days before, the regular 'diligence' service on all the lines had been suspended, and the post ordered for the future to travel in 'telegas.' Further researches resulted in the discovery, that St. Petersburg by way of Vitebsk was more accessible than Moscow by way of Koursk, the advantage in distance (sixty miles) of the latter being more than counterbalanced by the better road to Vitebsk. On calling at the carriage bureau, at the post, we were lucky enough to see a high official, who spoke excellent French, and was very civil and glad to see us, for he had a lady in charge desirous of making the same journey. We quickly agreed to take a 'diligence' between us, and start on the morning of Monday the 28th; before the time came two other people were found to occupy the vacant places. Tucker and I took the 'banquette,' and with pleasant weather and no dust, the long drive was by no means disagreeable.

The road was broad and metalled, and the stations, mostly tenanted by Jews, well-fitted-up, at least compared to those in the Caucasus. Our drivers were a constant source of amusement. Sometimes a postilion rode one of the leaders, sometimes two peasants sat side by side on the box—one driving the leaders, the other the four wheelers, harnessed abreast after the usual Russian fashion; for two stages only, a man bolder than usual gathered up the mass of reins, and drove the whole team single-handed. This part of the interior of Russia is not so ugly as that country is popularly supposed to be; where flat, it is generally well-wooded with pine and birch, and between Mohilef and Vitebsk, where the watershed between the Dnieper and Dwina is crossed, the country becomes really hilly. The autumn tints on the foliage were glorious. We accomplished the 360 miles in seventy hours, and reached Vitebsk early on Thursday morning. There we came upon one of the yet-unfinished threads of the web of European railways which will soon spread itself over the whole of Russia. The train carried us in twenty-four hours to St. Petersburg, where we arrived ten days after leaving Odessa, having, with the exception of our halts of one day at Birzoula, and two at Kieff, travelled day and night.

Anything connected with the Caucasus, a country associated in the national mind with a long and victorious struggle, is sure to attract attention at St. Petersburg. The notices of our ascents of Kazbek and Elbruz, which had appeared during the summer in the newspapers, had created considerable interest, and we received, on our arrival in the capital, many kind invitations, most of which, owing to the shortness of our stay, we were compelled, much against our wish, to decline. A matter of more serious regret to us was the miscarriage of an invitation to be present at a review of the Czarewitch's regiment of Cossacks, held in the Great Riding School. We did not receive the Imperial commands until it was too late to obey them, and were thus deprived of an opportunity of presenting the Czar with a portion of the highest rock in his European dominions.

It is now time to cut a long story short. Amongst the sights of Moscow and St. Petersburg, and on the homeward journey, I need not detain my readers, to whom I here bid a hearty farewell, trusting that I may persuade some of them to follow in our footsteps, and to learn for themselves, on the slopes of Elbruz and at Russian post-stations, the force of Shakspeare's questions:—

> Oh, who can hold a fire in his hand
> By thinking on the frosty Caucasus?
> Or cloy the hungry edge of appetite
> By bare imagination of a feast?

APPENDIX I.

THE ELBRUZ EXPEDITION OF 1829.

THE Expedition of 1829, led by General Emmanuel, was a sort of politico-geographical progress through some of the northern valleys of the Caucasus, with the ultimate object of ascending Elbruz. It was accompanied by several German savants, one of whom, Herr Kuppfer, has given an account of it in his 'Voyage dans les Environs du Mont Elbrouz dans le Caucase,' published at St. Petersburg, 1830.

After experiencing many difficulties on the road, the expedition, escorted by Cossacks and several cannon, reached in safety the headwaters of the Malka, 8,000 French feet above the sea. On the morning of the 21st July a portion of the party set out at 10 A.M., and at 4 P.M. attained the edge of the snow, at a height which they assumed to be not far from 11,000 ft. Here they encamped for the night, and at 3 the next morning started with some native (Circassian) mountaineers and a few Cossacks. At first all went smoothly, but as the steepness of the slopes and the heat of the sun increased, their progress became more laborious, until —at a point which was determined to be 14,000 French (14,921 English) feet above the sea, and therefore *really* 3,600 English feet, though estimated by them to be 1,492 English feet below the summit—M. Kuppfer and three of his companions fairly knocked up. In spite of this, with strange looseness of expression, he proceeds to add: 'However, this first attempt had succeeded beyond our hopes. On entering the Caucasus we had believed Elbruz inaccessible, and in a fortnight *we were on its summit.*' Meanwhile M. Lenz, who, accompanied by two Circassians and a

Cossack, had preceded his friends, got as far as the top of a ridge of rocks in the direction of the summit by 1 P.M., and *then turned back*, as time ran short and the snow was soft.

While his companions were engaged in assaulting the mountain, General Emmanuel, seated before his tent in the valley, watched their progress through a telescope. He suddenly observed a single man far in advance of the rest. We are rather superfluously informed that the features of the solitary climber were indistinguishable, but the General could tell from his dress that he was a Tcherkess. The figure advanced steadily towards a scarped crag, which appeared from the camp to be the summit, walked round its base, and then vanished behind the mists which cut off all further view of the mountain.

What he had thus seen satisfied the General that the object of his expedition was fulfilled, and that the highest summit of Elbrus had been trodden by human foot. He ordered the news to be proclaimed in camp, and gave notice that the successful climber should receive the promised reward of 400 roubles as soon as he appeared to claim it. Few of my readers will be surprised to hear that in the course of the evening a Tcherkess named Killar presented himself and received the money.

If, as the loosely-worded narrative seems to show, neither Mons. Lenz nor any of the German savants saw or heard anything of their more fortunate rival until they returned to the camp, Killar's claim to the honour of the first ascent rests entirely on General Emmanuel's account of what he saw through his telescope, under circumstances which render his testimony, to say the least, very questionable. It is difficult even for practised eyes to distinguish a solitary man on a snowslope broken by crags 10,000 feet in vertical height above the observer, and in such cases men often see what they both wish and look for. Moreover, in the present instance, General Emmanuel's credit was involved in the success of an expedition which had been organised with much care and expense, and he had every motive to make a discovery which would justify him in asserting officially that the top of Elbrus had been gained by one of the men under his command. Even Russians treat the official statements of their countrymen with a certain amount of reserve, and require external confirmation before believing them. If, however, both the General's good faith and his telescope are thought above suspicion, the only fact proved is that a Tcherkess reached the foot of rocks, which looked from below like the top, and was then lost in clouds.

In default of better evidence we can scarcely be expected to regard Killar as the Jacques Balmat of Elbruz.

APPENDIX II.

HEIGHTS OF PEAKS, PASSES, TOWNS, AND VILLAGES IN THE CAUCASIAN PROVINCES.

[*N.B.*—The barometrical heights are untrustworthy, and can only be considered as roughly approximative.]

PEAKS.

	Position	Russian Survey	Barometer	Estimate
Ararat	} S. of Erivan	16,916		
Little Ararat		12,840		
Alagoz	NW. of Etchmiadzin	13,436		
Kazbek	W. of Kazbek village on Dariel road	16,546	16,675	
Gimaran Khokh	} In the Kazbek group, lying off the watershed	15,672		
Tau Teply		14,510		
Zilga Khokh	Above the source of the Terek	12,645		
Adai Khokh	NW. of Mamison Pass	15,244		
Tau Bardisula	Between Gurdzieversk and Karagam Passes			16,000
Tau Rehoda	S. of Gebi	11,128		
Dychtau	} Between the two branches of the Tcherek	16,925		
Kuschtantau		17,098		
Tau Tiutinal	} N. of Suanetia			15,800
Uschba				16,500
Tungzorun	S. of the Baksan sources			16,000
Elbruz	Between the sources of the Kuban, Malka, and Baksan	18,526		
Beschtau	N. of Pätigorsk	4,594		

HEIGHTS OF PEAKS, PASSES, ETC.

PASSES

	Position	Russian Survey	Radde	Barometer
Suram Pass	On Tiflis-Kutais road	3,027		
Krestowaja Gora	On Tiflis-Vladikafkas road	7,977		7,770
Pass to Zneem	Between Terek and Ardon			10,690
Mamisson Pass	Between Ardon and Rion	9,390	9,421	9,450
Garbievcsek Pass	Between Rion and Urach			11,250
Karagam Pass				12,250
Goribolo Pass	Between Rion and Zenes-Squali		9,598	9,790
Nachha Pass	Between E. and W. Zenes-Squali		9,400	9,400
Nakengar Pass	Between W. Zenes-Squali and Ingur		8,831	8,780
Dashkjinsey Pass	Between Kalde and Adisch glens			9,074
Nakra Pass	Between Ingur and Baksan			10,830
Stulovesak Pass	Between Tcherek and Urach			10,500
Sikar Pass	Between Abastuman and Kutais	7,104		

TOWNS AND VILLAGES

	Position	Russian Survey	Radde	Barometer
Kutais	Capital of Mingrelia	670		
Tiflis	Capital of Georgia	1,350		
Duschet		2,918		
Kobi	On Dariel road	6,590		6,400
Kasbek		5,710		5,720
Alauo	Upper Terek Valley	7,600		7,400
Kazei Kom	Ardon Valley			7,100
Tseh				5,400
Gurschari				6,600
Gtola	Rion Valley	4,344	4,544	4,350
Chiora				4,390
Gebi		4,030	4,327	4,000
Ssanagunelli				6,160
Jibiani	Suanetia		7,004	7,130
Adisch				6,500
Sani				4,305
Latai				4,450
Pari		4,833		4,340
Uruspieh	Baksan Valley	6,156		6,100
Atschkutan				
Pitigorsk	On the Podkumok	1,800		
Kislowdsk		2,700		
Vladikafkas	On the Terek	2,345		
Borjom	On the Kur	2,212		
Achaltsich		3,376		
Abastuman	NW. of Achaltsich	4,113		

APPENDIX III.

CATALOGUE OF PLANTS

Collected in the summers of 1864–65 by Herr Radde, and arranged by Herr V. Trautvetter

Borjom, June, 1865.   2,600—3,400 ft.

Scrophularia lucida *L.*
Scrophularia variegata *M. B.*
Ziziphora clinopodioides *Lam.*, vart. canescens *Led.*
Melica ciliata *L.*
Dactylis glomerata *L.*
Centaurea leucolepis *Dec.* (*Ledeb.*)
Carduus hamulosus *Ehrh.*
Adonis aestivalis *L.*
Onobrychis sativa *Lam.*
Pyrus salicifolia *L.*
Centaurea bella *Trautv.*
Medicago falcata *L.*
Rhus Cotinus *L.*
Leontodon biscutellaefolius *Dec.*
Onobrychis petraea *Dec.*
Medicago minima *Lam.*
Crataegus melanocarpa *M. B.* vart. glabrata *Trautv.*
Cerastium grandiflorum *W. et Kit.* vart. glabra *Koch.*
Arabis hirsuta *Scop.*
Convolvulus lineatus *L.*
Ribes Grossularia *L.*
Alsine setacea *M. et Koch.*
Scrophularia rupestris *M. B.*

Salvia sylvestris *L.*
Mulgedium albanum *Dec.*
Helianthemum oelandicum *Wahlenb.* vart. hirta *Ledeb.*
Thalictrum minus *L.* vart. stipellata.
Campanula Raddeana *Trautv.*
Crupina vulgaris *Cass.*
Valerianella Morisonii *Dec.* vart. dasycarpa *Trautv.*
Saponaria alocioides *Boiss.*
Briza media *L.*
Anacamptis pyramidalis *Rich.*
Verbascum, sp.
Ervum Ervilia *L.*
Centranthus longiflora *Stev.*
Polygonatum latifolium *Desf.*
Lathyrus rotundifolius *W.*
Hablitzia tamnoides *M. B.*
Onosma microcarpum *Stev.*
Asperula azurea *Jaub et Spach.*
Vincetoxicum medium *Decaisn.*, vart. latifolia *Trautv.*
Onobrychis Michauxii *Dec.* vart. glabra *Regel.*
Rhamnus grandifolia *F. et M.* vart. umbellis sessilibus.

CATALOGUE OF PLANTS.

Valerianella carinata *Lois.*
Lysimachia punctata *L.*
Odontarrhena argentea *Ledb.*
Stachys pubescens *Ten.*
Silene saxatilis *Sims.*
Rhamnus Pallasii *F. et M.*
Scabiosa Columbaria *L.* vart.
Pyrethrum parthenifolium *L.*
Galium Aparine *L.*
Geranium robertianum *L.*
Vincetoxicum nigrum *Mönch.*
Lampsana intermedia *M. B.*
Philadelphus coronarius *L.*
Valeriana officinalis *L.*
Epilobium montanum *L.*
Astragalus galegæformis *L.*
Lactuca muralis *Dec.*
Papaver caucasicum *M. B.*
Geranium lucidum *L.*
Tamus communis *L.*
Orobus roseus *Ledeb.*
Clinopodium vulgare *L.*
Spiræa Aruncus *L.*
Geranium sanguineum *L.*
Echenais carlinoides *Cass.*
Orobus aurantiacus *Stev.*
Solanum Dulcamara *L.*, vart. persica (Solan. persicum *W.*).
Mœhringia trinervia *Clairv.*
Carex remota *L.*
Festuca Drymeja *Mert. et Koch.*
Euphorbia glareosa *M. B.*
Cuscuta cupulata *Engelm.*
Veronica orbicularis *Fisch.*
Rosa canina, *L.* var. collina *Koch.* forma 1, sempervirens *Rau* (*Ledb.*).
Juniperus communis *L.*
Knautia montana *Dec.*
Scutellaria altissima *L.*
Gymnadenia conopsea *R. Br.*
Alnus glutinosa *W.* typica.

Veronica Anagallis *L.* typica.
Pimpinella rotundifolia *M. B.*
Saxifraga cartilaginea *W.*
Leontodon hastilis *L.*, vart. glabrata *Koch.*
Cephalanthera rubra *Rich.*
Epipactis Helleborine *Crantz.* vart.
Rhaponticum pulchrum *F. et M.*
Pæonia corallina *Retz.*
Euphorbia aspera *M. B.*
Silene nemoralis *W.* et *Kit.*
Veronica officinalis *L.*
Reseda lutea *L.*
Cardamine impatiens *L.*
Genista tinctoria *L.*
Lathyrus pratensis *L.*
Saxifraga rotundifolia *L.*
Fragaria vesca *L.*
Orobus hirsutus *L.*
Farsetia clypeata *R. Br.*
Dianthus Carthusianorum *L.*
Silene chlorœfolia *Sm.*
Tragopogon pusillus *M. B.*
Acantholimon Kotschyi *Boiss*, vart. pontica *Trautv.*
Onosma sericeum *W.*
Achillea pubescens *L.* (Ach. micrantha *M. B.*).
Astragalus denudatus *Stev.*
Oxytropis pilosa *Dec.*
Lathyrus Nissolia *L.*
Thesium ramosum *Hayne.*
Crucianella glomerata *M. B.*
Alopecuri sp.?
Anthyllis Vulneraria *L.*
Blitum virgatum *L.*
Phleum alpinum *L.*
Evonymus latifolius *Scop.*
Rubus cæsius *L.*
Ostrya carpinifolia *Scop.*
Thlaspi macrophyllum *Hoffm.*

Doronicum caucasicum *M. B.*
Paris incompleta *M. B.*
Luzula pilosa *W.*
Oxalis Acetosella *L.*
Anemone ranunculoides *L.*
Symphytum tauricum *W.*
Viola canina *L.,* vart. sylvestris *Koch.*
Quercus Robur *L.,* iberica *Stev.*
Acer campestre *L.*
Orobus hirsutus *L.*
Pterotheca bifida *F. et M.*
Sideritis montana *L.*
Pastinaca intermedia *F. et M.*
Centaurea dealbata *W.*
Cerastium grandiflorum *W. et Kit.* vart. glabra *Koch.*
Cotoneaster Nummularia *F. et Mey.*
Alyssum campestre *L.,* vart. hirsuta *Trautv.*
Coronilla iberica *Stev.*
Pedicularis comosa *L.*
Thlaspi orbiculatum *Stev.*
Fumaria parviflora *Lam.*
Carpinus duinensis *Scop.*
Potentilla recta *L.*
Stellaria bolostea *L.*
Medicago falcata *L.*
Convolvulus Cantabrica *L.*
Anthriscus trichosperma *Schult.*
Scleranthus annuus *L.*
Hieracium praealtum *Koch.*
Vicia tenuifolia *Roth.*
Lathyrus rotundifolius *W.*
Polygala major *Jacq.*
Veronica austriaca *L.,* vart. pinnatifida *Koch.*
Cornus mascula *L.*
Melica ciliata *L.*
Ziziphora capitata *L.*
Campanula sibirica *L.*

Dactylus glomerata *L.*
Poa trivialis *L.*
Astragalus Raddeanus *Regel.*
Cytisus ratisbonensis *Schaff.*
Cerinthe minor *L.,* vart. maculata *C. A. Meyer.*
Picridium dichotomum *F. et Mey.*
Aethionema Buxbaumii *Dec.*
Melilotus arvensis *Wallr.*
Onosma rupestre *M. B.*
Marrubium catariaefolium *Desv.*
Daucus pulcherrimus *Koch.*
Salvia grandiflora *Ettl.* affinis.
Coronilla varia *L.*
Leonurus Cardiaca *L.*
Campanula ranunculoides *L.*
Lathyrus pratensis *L.*
Coronilla coronata *L.*
Teucrium orientale *L.*
Sophora alopecuroides *L.*
Pterotheca bifida *F. et M.* vart.
Linaria armeniaca *Chav.*
Cleome virgata *Stev.,* vart. macropoda *Trautv.*

Schambobell, 6,000 ft.
Lomatocarum alpinum *F. et M.*

Kutais, July. 600 ft.
Zolkowa crenata *Spach.*

Schambobell, 4–5,000 ft.
Trifolium alpestre *L.*
Echium rubrum *Jacq.*

Abastuman, July. 4,500 ft.
Hypopitys multiflora *Scop.,* vart. hirsuta *Koch.*
Dianthus recticaulis *Ledeb.*
Rubus fruticosus *L.*

Schambobell, south of Achaltzich,
  July. 5–7,000 ft.
Juncus alpigenus *C. Koch.*
Scirpus sylvaticus *L.*
Campanula collina *M. M.* et
  Ledeb.
Campanula Rapunculus *L.*
Aquilegia Wittmanniana *Stev.*
Campanula Saxifraga *M. B.*
Scabiosa caucasica *M. B.*, vart.
  heterophylla *Ledb.*
Silene saxatilis *Sims.*
Cerastium purpurascens *Adam.*
Hypericum hyssopifolium *Will,*
  vart. abbreviata *Ledb.*
Lotus corniculatus *L.*, vart.
  hirsutissima *Ledb.*
Trifolium ochroleucum *L.*
Chamæsciadium flavescens *C.
  A. Meyer.*

Sikar Pass, north of Abastuman, July. 6–7,000 ft.
Pimpinella magna *L.*, vart.
  rosea *Stev.*
Epilobium trigonum *Schrank.
  Ledb.*
Scrophularia macrobotrys *Ledb.*
Ranunculus caucasicus *M. B.*
Viburnum Lantana *L.*
Geranium psilostemon *Ledb.*
Rosa canina *L.*, vart. dumetorum *Koch.*
Cardamine impatiens *L.*
Arnebia echioides *Dec.*
Rumex scutatus *L.* β hastifolius
  C. A. Meyer.

Schambobell, July. 5–7,000 ft.
Ranunculus Villarsii *Dec.*
Papaver monanthum *Trautv*
Pimpinella Saxifraga *L.*

Betonica grandiflora *Staph.*
Pedicularis condensata *M. B.,*
  vart. minor *Trautv.*
Spiræa Filipendula *L.*
Orchis maculata *L.*
Gymnadenia conopsea, *R. Br.*
Centaurea montana *L.* vart.
  purpurascens *Dec.* et vart. albida *Dec.*
Alsine hirsuta *Fenzl.*
Linum hirsutum *L.*
Crucianella aspera *M. B.*
Dianthus Seguierii *Vill.* vart. ?

Sikar Pass. 6,000 ft.
Lonicera caucasica *Pall.*

Schambobell. 4–5,000 ft.
Tragopogon pusillus *M. B.*

Sikar Pass. 6–7,000 ft.
Scrophularia congesta *Stev.*
Nonnea intermedia *Ledb.*

Foot of Elbruz, August 9.
  5,000 ft.
Nepeta cyanea *Stev.*

Nachar Pass, south side, August.
  6–7,000 ft.
Saxifraga exarata *Vill.*
Vicia variegata *W.*
Hedysarum caucasicum *M. B.*

Foot of Elbruz, August 9.
  5–6,000 ft.
Salvia canescens *C. A. Meyer.*

Nachar Pass, south side, August.
  5–7,000 ft.
Myosotis sylvatica *Hoffm.*
Veronica monticola *Trautv.*

Campanula Saxifraga *M. B.*
Gentiana auriculata *Pall.*
Ranunculus subtilis *Trautv.*

Chursuk Valley, west foot of Elbrus, August 9. 4,000 ft.

Gypsophila elegans *M. B.*

Nachar Pass, south side. Anfang, August.

Trifolium polyphyllum *C. A. Meyer.*

West side of Elbrus. 8,000 ft.

Sedum tenellum *M. B.*
Senecio pyroglossus *Kar et Kir.*

Nachar Pass, August 6. 6,500 ft.
Scophularia Scopolii *Hoppe*

West of Elbrus, August 10. 10,000 ft.

Eritrichium nanum *Schrad.*
Draba scabra *C. A. Mey.*

East of Elbrus, August 10. 9,000 ft.

Delphinium caucasicum *C. A. Meyer.*

West side of Elbrus, August 10. 8–10,000 ft.

Saxifraga flagellaris *W.*
Anthemis Marschalliana *W.*, vart. Rudolphiana *C. A. Meyer.*

Nachar Pass, south side, Aug. 6. 5–7,000 ft.

Arenaria rotundifolia *M. B.*
Epilobium origanifolium *Lam.*

Saxifraga sibirica *L.*
Scrophularia pyrrholopha *Boiss.* et *Kotschy*, vart. pinnatifida *Trautv.*

West of Elbrus, August 10. 8–9,000 ft.

Ranunculus arachnoideus *C. A. Meyer.*

North side of Elbrus, August 10. 10–12,000 ft.

Lamium tomentosum *W.*
Hypericum nummularioides *Trautv.*

West and East sides of Elbrus, August 10. 6–8,000 ft.
Pedicularis crassirostris *Bunge*
Pedicularis Nordmanniana *Bunge.*

Minitaues Valley, August 10. 6,000 ft.

Aconitum Anthora *L.*

Elbrus, Aug. 10. 7–9,000 ft.
Luzula spicata *Dec.*
Cerastium latifolium *L. (C. A. Meyer.)*

Elbrus, Aug. 10. 8–10,000 ft.
Arenaria lychnidea *M. B.*

Nachar Pass, August 6. 8–9,000 ft.
Myosotis sylvatica *Hoffm.*

West and North sides of Elbrus. 9–12,000 ft.
Cerastium purpurascens *Adam.*

South side of Nachar Pass, August 6. 6,000 ft.
Gnaphalium sylvaticum *L.*

CATALOGUE OF PLANTS. 507

South side of Nachar Pass,
August 6. 9,500 ft.
Saxifraga exarata Vill.

North and West sides of Elbrus,
August 10. 10-12,000 ft.
Veronica repens Clar.
Veronica minuta C. A. Meyer.
Eunomia rotundifolia C. A.
Meyer.
Alsine imbricata C. A. Meyer.

West side of Elbrus, August 10.
9-10,000 ft.
Saxifraga sibirica L.
Taraxacum Stevenii Dec.
Potentilla gelida C. A. Meyer.

North side of Nachar Pass,
August 6. 9,500 ft.
Veronica gentianoides Vahl.
Campanulae sp.
Saxifraga muscoides Wulf.

From Muri to Lentechi, June
16-19. 1,600-2,600 ft.
Staphylea colchica Stev.
Myosotis sparsiflora Mikan.
Valeriana saxicola C. A. Meyer.,
var. lyrata Trautv.
Cystopteris fragilis Bernh.
Asplenium septentrionale Sw.
Scrophularia latoriflora Trautv.
Pyrethrum macrophyllum W.
Galium valantioides M. B.
Aspidium aculeatum Sw.
Saxifraga orientalis Jacq.
Andromaenum officinale All.
Orobanche alba Stev.
Hypericum montanum L.
Poa nemoralis L.

Dadiasch, June 23. 7-9,000 ft.
Gentiana verna L., var. alata
Griseb.
Draba tridentata Dec.
Eleocharis palustris R. Br.
Rhododendron caucasicum
Pall.
Daphne glomerata Lam.
Veronica Chamædrys L., var.
peduncularis Led.
Sibbaldia procumbens L.
Primula amœna M. B.
Potentilla Nordmanniana Ledb.?
Carex leporina L.
Primula farinosa L., var. xan-
thophylla Trautv. et Meyer =
Pr. algida Ad., var. luteo-
farinosa Rupr.

Dadiasch, June 23. 6,000 ft.
Acer hyrcanum F. et Meyer.

Laschketi, June 20. 3-4,000 ft.
Rhynchocoris Elephas Grisb.
Coronilla iberica Stev.

Dadiasch, June 23. 5-6,000 ft.
Andromace albana Stev.
Pedicularis comosa L.
Astrantia helleborifolia Salisb.

Dadiasch, June 23. 7-9,000 ft.
Alchemilla sericea W.
Campanula Biebersteiniana R.
et Sch.

Laschketi, June 20. 4,000 ft.
Psoralea acaulis Stev.

Dadiasch, June 23. 7-8,000 ft.
Alsine hirsuta Fenzl.
Jurinea subacaulis F. et Meyer.

Anemone alpina L. β. sulphurea Ledb.
Primula grandis *Trautv.*

Pari, July 11. 7–8,000 ft.
Cnidium meifolium M. B.

Pari, July 11. 4–5,000 ft.
Bupleurum falcatum L., vart. oblongifolia *Trautv.*
Hypericum Richeri Vill.

Pari, July 11. 7–8,000 ft.
Ranunculus montanus W., vart. glabrata *Trautv.*

Kalde-tskalai, July 5. 6,000 ft.
Saxifraga Kolenatiana Regel n. sp.

Karel Pass, July 5. 9,000 ft.
Orobanche sp. n.

Kaksagar Pass, June 29. 7,000 ft.
Salix apus *Trautv.* n. sp.

Pari, July 11. 6–8,000 ft.
Solidago Virgaurea L.
Corydalis spc.

Pari, July 11. 5,500 ft.
Scrophularia divaricata Ledb.

Karel Pass, July 5. 9,000 ft.
Primula Meyeri Rupr., vart. hypoleuca *Trautv.*

Pari, July 11. 7,000 ft.
Valeriana dubia *Dunge* affinis.

Pari, July 11. 4,600 ft.
Stachys persica S. G. Gmel.

Pari, July 11. 7,000 ft.
Ranunculus Villarsii Dec.
Digitalis ciliata *Trautv.*

Tibiani, July 4. 7,500 ft.
Scutellaria orientalis L., vart. chamaedryfolia Reichb.

Pari, July 11. 8,500 ft.
Gagea Liottardi Schult. Ledb.
Phleum alpinum L.
Nonnea ? (intermedia Ledeb.)

Laschketi, June 23. 4,000 ft.
Hypericum ramosissimum Ledb.

Kutais, May. 700 ft.
Euphorbia Lathyris L.
Dorycnium latifolium W.
Fragaria indica Audr.

North side of Nakerala, 3,500 ft.
Azalea pontica L.

Nakerala, south side, June.
3–4,000 ft.

Vaccinium Arctostaphylos L.
Scolopendrium officinarum Sw.
Veronica chamaedrys L., vart. peduncularis Led.
Gentiana asclepiadea L.
Mulgedium petiolatum Koch.
Cirsium fimbriatum Dec.
Orobus roseus Ledeb.

Tschitcharo, June, Alpine Region. 6–8,000 ft.

Saxifraga laevis M. B.
Podospermum Meyeri C. Koch.
Campanula Biebersteiniana B. et Sch.

Pedicularis crassirostris *Bunge.*
Salix arbuscula *L.*
Saxifraga exarata *Vill.*
Saxifraga rotundifolia *L.*
Corydalis angustifolia *Dec.*
Cnidium carvifolium *M. B.*
Primula pycnorhiza *Ledb.*
Oxytropis caucasica *Regel.*
Piristylus viridis *Lindl.*
Viola grandiflora *L.* (V. orcades *M. B.*)
Galanthus plicatus *M. B.*
Arenaria lychnidea *M. B.*

Nöschko Pass, June. 6–7,000 ft.

Salix apus *Trautv.*
Andromace villosa *L.* β. latifolia *Ledb.*
Arnebia echioides *Dec.*
Trichasma cylycianm *Walpers.*
Silene lacera *Sims.*

Zenes-Squali sources. 4,500 ft.

Ribes Dieberstenii *Berl.* (R. petraeum *Fl. ross.*).

Laschketi, June. 4,500–5,000 ft.

Ranunculus arvensis *L.* β. tuberculatus *Ledb.*
Hypericum hirsutum *L.*
Hypericum orientale *L.*
Circaea alpina *L.*
Trifolium elegans *Fl. germ.*
Tamus communis *L.*
Scrophularia Scopolii *Hoppe.*
Euphorbia micrantha *Steph.*
Mulgedium albanum *Dec.*
Datisca cannabina *L.*

Laschketi, June. 4,000 ft.

Lampsana grandiflora *M. B.*
Nonnea versicolor *Sweet.*
Scrophularia lucida *L.*
Sanicula europaea *L.*
Genista tinctoria *L.*
Agrostis vulgaris *With.*
Scandix Pecten *L.*, vart. trachycarpa *Trautv.*
Gypsophila elegans *M. B.*
Valerianella Morisonii *Dec.*, vart. leiocarpa *Trautv.*

Rion sources, Mamisson Pass, Goribolo, Aug. and *Sept.* 6–7,000 ft.

Phleum alpinum *L.*
Agrostis calamagrostoides *Regel.* n. sp.
Senecio longiradiatus *Trautv.*
Delphinium speciosum *M. B.*, var. dasycarpa *Trautv.*
Crocus Saworovianus *C. Koch.*
Potentilla elatior *Schlechtend.*
Aconitum variegatum *L.*
Draba tridentata *Dec.*
Gentiana septemfida *Pall.*
Cirsium munitum *M. B.*
Poa alpina *L.*
Ranunculus caucasicus *M. B.*
Briza media *L.*
Colchicum speciosum *Stev.*
Senecio nemorensis *L.*
Phyteuma campanuloides *M. B.*
Campanula collina *M. B.* subuniflora *Ledb.*
Knautia montana *Dec.* vart.
Cirsium simplex *C. A. Meyer.*
Swertia iberica.

www.ingramcontent.com/pod-product-compliance
Lightning Source LLC
Chambersburg PA
CBHW031947290426
44108CB00011B/714